Biosorption for Wastewater Contaminants

Biosorption for Wastewater Contaminants

Edited by

Rangabhashiyam Selvasembian
SASTRA University, School of Chemical and Biotechnology,
Thanjavur, India

and

Pardeep Singh
PGDAV College, University of Delhi, New Delhi, India

WILEY Blackwell

This edition first published 2022

Registered Offices
John Wiley & Sons, Inc., 111 River Street, Hoboken, NJ 07030, USA
John Wiley & Sons Ltd, The Atrium, Southern Gate, Chichester, West Sussex, PO19 8SQ, UK

Editorial Office
9600 Garsington Road, Oxford, OX4 2DQ, UK

For details of our global editorial offices, customer services, and more information about Wiley products visit us at www.wiley.com.

Wiley also publishes its books in a variety of electronic formats and by print-on-demand. Some content that appears in standard print versions of this book may not be available in other formats.

Library of Congress Cataloging-in-Publication data applied for

HB ISBN: 9781119737599

Cover Design: Wiley
Cover Images: © Chanin Wardkhian/Getty images; krystiannawrocki/Getty Images; Image Source/Getty Images

Set in 9.5/12.5pt STIXTwoText by Straive, Pondicherry, India

10 9 8 7 6 5 4 3 2 1

Contents

List of Contributors

Abhishek Agrawal
School of Pharmaceutical Sciences
Pratap University
Jaipur
Rajasthan
India

Pinki Rani Agrawal
CSIR National Physical Laboratory
New Delhi
India
and
Academy of Scientific and Innovative Research (AcSIR)
Ghaziabad
Uttar Pradesh
India

R. Balaji
Department of Biotechnology
Aarupadai Veedu Institute of Technology
Chennai
Tamilnadu
India

João Paulo Bassin
Chemical Engineering Program
COPPE
Federal University of Rio de Janeiro
Rio de Janeiro
Brazil
and
Civil Engineering Program
COPPE
Federal University of Rio de Janeiro
Rio de Janeiro
Brazil

Subramanyam Busetty
School of Civil Engineering
SASTRA Deemed University
Thanjavur
Tamilnadu
India

Ramprasad Chandrasekaran
School of Civil Engineering
SASTRA Deemed University
Thanjavur
Tamilnadu
India

B. Deepanraj
Department of Mechanical Engineering
Jyothi Engineering College
Thrissur
Kerala
India

R. Devika
Department of Biotechnology
Aarupadai Veedu
Institute of Technology
Chennai
Tamilnadu
India

João Pedro Neves Goldenstein
Civil Engineering Program
COPPE
Federal University of Rio de Janeiro
Rio de Janeiro
Brazil

Gaurav Gupta
Department of Physics and Astrophysics
University of Delhi
New Delhi
India

Ittishree
G.V.M College of Pharmacy
Pt. B.D. Sharma University of Health Sciences
Sonipat
Haryana
India

Jay Mant Jha
Department of Chemical Engineering
Maulana Azad National Institute of Technology
Bhopal
Madhya Pradesh
India

Noel Jacob Kaleekkal
Membrane Separation Group
Department of Chemical Engineering
National Institute of Technology Calicut
Kozhikode
Kerala
India

J. Kanimozhi
Department of Biotechnology
Kalasalingam Academy of Research and Education
Krishnankoil
Tamilnadu
India

Ravi Kumar
CSIR National Physical Laboratory
New Delhi
India
and
Academy of Scientific and Innovative Research (AcSIR)
Ghaziabad
Uttar Pradesh
India

Vishal Mishra
School of Biochemical Engineering
IIT (BHU)
Varanasi
Uttar Pradesh
India

G Nandu
Membrane Separation Group
Department of Chemical Engineering
National Institute of Technology Calicut
Kozhikode
Kerala
India

Rajarathinam Nithya
Department of Industrial Biotechnology
Government College of Technology
Coimbatore
Tamilnadu
India

Ramesh Oraon
Department of Nanoscience and Technology
Central University of Jharkhand
Brambe
Ranchi
Jharkhand
India

Dan Bahadur Pal
Department of Chemical Engineering
Birla Institute of Technology
Mesra
Ranchi
Jharkhand
India

Camila Pesci Pereira
Chemical Engineering Program
COPPE
Federal University of Rio de Janeiro
Rio de Janeiro
Brazil

Maheswari Purushothaman
Department of Chemistry
SRM Valliammai Engineering College
SRM Nagar
Kattankulathur
Tamilnadu
India

Bidhan Chandra Ruidas
Department of Chemical Engineering
Birla Institute of Technology
Mesra
Ranchi
Jharkhand
India

Guru Charan Sahu
Department of Chemistry
Ranchi University
Ranchi
Jharkhand
India

A.B. Sathya
Department of Biotechnology
Aarupadai Veedu Institute of Technology
Chennai
Tamilnadu
India

Rahul Sharma
Department of Chemistry
Sabarmati University
Ahmedabad
Gujarat
India

Rahul Sharma
AcSIR
CSIR-National Physical Laboratory
New Delhi
India

Nidhi Singh
Centre of Bioinformatics
University of Allahabad
Prayagraj
Uttar Pradesh
India

Veer Singh
School of Biochemical Engineering
IIT (BHU)
Varanasi
Uttar Pradesh
India

R. Sivashankar
Department of Chemical Engineering
Hindustan Institute of Technology and Science
Chennai
Tamilnadu
India

Arunachalam Thirunavukkarasu
Department of Industrial Biotechnology
Government College of Technology
Coimbatore
Tamilnadu
India

Amit Kumar Tiwari
Department of Chemical Engineering
Birla Institute of Technology
Mesra
Ranchi
Jharkhand
India

Srihari Vedartham
NICMAR
Hyderabad Campus at Shamirpet
Hyderabad
Telangana
India

Shashikant Shivaji Vhatkar
Department of Nanoscience and Technology
Central University of Jharkhand
Brambe
Ranchi
Jharkhand
India

Priyanka Yadav
School of Biochemical Engineering
IIT (BHU)
Varanasi
Uttar Pradesh
India

Preface

Today, pollution due to various anthropogenic activities has increased many times. Organic and inorganic pollutants are the most significant problem that humanity currently faces in the broad categories of water pollutants. Although several measures have been proposed and implemented to reduce these pollutants, their increasing concentration in bodies of water has created serious concerns. Over the years, the problem has been aggravated by uncontrolled industrialization and urbanization and the consequent alteration of humanity's natural resources. The direct discharge of wastewater contaminants and their geographical mobilization have led to increased concentrations in ground, surface, and residual waters. Therefore, it is essential to understand this problem today and to control the source. In addition, extensive knowledge of detection and disposal methods is needed to develop technological solutions for various environments, including urban, rural, and urban areas.

The purpose of this book is to provide an information platform about wastewater contaminants in the current context, where researchers, engineers, and technologists working in this field face various challenges. Conventional physicochemical techniques used to remove contaminants from wastewater include ion exchange, precipitation, degradation, coagulation, coating, membrane processes, and adsorption. However, these applications have technological and economic limitations. Biomass-related precursors for the preparation of biosorbents are attracting increasing attention from researchers. Different activation approaches are used in addition to native biomass utilizations. This book deliberately minimizes basic information and focuses on sources of water pollution, biomass for biosorbent preparation, characterization of biosorbents, understanding associated biosorption mechanisms, modeling analyses of biosorption, sustainable approaches for biosorption applications, and possibilities for recovering precious metals. This book will be a key guide for environmental engineers, researchers, water authorities, and students in these fields.

Rangabhashiyam Selvasembian
Pardeep Singh

Foreword

I am extremely pleased to provide the foreword to a book on biosorption that deals with the fundamentals, advancements, and future possibilities of this exciting technology. Biosorption employs the ability of biomass to adsorb different contaminants through various physicochemical mechanisms. A wide variety of biosorbents of different origins have been recognized to have high sorption capacity toward various contaminants. Biosorption mechanisms have been explored by scientists to obtain a basic understanding of the process. Nevertheless, biosorption has seldom been used for real industrial applications. The reasons are many and include lack of experimentation on real industrial effluents, multicomponent studies, and surface functionalization. Thus, a thorough compilation of the biosorption literature and a solid understanding of the biosorption technology's essentials, progress, and potential are needed. This book, *Biosorption for Wastewater Contaminants,* offers the complete package.

The book comprises 14 chapters contributed by various experts in the field. Each chapter is edited carefully to provide the required information for a thorough understanding of different aspects of biosorption technology. I congratulate the editors and other contributors of this book for providing us with a complete guide to biosorption for the welfare of the research community.

Dr. K. Vijayaraghavan
Senior Scientist
Department of Civil and Environmental Engineering
National University of Singapore, Singapore

1

Industrial Wastewater Contaminants and Their Hazardous Impacts

Camila Pesci Pereira[1], João Pedro Neves Goldenstein[2], and João Paulo Bassin[1,2]

[1] *Chemical Engineering Program, COPPE, Federal University of Rio de Janeiro, Rio de Janeiro, Brazil*
[2] *Civil Engineering Program, COPPE, Federal University of Rio de Janeiro, Rio de Janeiro, Brazil*

List of Abbreviations

AOP:	Advanced oxidation process
ASW:	Artificial sweetener
BIT:	Benzisothiazolinone
BOD:	Biochemical oxygen demand
BPA:	Bisphenol A
BPD:	Biocidal Products Directive
CD:	Carbendazim
CEC:	Contaminant of emerging concern
CO:	Carbon monoxide
COD:	Chemical oxygen demand
DBD:	Dielectric barrier discharge
DCOIT:	Dichloroctylisothiazolinone
DDT:	Dichlorodiphenyltrichloroethane
DEET:	N, N-Diethyl-meta-toluamide
DR:	Diuron
EDC:	Endocrine disrupting compound
EPA:	US Environmental Protection Agency
FR:	Flame retardant
HCB:	Hexachlorobenzene
IARC:	International Agency for Research on Cancer
IPBC:	Iodocarb
IP:	Isoproturon
IRG:	Cybutryn, Irgarol 1051
MBR:	Membrane bioreactor
MCI:	Methyl chloroisothiazolinone

Biosorption for Wastewater Contaminants, First Edition.
Edited by Rangabhashiyam Selvasembian and Pardeep Singh.

MCPP: Mecoprop
MI: Methylisothiazolinone
NO_x: Nitrogen oxides
OIT: Octylisothiazolone
PAH: Polycyclic aromatic hydrocarbon
PBB: Polybrominated biphenyl
PCB: Polychlorinated biphenyl
PCP: Pentachlorophenol
PeCB: Pentachlorobenzene
PFOS: Perfluorooctanesulfonic acid
POP: Persistent organic pollutant
PPCPs: Pharmaceutical and personal care products
PPZ: Propiconazole
PT: Product type
SBR: Sequencing batch reactor
SVOC: Semi volatile organic compound
TB: Terbutryn
TBU: Tebuconazole
TOC: Total organic carbon
TOD: Total oxygen demand
VOC: Volatile organic compound
VVOC: Very volatile organic compound
WWTP: Wastewater treatment plant

Introduction

Industrial and technological development and population growth have provided countless advances in society and improved quality of life. However, the increase in production and consumption capacity also negatively impacts the planet due to emission of harmful gases and the increasing production and consequent disposal of solid wastes and wastewaters from domestic and industrial sources. The volume and concentration of contaminants present in waste streams depend on several factors such as population density, level of technological development in the country or region, climate, seasonality, etc.

Contamination of water and soil by industrial waste is a global concern. For most industries, providing access to clean, affordable water is one of the biggest challenges. In addition, industrial wastewater contains several polluting compounds (Hashemi et al., 2018; Barak et al., 2020). Thus, there is a need to establish the management of and industrial wastewaters based on strict environmental protocols, with the objective of eliminating these contaminants from the environment as much as possible. As a result, through the use of recent technologies, research has been carried out to provide intelligent solutions not only from an environmental point of view but also from an economic standpoint. Therefore, several technologies based on physical, chemical, and biological methods are investigated (Hashemi et al., 2018).

This chapter describes some of the contaminants present in industrial wastewaters, their environmental impacts, and their risks to human health. In addition, the efforts made to evaluate and define the best ways to control their dissemination in the environment are presented, as well as the treatment processes used and studied for eliminating contaminants.

Toxic Heavy Metals

Heavy metals are a classic group of metallic elements that cannot be overlooked as industrial wastewater contaminants due to their toxicity. According to the World Health Organization (WHO), some of these pollutants (e.g. arsenic, lead, mercury, and cadmium) are included in the top 10 issues of primary concern (air pollution, arsenic, asbestos, benzene, cadmium, dioxin and dioxin-like substances, inadequate or excess fluoride, lead, mercury, and highly hazardous pesticides) because they are widely spread worldwide, they are highly hazardous, and negatively affect human health and natural ecosystems (WHO, 2020). Moreover, such substances are significantly dense and persistent and have shown toxicity even at part per billion (ppb) levels (Yadav et al., 2019). Compared to other contaminants, heavy metals can be neither degraded nor destroyed and can easily accumulate along the food chain. This phenomenon, known as *bioaccumulation*, increases the concern about suitable methods to treat industrial wastewaters.

Heavy metals are natural constituents of the Earth's crust, and heavy metal contamination in the environment results from natural and anthropogenic sources. Human activities are the leading causes of pollution. First, these metals are found in the form of chemical compounds or elemental state in rocks as ore deposits, which are explored through mining techniques. Environmental contamination by these compounds is not only a result of the business of mining – the largest contributor to heavy metal pollution – but may also be caused by improperly treated industrial wastewaters that reach natural ecosystems. The most common heavy metals found in industrial wastewaters are lead, cadmium, mercury, arsenic, copper, zinc, nickel, and chromium, which come from the paint, mining, metal manufacturing (electroplating, smelting, finishing, etc.), and textile industries (Akpor et al., 2014). From natural sources, the main contributions come from volcanic activity, soil erosion, and run-off. These hazardous waters deteriorate aquatic systems and terrestrial environments and retard biota growth (i.e., seed germination, chlorophyll production, and enzyme activity), with reported cases of carcinogenic, teratogenic, mutagenic, and neurotoxic effects.

There are multiple criteria to define an element as a heavy metal; however, one of the most frequently adopted classifications is based on the element's density. Some authors define *heavy metals* as metals with a density greater than 4 g/cm^3, while others say that the density needs to be greater than 7 g/cm^3, or five times denser than water. Clearly, there is no consistency in the literature (Duffus, 2002).

Regardless of the definition, the term *heavy metals* is considered inadequate and obsolete, since the majority of the population misuse and misunderstand it and the term sometimes leads to confused policies and regulations. In this chapter and most discussions, such as the WHO's list, when people refer to *heavy metals*, they relate *heaviness* to *toxicity* and are actually referring exclusively to toxic metallic elements. However, in appropriate concentrations, some of these elements are vital in our daily diets, such as calcium, iron, magnesium, and zinc. Their roles are very diverse. Calcium is the primary constituent of human bones and teeth, providing structure and hardness, but it also is necessary for muscles, metabolism, and nerve responses, together with magnesium and zinc. While iron and magnesium are essential for blood production, especially hemoglobin and myoglobin, zinc supports the human immune system and male reproductive activity, and magnesium also aids in muscle recovery and sleep. Signs of illness (diarrhea, anemia, liver and kidney failure, bloody urine, vomiting) have been reported when inadequate zinc concentrations are present, which can lead to a misdiagnosis of lead poisoning (Duruibe et al., 2007). Since the main concern is not the vital concentrations of heavy metals in humans, it was decided to name this section of the chapter "Toxic Heavy Metals" to avoid any room for misinterpretation.

That being said, toxic heavy metals are a subgroup with harmful effects leading to poisoning and bio-intoxication. Some of them are dangerous at extremely low levels. For instance, the general effects of cadmium exposure in humans have been reported as renal and gastrointestinal symptoms since the body tries metabolize and eliminate the substance. However, fumes and dust can cause high levels of cadmium exposure, which compromises the respiratory tract and causes coughing and chest pain because the lungs may fill with watery fluids; the result is a high risk of death. On the other hand, long-term, low-dose exposure can be associated with osteoporosis, myocardial dysfunctions, proteinuria, and higher blood pressure (Hallenbeck, 1986; Duruibe et al., 2007). This metal is commonly associated with high-tech and power-generation components, such as batteries, alloys, solar cells, pigments, and neutron absorbers in nuclear reactors. It is also used in the automotive industry due to its corrosion resistance. Since it inhibits organism growth, like many heavy metals, it is found in phosphate fertilizers as well (Sharma et al., 2015).

Arsenic is a harmful metalloid widely distributed around the world through natural and human activities. In reported long-term, low-dose exposure, victims suffer from skin lesions, neuropathy, diabetes, cardiovascular diseases, and cancer of the skin and internal organs (WHO 2019). Arsenic can be found in drinking water, industrial processes, food, and various tobacco products. In the industrial sector, wastewaters from pesticide, glass, and pharmaceutical manufacturing may contain the metalloid, but microelectronics and optical industries also handle arsenic compounds. Arsenic has been historically used in the agricultural sector as a pesticide, an insecticide, a defoliant, and a desiccant. However, due to its high biotoxicity, the United States Environmental Protection Agency (EPA) has banned its application in crops. Furthermore, elemental arsenic is used in alloys with copper and lead, while gallium arsenide and arsine have high electron mobility and are used in superconductors, magnets, light-emitting and chips, and many other applications in the electronic industries.

On the other hand, lead is not a good electric conductor. It is used in construction, ammunition, metallic alloys, cable sheaths, lead-acid batteries, pigments, fuses, and X-ray and radiation protection. Lead-contaminated potable water causes various undesirable effects, such as anemia, kidney and brain damage, infertility, miscarriage, and changes in behavior (hypersensibility, impulsivity, and aggressiveness). Despite oral exposure being the most common, anal, dermal, respiratory, and genital routes are also possible and result in the same kind of effects. This metal is well known for its chronic toxicity, easily damaging the central nervous system even at very low doses. In acute poisoning, the person shows signs of muscle weakness, pain, diarrhea, constipation, weight loss, bloody urine, etc. In industrial wastewaters, lead concentration ranges from 200 to 500 mg/L. According to WHO and EPA standards, its levels must be critically reduced to achieve 0.05–0.10 mg/L before discharge (Arbabi et al., 2015). In order to mitigate lead pollution, several methods for industrial wastewater treatment have been adopted, such as precipitation, adsorption, ion exchange, reverse osmosis, coagulation, and electro-coagulation (Arbabi et al., 2015).

Mercury, the only toxic metal present in liquid form at ambient temperature, is found in all living things and plays a crucial biological role. Nevertheless, it must remain at low levels, otherwise leads to toxic effects. The usual symptoms for acute exposure are fever, diarrhea, and vomiting, while chronic exposure causes tremors, nausea, stomatitis, pink disease, parageusia, and kidney malfunctions. Historically, gold amalgam (mercury alloy) was commonly used for gilding in architecture and fine arts. For example, in Saint Petersburg, Russia, at least 60 workers died from gilding the main dome of Saint Isaac's Cathedral (Emsley, 2011). Nowadays, industrial waters containing mercury are usually associated with dental care, paints, and fertilizers, but landfill leachate must also be considered mercury-containing waste (Hargreaves et al., 2016).

There are other toxic metal compounds of great concern, such as hexavalent chromium and copper. However, these chemicals were not included in the WHO list due to fewer cases compared to the ones previously mentioned. Many of them are also genotoxic gastrointestinal disruptors and are found in fertilizers and biocide compositions. Consequently, industrial wastewater containing toxic metals must be properly treated in order to not expose humans and other animals to harmful health impacts.

Toxic heavy metals have been part of human life for quite some time; hence there have been several attempts to find the best method to effectively remove them from industrial wastewater. Conventional methods including chemical, physical, and biological mechanisms have been adopted. For inorganic wastewater, chemical precipitation and ion exchange have been most widely used, but more recent advances have shown promising results using adsorption, electrodialysis, membrane filtration, and photocatalysis (Barakat, 2011). It is noteworthy that biosorption has also shown high efficiency for removing and recovering metals from wastewater, in addition to being a much cleaner and more cost-effective alternative (Kurniawan et al., 2006; Calderón et al., 2020).

Taking note of the facts and trends regarding toxic heavy metals, proper management of industrial wastewater must be taken into consideration. Each type of industry brings its own complexity and selected metals due to its manufacturing processes, as well as other constituents that must be treated before effluent discharge.

Dyes

Dyes are strongly related to toxic metals, given that they may contain such chemicals, and must be addressed. Some of their negative effects are inhibiting photosynthesis and plant growth, toxicity, carcinogenicity, mutagenicity, persistence, and recalcitrance. In addition, they increase biochemical oxygen demand (BOD) and chemical oxygen demand (COD) in industrial wastewaters (Lellis et al., 2019), which leads to further treatment expenditures. These wastewater contaminants are soluble organic compounds; because of the presence of chromophoric groups (azo, carbonyl, nitro, methine, etc.) in their molecules, it is hard to remove them from solution by conventional methods due to the presence of hydrophilic groups such as auxochromes.

Since such substances are used to color solid structures, when disposed in water bodies, they block the penetration of solar rays and reduce photosynthesis activity, which leads to lower dissolved oxygen levels and compromises aquatic life and self-purification processes. As a result, dye-containing industrial wastewater requires efficient methods to eliminate this chain of negative impacts throughout ecosystems.

Dyes can be manufactured from artificial and organic materials. Synthetic dyes are usually derived from petroleum and contain various toxic chemicals. Despite natural dyes being much more biodegradable and environmentally friendly, they must be mixed with a *mordant*, a kind of binding agent that helps attach dyes to fabrics. According to Katheresan et al. (2018), such binding agents may be more dangerous than synthetic dyes alone.

Over 800,000 tons of synthetic dyes are produced annually worldwide and generate around 1 trillion US dollars (Hassaan and El Nemr, 2017; Lellis et al., 2019). However, the manufacturing processes make them recalcitrant (resistant to biological degradation), and this is one of the most water-consuming sectors. For example, 1 kg of dye produced normally requires 0.15 m^3 of water. Unfortunately, over 200,000 tons of these dyes are lost during coloring processes and are carried in wastewater (Chequer et al., 2012), implying an unsustainable situation. The primary use of dyes is in the textile industries, but they are also present in the paper, tannery, food, pharmaceutical, and cosmetic industries, as shown in Figure 1.1.

In the textile industry, most dyes are used for fabric coloring, especially synthetic fabrics such as polyester, polyamide, and other polymers. In this case, dyes may impact both the environment and consumers because some dyes cause allergic reactions and skin and respiratory tract irritation. When dyes are inhaled, they attack the immune system and may

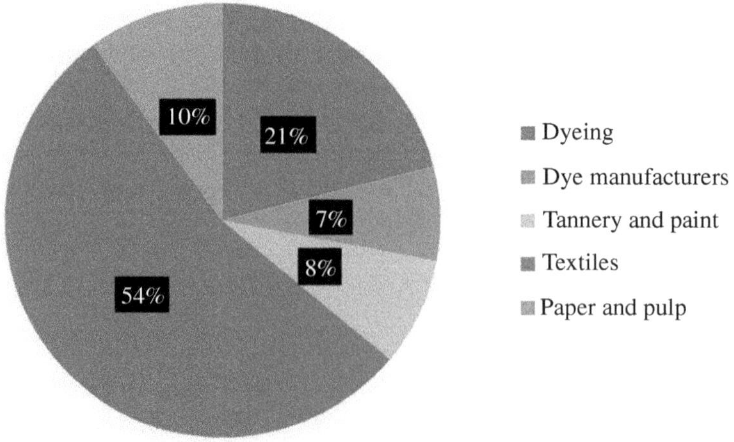

Figure 1.1 Main industries dealing with dyes. *Source:* Modified from Katheresan et al., 2018.

cause itchy, blocked noses, sneezing, and sore eyes. Furthermore, textile dyes are usually made of azo compounds, which are strong and complex structures made of nitrogen-nitrogen bonds that can be harmful even at concentrations as low as 1 part per million (ppm). To destroy such components, advanced oxidation processes (AOPs) have shown promising results, despite their high cost and sophisticated equipment requirements.

In the food industry, dyes are divided into two groups: those that require certification and those that do not. Based on the previous explanation, synthetic dyes such as azo, xanthene, triphenylmethane, and indigoid dyes are subject to a certification process (Barrows et al., 2014) due to their greater risk of toxicity. On the other hand, natural-based dyes commonly made from plants, insects, or minerals may be exempt from licensing, but their usage still needs concentration limits and attendance to levels of purity. It is worth noting that such licensing practices vary from country to country; hence, deeper understanding is necessary for any individual case. Besides the usual adverse effects listed so far, in rare cases, consuming colored foods may trigger anaphylactic shock that can cause death in minutes if proper treatment is not received, due to hypersensitivity to a dye. Any harmful effect of dye-containing food may occur due to an individual's intolerance or overdose of such chemicals when the food was not carefully manufactured.

In a completely different approach from how they are usually applied, dyes can be used with drugs in the pharmaceutical industry for the same purpose. Dyes are gaining credence in therapy due to their ability to engage with drug-resistant organisms, creating intolerable toxicity for microbes but not humans. A well-known example is Prontosil Rubrum® from Bayer (IG Farben Conglomerate, Germany), which behaves like a pro-drug since its reduction products are highly effective against bacterial illness, as shown in Figure 1.2. Consequently, pharmaceutical wastewaters contain, among a variety of complex, toxic, and recalcitrant compounds, dye waste from manufacturing processes and must be treated adequately before discharge.

There is a lack of complete understanding about the most effective technique for removing dyes from wastewater, and some approaches can generate by-products of secondary pollution. One of the first methods was an activated sludge process. However, although this biological process is relatively cheap and widely implemented worldwide, it is insufficient to remove such hazardous compounds (Katheresan et al., 2018). Another biological method that has achieved better removal is adsorption in biomass simultaneously with biodegradation by enzymes. Further research must take place since enzyme degradation seems to be a cheaper and safer method for dye wastewater treatment (Katheresan et al., 2018).

More expensive methods are available to treat dye wastewater using chemical and physical processes. Among the chemical methods are AOP, electrochemical destruction, ozone oxidation, ozone plus ultraviolet (UV) irradiation, etc. Unfortunately, such methods remain unattractive due to their higher investment and operational costs. On the flip side, physical methods are straightforward approaches relying on mass-transfer phenomena. These methods encompass adsorption, precipitation, cementation, coagulation, filtration, ion exchange, and reverse osmosis. Such technologies also have advantages and disadvantages and must be carefully analyzed before implementation.

After this brief review of dye wastewater and its hazardous impacts, it should be noted that the structure of dyes turns them into significant threats to the environment. With hydrophilic groups and properties that allow them to add or change color, they can

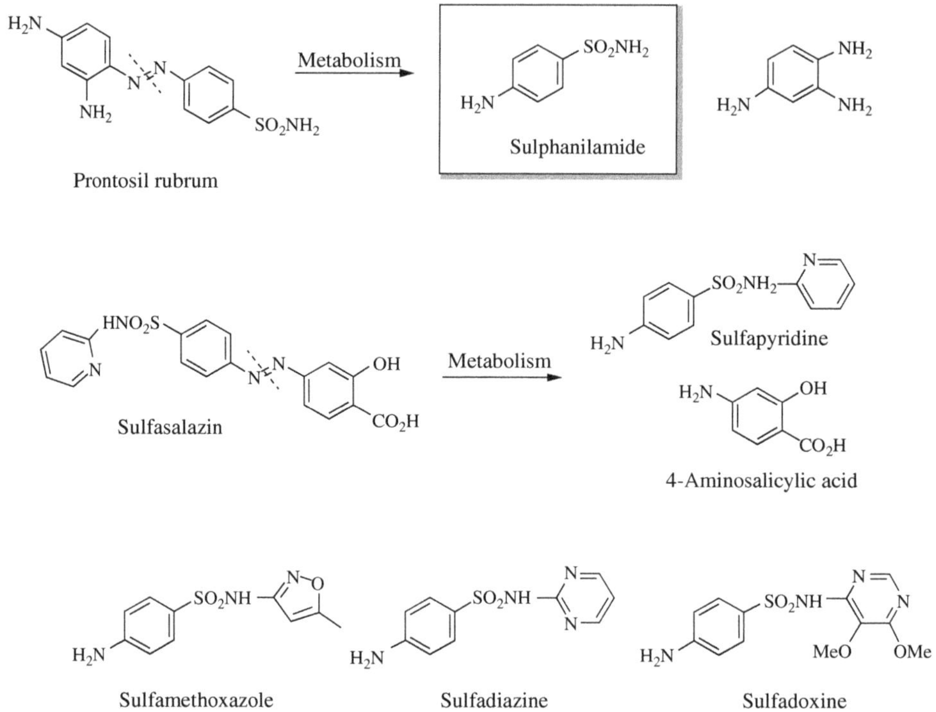

Figure 1.2 Prontosil and its reduction products. *Source:* Wainwright, 2008.

seriously modify the environment and affect all dynamics of ecosystems. In light of this, every industry using dyes during manufacturing processes must be aware of their negative environmental impacts and properly treat wastewater before discharging it into water course.

Oil and Grease

Oil and grease (O&G) form a well-known group of wastewater contaminants including oils, waxes, grease, fats, and other compounds. Their hydrophobic property, together with lower densities compared to water, allows them to form a top layer in bodies of water that interferes with the ecosystem. Moreover, some substances in this pollutant category were found to be plant and animal growth inhibitors and may also cause mutations and cancers among exposed individuals (Odeigah et al., 1997). Similar to dye wastewater, oily wastewater normally has high COD due to its source of generation.

At first, the oily top layer may seem to be a consequence of the density difference and has only an aesthetic or recreational impact. However, this sign of contamination hinders oxygen transfer between the atmosphere and the water body, compromising the dissolved oxygen levels and, therefore, all aquatic life. Besides that, to degrade such compounds, many bacteria consume the available oxygen during their metabolization, further increasing the impact (Ekpu, 1995). If the oil comes from the petroleum industry, it may also be inflammable and contain toxic metal and polycyclic aromatic hydrocarbons (PAHs). O&G also increase the water temperature due to their capacity to absorb solar radiation.

O&G directly and indirectly affect humans. Dermal exposure to oily wastewater can cause skin irritation and strong allergic reactions; when ingested, O&G accumulate in the liver and may cause vomiting, diarrhea, etc. But they also act indirectly, with adverse effects in ecosystems that compromise the basis of the food chain, starting from intoxicating plankton and phytoplankton and contributing to an imbalance that finally reaches the human diet. Furthermore, oily water may enter wastewater treatment plants (WWTPs) and damage pumping systems due to its viscosity, which leads to treatment failure and water and sanitation problems (Diphare et al., 2013).

Over the years, this type of contaminant is increasing in wastewater streams because of greater demand in the industrial sector and population growth. O&G are present in wastewaters from crude oil production, refineries, petrochemical industries, lubricant and cooling materials, compressor condensate, and metal processing. Domestic sewage is also a source of oily substances related to the human diet, restaurants, and car washing.

Oil in wastewater can be classified as a free oil phase, oil-water mixture, or emulsified oil mixture. As a free oil phase, droplets of oil must have a size greater than or equal to 150 microns; in oil-water mixtures, the size of oil droplets ranges from 20 to 150 microns; and in emulsified mixtures, their size is smaller than 20 microns. When the droplets are smaller than 5 microns, the oil is considered solubilized (Alade et al., 2011).

In relation to oil production, especially vegetable oil, consumption has increased much more than the population. While Brazil, the United States, the Philippines, Malaysia, and Argentina are some of the biggest producers, China consumes an enormous amount of oil by importation. The most important step in vegetable oil manufacturing and the primary cause of wastewater generation is the extraction process, which can be done chemically or mechanically. When oil is extracted mechanically, the industry uses oil mills, ram presses, or screw presses to pressurize and crush plants and seeds and obtain the desired product. Chemical extraction uses solvents such as hexane (Azad, 2018).

Despite the increasing consumption of vegetable oil, the largest producer of O&G contaminants is the petroleum industry and its refineries. Since petroleum is the world's primary energy source and will continue to be for many years, the production of wastewaters containing O&G will remain a concern. The oily wastewater comes from upstream and downstream operations with high BOD and COD derived from fatty acids. The oil and gas industry generates significant waste streams, such as water from drilling and extraction operations in oil reservoirs, and uses water in washing, cooling, and processing operations (Odeigah et al., 1997). When a refinery has vapocracking and fluid catalytic cracking units, it generates higher volumes of wastewater. So, when refining oil, there is process water and oil-free water. The first comes from petroleum distillation, cracking, coking and viscosity reduction, and desalting, while oil-free water includes purge water from boilers, cooling water from circuits, accidentally contaminated water, stormwater, and some smaller units (Suez, n.d.).

Another kind of pollutant is lubricating grease, deriving from petroleum oils and thickened with additives. It is used in many industries to lubricate moving machinery and other heavy equipment. As a result, friction, heat, and combustion activities may transfer metals to the grease, increasing its chemical complexity or toxicity. For instance, the combustion of heavy fuels commonly results in waste lubricants containing lead, a poisonous contaminant. From time to time, this viscous fluid needs to be replaced. Around 37 million tons of lubricants are consumed every year (Sönnichsen, 2021). Therefore, a collection system and treatment for these pollutants are necessary.

Unlike treatment processes for dye wastewater, the available methods for oily wastewater have demonstrated lower cost and complexity. There are many ways to eliminate O&G from water, including conventional, combined, and advanced technologies. According to Yu et al. (2017), conventional methods, which differ based on their separation mechanisms, are flotation, coagulation, biological treatment, and membrane filtration. On the other hand, combined technologies involve multistage processes and vary according to their studies and design. Lastly, advanced processes such as electrochemical catalysis, supercritical water oxidation, the Fenton process, ozonation, and photochemical reduction may be applied to treat wastewater containing O&G.

Biocides

A *biocide* is a substance or mixture that contains or generates one or more active substances to destroy, deter, render harmless, prevent the action of, or exercise control over any harmful organism by any means other than mere physical or mechanical action (ECHA, 2017).

Biocides can be used in different applications such as in-can preservatives, film preservatives, and building materials, such as coatings and wood protection products. In addition, most of these compounds have a broad spectrum of uses and application, as shown in Table 1.1 (Bollmann et al., 2014).

The European Parliament and Council directive 2013/39/EU incorporated four biocidal compounds as priority substances: cybutryne and terbutryn, diuron, and isoproturon (mentioned in Table 1.1). Diuron and isoproturon belong to the phenylureas group. Isoproturon is a component used mainly for the protection of cereals. Diuron is a herbicide present in preparations used to prevent algae growth and is also used as an additive and antifouling agent (Durak et al., 2020).

Another very relevant group of biocides is the isothiazolinones. These substances can be present in cosmetics and as chemical additives for industrial use, as they have bacteriostatic and fungistatic activities or properties. However, they may represent ecotoxicological risks because they can cause irritation and allergies. This category of biocides has aroused interest in drug development due to its biological effects, resulting in an increase in patent applications related to these substances. Isothiazolinones most commonly found in commercial applications, alone or in combination, are methylisothiazolinone (MI), methyl chloroisothiazolinone (MCI), benzisothiazolinone (BIT), octylisothiazolone (OIT) and dichloroctylisothiazolinone (DCOIT) (Silva et al., 2020).

Industrial development provides greater exposure to biocides and other contaminants, which may be discharged in many environmental matrices. The occurrence of these substances in the environment is determined by environmental conditions, physicochemical properties, and anthropogenic factors (Durak et al., 2020) and has consequences in the form of acute toxicity to aquatic organisms and accumulation of toxic substances in

Table 1.1 Examples of biocides, activities where they are employed, and their types of products (BPD).

Group ***Compound (abbreviation)*** ***CAS number***	**Activity** ***Product types (BPD)***
Triazines	
Terbutryn (TB) 886-50-0	Algaecide Biocide: PT 7, 9, 10
Cybutryn, Irgarol 1051 (IRG) 28159-98-0	Algaecide Biocide: PT 21
Carbamates	
Carbendazim (CD) 10605-21-7	Fungicide Biocide: PT 7, 9, 10
Iodocarb (IPBC) 55406-53-6	Fungicide Biocide. PT 6, 7, 8, 9, 10, 13 Cosmetics
Isothiazolinones	
Methylisothiazolinone (MI) 2682-20-4	Bactericide/Fungicide Biocide: PT 6, 11, 12, 13 Cosmetics
Benzisothiazolinone (BIT) 2634-33-5	Bactericide/Fungicide Biocide: PT 2, 6, 9, 11, 12, 13
Octylisothiazolinone (OIT) 26530-20-1	Bactericide/Fungicide Biocide: PT 6, 7, 9, 10, 13
Dichlorooctylisothiazolinone (DCOIT) 64359-81-5	Bactericide/Fungicide Biocide: PT 7, 8, 9, 10, 11, 21
Phenylureas	
Isoproturon (IP) 34123-59-6	Algaecide Biocide: PT 7, 10
Diuron (DR) 330-54-1	Algaecide Biocide: PT 7, 10
Triazoles	
Tebuconazole (TBU) 107534-96-3	Fungicide Biocide: PT 7, 8, 9, 10
Propiconazole (PPZ) 60207-90-1	Fungicide Biocide: PT 7, 8, 9
Miscellaneous	
Mecoprop (MCPP) 93-65-2	Algaecide Roof protection (not registered under BPD, since higher plants)

BPD, Biocidal Products Directive; PT, product type.
Source: Adapted from Bollmann et al., 2014.

the ecosystem, leading to decreased biodiversity and risks to human health (Mohr et al., 2008; Bollmann et al., 2014; Durak et al., 2020).

In response to these problems, environmental agencies have made an effort to restrict and/or regulate the use and disposal of biocides. At the same time, the scientific community aims to study efficient treatment processes to remove them from wastewaters, whether domestic or industrial.

Different processes have been tested to evaluate an efficient treatment approach for industrial wastewaters. AOPs have been tried to oxidize organic compounds, which are difficult to convert into biologically less-harmful final products (Pirilä et al., 2015). Pirilä et al. (2015) studied the photocatalysis process to treat four biocide compounds, including diuron, which is commonly used as a herbicide in agriculture, leading to contamination of the aquatic environment due to leaching from the soil. The elimination of this substance from wastewater is important due to its toxicity to photosynthetic organisms. This study revealed that diuron was removed more efficiently and with minimum energy consumption (Pirilä et al., 2015).

AOPs can also be applied in combination with other processes to treat biocides, as in the study carried out by Vanraes et al. (2018), who developed and optimized a reactor that combines oxidative treatment by dielectric barrier discharge (DBD) with adsorption of micropollutants on activated carbon and with additional plasma gas. The objective was to minimize the formation of dangerous oxidation byproducts from the treatment of persistent pesticides, including diuron and isoproturon. The authors observed that activated carbon in DBD reactors causes a synergistic effect in removing pollutants and that DBD reactors allow the regeneration of activated carbon, increasing the life of the reactor (Vanraes et al., 2017; Vanraes et al., 2018).

Organic Compounds

Industrial wastewater contains multiple polluting compounds, with the most toxic contaminants being organic compounds (Hashemi et al., 2018; Barak et al., 2020). Organic pollutants include phenols, chlorinated phenols, endocrine disrupting compounds (EDCs), azo dyes, polycyclic aromatic hydrocarbons (PAHs), polychlorinated biphenyls (PCBs), dichlorodiphenyltrichloroethane (DDT), hexachlorobenzene (HCB), pesticides, etc. (Das and Dash, 2019; Trojanowicz, 2020). Many of these compounds, such as some PAH, PCBs, EDCs, most phenols, DDT, etc., are recalcitrant: i.e. they can be more resistant to biological degradation (Knapp and Bromley-Challoner, 2003). Among the pollutants of greatest concern are volatile organic compounds (VOCs) and persistent organic pollutants (POPs).

VOCs are generally defined as substances whose composition allows evaporation under normal internal temperature and pressure conditions. Thus, the lower the boiling point of a substance, the greater its volatility. Therefore, VOCs can be defined and classified in some cases according to their boiling point in indoor environments. The WHO classification is presented in Table 1.2 for illustration (WHO, 1989; EPA, 2017).

In outdoor environments, the emission of VOCs is regulated to prevent the formation of ozone (a constituent of photochemical pollution) and other pollutants. VOCs that are considered reactive can interact with nitrogen oxides (NO_x) and carbon monoxide (CO) in

Table 1.2 Classification of organic pollutants by WHO.

Description	Abbreviation	Boiling Point Range (°C)	Example of compounds
Very volatile (gaseous) organic compounds	VVOC	<0 to 50–100	Propane, butane, methyl chloride
Volatile organic compounds	VOC	50–100 to 240–260	Formaldehyde, d-Limonene, toluene, acetone, ethanol (ethyl alcohol), 2-propanol (isopropyl alcohol), hexanal
Semi volatile organic compounds	SVOC	240–260 to 380–400	Pesticides (DDT, chlordane, plasticizers (phthalates), fire retardants (PCBs, PBB))

Source: Adapted from EPA, 2017.

the presence of sunlight, producing ozone. However, compounds considered nonreactive can also cause damage to human health if emitted in indoor environments. For example, methylene chloride, used as a paint remover, is exempt from outdoor regulation according to the EPA but is considered a substance with carcinogenic potential by the International Agency for Research on Cancer (IARC) (EPA, 2017).

In addition to the environmental and carcinogenic risks mentioned, VOCs can produce teratogenic and mutagenic effects; therefore, their emission must be controlled (Yeh et al., 2011; Zhang et al., 2019). Industrial wastewaters are important sources of VOCs. Significant amounts of benzene, toluene, xylenes, and other compounds are released during wastewater treatment and can harm the health of staff working at treatment plants and people living in nearby communities (Zhang et al., 2019).

Studies of VOC emissions focus mainly on municipal wastewater treatment plants. However, the treatment of industrial wastewaters has attracted attention. In the research conducted by Zhang et al. (2019), a full-scale plant for coke-oven wastewater treatment was evaluated, estimating the VOC emissions in each treatment unit and discussing factors that influence these emissions.

Another category of organic compounds that cause concern is POPs. They are defined as carbon compounds that remain immutable for a long time in the environment. In living organisms, they can accumulate in adipose tissues and are toxic to both animals and humans: they can cause different forms of cancer, reproductive problems, and changes in the nervous system (Stockholm Convention, 2001; Trojanowicz, 2020). Table 1.3 presents a list of POPs according to the Stockholm Convention.

The organic content of wastewaters can be estimated by the BOD, COD, total organic carbon (TOC), or total oxygen demand (TOD). The BOD test measures biodegradable organic carbon, and the COD test measures total organic carbon with the exception of certain aromatics, such as benzene, that are not completely oxidized in the reaction. Because it is an oxidation-reduction reaction, other non-organic reduced substances, such as sulfides, sulfites, and ferrous iron, will also be oxidized and reported as COD, except NH_3-N. The TOC test measures all carbon as CO_2, and hence inorganic carbon (CO_2, HCO_3^-, etc.) present in wastewater must be removed before the analysis or corrected for in the calculation. The

Table 1.3 Listing of persistent organic compounds according to the Stockholm Convention.

Category of compounds	Examples of compounds
Compounds for elimination (Should be eliminated from the production, use, and trade, although their import and export can occur under specific and restrictive conditions)	Aldrin a- and b-hexachlorocyclohexane Chlordane Chlordecone Dieldrin Endrin Heptachlor Hexabromobiphenyl Hexabromodiphenyl ether and Heptabromodiphenyl ether Hexachlorobenzene Lindane Mirex Pentachlorobenzene Polychlorinated biphenyls (PCB) Tetrabromodiphenyl ether and pentabromodiphenyl ether Toxaphene
Compounds whose production should be restricted (Subject to restrictions concerning their production and use)	DDT (1,1,1-trichloro-2,2-bis(4-chlorophenyl) ethane Perfluorooctanesulfonic acid (PFOS), its salts, and perfluorooctane sulfonyl fluoride
Compounds that are unintentionally produced (Should reduce and eliminate their release from unintentionally produced POPs)	Hexachlorobenzene (HCB) Pentachlorobenzene (PeCB) Polychlorinated biphenyls (PCB) Polychlorinated dibenzo p-dioxins and dibenzofurans

Source: Adapted from Trojanowicz, 2020.

TOD test measures organic carbon and unoxidized nitrogen and sulfur (Eckenfelder et al., 2008). TOC measures are used to monitor organic compounds in complex industrial wastewaters because they are more reliable than conventional BOD or COD measures, although the latter parameters are still widely implement (Dubber and Gray, 2010; Barak et al., 2020).

Several processes are used in industry to treat organic compounds. Biological treatment is one of the most-used processes since it is considered a reliable and economical technology in comparison to chemical (coagulation-flocculation, oxidation) and thermal (incineration) treatment (Barak et al., 2020; Chávez et al., 2019). Barak et al. (2020) studied full-scale membrane bioreactors (MBRs) for the treatment of wastewater from one of the major factories in an industrial park and observed good results in relation to the microbial diversity of the bioreactor and the removal of TOC.

However, some industrial wastewaters contain sufficient toxic and refractory compounds to affect the COD and removal of other parameters, not allowing them to reach the expected discharge standards. Thus, AOPs can be an alternative to achieve a lesser impact of the final effluent on the receiving ecosystem (Chávez et al., 2019; Li et al., 2019). AOPs have been extensively studied for treating industrial wastewaters such as those from the textile industry, not only for the removal of COD and TOC but also for the effective removal of colors and other toxic byproducts (Bilińska et al., 2020). They are usually implemented independently or can be combined with other processes (Bilińska et al., 2020; Buthiyappan and Raman, 2019; Chávez et al., 2019).

Catalytic ozonation has also occupied space for the treatment of industrial effluents. Bilińska et al. (2020) used catalytic ozonation to remove organic matter and toxicity in raw effluent from the textile industry, evaluating three types of catalysts. As a result, they observed that, with the best-performing catalyst (activated carbon), 35% of TOC and 40% of COD were removed, and toxicity decreased by 30%. Chávez et al. (2019) evaluated the feasibility of a photocatalytic ozonation treatment preceded by aerobic biodegradation in a sequencing batch reactor (SBR). The system was assembled on a bench scale to treat wastewaters from petrochemical and cosmetic products with a high organic load (TOC > 3 g/L, COD > 12 g/L, BOD_5 > 2 g/L), other toxic compounds, and metals. Due to the quality of the effluent, it was diluted with urban wastewater (1:5) for treatment in the SBR. The implemented approach achieved a final effluent suitable for disposal according to environmental regulations (COD < 125 mg/L, BOD_5 < 25 mg/L).

Zeolites are also widely used as an adsorbent to remove organic contaminants in the treatment of industrial wastewaters. Zeolites are porous materials with a three-dimensional structure used as an adsorbent base during treatment (Hashemi et al., 2018). Using a surfactant-modified zeolite, Hashemi et al. (2018) reported that the TOC content of olefin plant wastewater was reduced up to 89%.

Contaminants of Emerging Concern (CECs)

CECs are any substances originating from human activity or natural occurrence not generally monitored in the environment (Nawaz and Sengupta, 2019). They mainly result from the discharge of wastewaters (industrial or domestic), and conventional treatment processes are not capable of degrading these compounds. They can also be present in surface and drinking water. Even in very low concentrations (μg/L or ng/L), they can be bioaccumulative and become potentially dangerous to the ecosystem and human health (Prada-Vásquez et al., 2020).

CECs include pharmaceutical and personal care products (PPCPs), pesticides, POPs, EDCs, flame retardants (FRs), artificial sweeteners (ASWs), various industrial chemicals, etc. (Salimi et al., 2017). Two criteria for the long-term classification of a compound as a CEC are persistence in the environment and/or potential ecotoxicological and harmful effects to humans (Nawaz and Sengupta, 2019). Countless compounds fit in this classification, and over time, new substances and their effects are discovered, so the list of CEC has been constantly updated. Salimi et al. (2017) listed a series of CECs, some of which are shown in Table 1.4.

Table 1.4 Classification of some CECs such as PPCPs, EDCs, FRs, pesticides, and ASWs.

Classes	Used	Examples
PPCPs		
Analgesics	Pain reliever	Acetaminophen and acetylsalicylic acid
Anti-epileptic drugs	Anticonvulsant	Carbamazepine and primidone
Antihyperlipidemics	Lipid regulators	Gemfibrozil, clofibric acid, and fenofibric acid
Synthetic hormones	Hormone	Estrone, 17a-estradiol, 17a-ethinylestradiol, and estriol
Antimicrobials	Antibiotic	Erythromycin, sulfamethoxazole, and tetracycline
	Antiseptic	Triclosan, biphenylol, and chlorophene
Polycyclic	Musk fragrances	Hexahydrohexamethyl-cyclopentabenzopyran
Other	Insect repellant	N, N-Diethyl-meta-toluamide (DEET)
	Fragrances	Acetophenone
	Stimulant	Caffeine
EDCs		
Steroids	Natural human estrogen metabolite	17b-estradiol Estrone
Alkylphenols	Manufacture of household and industrial products	Nonylphenol and octylphenol
Polyaromatic compounds		Polychlorinated biphenyls and brominated flame retardants
Organic oxygen compounds	Plasticizers	Phthalates
	Industrial production of polycarbonates and epoxy resins	BPA
Pesticides	Insecticides, herbicides, fungicides	Atrazine, chlordane, and trifluralin
Others	By-products of various industrial and combustion processes	Dioxins and furans
Flame retardants		
Halogen-containing flame retardants (fluorine, chlorine, bromine, or iodine)	FR	Brominated bisphenols and phenols
Inorganic hydroxide FRs	FR	Aluminum hydroxide and magnesium hydroxide
Borate FRs	FR	Sodium borate and boric acid

Table 1.4 (Continued)

Classes	Used	Examples
Pesticides		
Carbamates	Herbicides, insecticides, and fungicide	Carbendazim, benomyl, and carbaryl
Organochlorines	Insecticides	DDT, dieldrin, endrin, and endosulfan
Organophosphates	Insecticides	Diazinon, malathion, and chlorpyrifos
Artificial sweeteners		
Artificial sweeteners	Sugar substitutes	Acesulfame, sucralose, saccharin, cyclama, aspartame, neotame, and neohesperidine dihydrochalcone

Source: Adapted from Salimi et al., 2017.

CECs can result from commercial, domestic, agricultural, and industrial origins, like any pollutant. They can be detected in remote and unlikely areas due to their persistence and long-distance mobility. Some of these substances can disturb the endocrine system, interfering with the hormonal action of animals and humans; they are referred to as EDCs and can cause reproductive effects, hormonal dysfunction, fertility problems, and diseases such as cancer, etc. (Nawaz and Sengupta, 2019; EPA, 2020).

EDCs include steroids, alkylphenols, polyaromatics, organic oxygen compounds, pesticides, and others (Salimi et al., 2017). They can be synthetic or natural. Examples of synthetic compounds are alkylphenols, phthalates, polychlorinated biphenyls, bisphenol A (BPA), and pharmaceutical agents such as estrogen 17α-ethinylestradiol (Salimi et al., 2017). Among natural EDCs are the estrogens estrone, estriol, and 17β-estradiol. Estrogens are hormones responsible for several female endocrine functions, and some of them can be used in the production of contraceptives and hormone replacement medicine (Hamid and Eskicioglu, 2012).

Few studies in the literature discuss the biodegradation, metabolization, and effects of CECs (Salimi et al., 2017). However, the advance of modern technologies, such as high-performance liquid chromatography, allow the detection of substances in lower concentrations (on the order of ng/L) in different environmental matrices. Thus, studies have been carried out to investigate and improve the removal of these compounds from both domestic and industrial wastewaters. Biological treatment (activated sludge and biofilm-based processes) and AOPs (ozonation, TiO_2 photocatalysis, Fenton process, and photo-Fenton process) have been extensively investigated for this purpose.

Conclusion

The primary contaminants found in industrial wastewater have been shown to commonly originate from human activities. Industrial wastes are becoming more recalcitrant, complex, and abundant due to the increasing development of society, and the list of pollutants.

This chapter addressed the effects of toxic heavy metals, oil and grease, biocides, dyes, recalcitrant and organic compounds, and contaminants of emerging concern.

Regarding toxic heavy metals, it is important to differentiate between heavy metals and toxic metals. The classification of many elements and compounds is largely based on their density. However, not every heavy element is toxic. With that in mind, copper, lead, chromium, cadmium, zinc, and arsenic are the most important. Bioaccumulation, vomiting, diarrhea, and inflammation are the usual symptoms of toxic heavy metal poisoning. Furthermore, such pollutants negatively impact humans and other species, hindering their reproductivity and metabolism, which unbalances the food chain as a whole.

Chemically speaking, dyes were developed to resist degradation since they have hydrophilic and chromophoric groups that provide color and strong persistence in aqueous phases. The primary consumer of dyes is the textile industry, which generates millions of dollars annually worldwide. Other sectors are also important in this context, such as the paper, tannery, food, pharmaceutical, and cosmetic industries. It must be noted that natural-based dyes are not the focus of this discussion since they are much more biodegradable than synthetic dyes and are less frequently used.

Oil and grease have always been one of the primary human contaminants in wastewaters. However, society is and will continue to be strongly dependent on petroleum and its derivatives. Thus, oil and grease from this industry will continue to be a significant threat to natural ecosystems. Fortunately, many methods are known to reduce their concentration in industrial waters.

Taking into account population growth, the need for food has significantly increased. To supply demand, the manufacturing and application of biocides have become increasingly necessary to guarantee crops and preserve them against stronger adverse agents (insects and other organisms). Such chemicals are made to be recalcitrant and toxic. Thus, they need to be properly removed from agricultural and industrial wastewaters so they do not affect human health through food.

Human activities have increased the consumption of organic and recalcitrant compounds such as phenols, VOCs, DDT, PCHs, PAHs, PCBs, POPs, etc., but these substances are proven to be carcinogenic, teratogenic, and mutagenic, besides being persistent in the environment. Commonly, industrial wastewater containing recalcitrant compounds contains other toxic materials that may affect biological treatment, given that advanced oxidation processes have been addressed and shown more suitable results.

Considering a more recent approach, contaminants of emerging concern must be highlighted. Even at small levels, they have shown harmful effects in humans and are capable of disrupting the endocrine system and causing other, as-yet-unknown impacts. Such pollutants are persistent in the environment and result from human activities. Hence, the list of these compounds is constantly growing and must be studied in depth.

The industrial sector has to acknowledge the impacts and complexity of its contaminants for humans and the environment. Therefore, suitable treatment methodologies must be developed and addressed. Once such understanding is adopted worldwide, hazardous impacts from these activities will not be a threat but instead can be properly controlled and made harmless.

References

Alade, A.O., Jameel, A.T., Muyubi, S.A. et al. (2011). Removal of oil and grease as emerging pollutants of concern (EPC) in wastewater stream. *IIUM Engineering Journal* 12 (4). doi:10.31436/iiumej.v12i4.218.

Akpor, O.B., Ohiobor, G.O., and Olaolu, T.D. (2014). Heavy metal pollutants in wastewater effluents: sources, effects and remediation. *Advances in Bioscience and Bioengineering* 2 (4): 37. doi:10.11648/j.abb.20140204.11.

Arbabi, M., Hemati, S., and Amiri, M. (2015). Removal of lead ions from industrial wastewater: A review of removal methods. *International Journal of Epidemiologic Research* 2 (2): 105–109.

Azad, A.K. (ed.) (2018). Advances in eco-fuels for a sustainable environment. Woodhead Publishing.

Barak, H., Brenner, A., Sivan, A., and Kushmaro, A. (2020). Temporal distribution of microbial community in an industrial wastewater treatment system following crash and during recovery periods. *Chemosphere* 258: 127–271.

Barakat, M.A. (2011). New trends in removing heavy metals from industrial wastewater. *Arabian Journal of Chemistry* 4 (4): 361–377.

Barrows, J.N., Lipman, A.L., and Belai, N. (2014). History of US regulation of color additives and colorants. In: *Food Additives and Packaging* (ed. V. Komolprasert and: Turowski), 57–71. American Chemical Society.

Bilińska, L., Blus, K., Bilińska, M. et al. (2020). Industrial textile wastewater ozone treatment: catalyst selection. *Catalysts* 10 (6): 611. doi:10.3390/catal10060611.

Bollmann, U.E., Tang, C., Eriksson, E. et al. (2014). Biocides in urban wastewater treatment plant influent at dry and wet weather: Concentrations, mass flows and possible sources. *Water Research* 60: 64–74. doi:10.1016/j.watres.2014.04.014.

Buthiyappan, A. and Raman, A.A.A. (2019). Energy intensified integrated advanced oxidation technology for the treatment of recalcitrant industrial wastewater. *Journal of Cleaner Production* 206: 1025–1040.

Calderón, O.A.R., Abdeldayem, O.M., Pugazhendhi, A., and Rene, E.R. (2020). Current updates and perspectives of biosorption technology: an alternative for the removal of heavy metals from wastewater. *Current Pollution Reports* 6 (1): 8–27.

Chávez, A.M., Gimeno, O., Rey, A. et al. (2019). Treatment of highly polluted industrial wastewater by means of sequential aerobic biological oxidation-ozone based AOPs. *Chemical Engineering Journal* 361: 89–98.

Chequer, F.D., De Oliveira, G.R., Ferraz, E.A. et al. (2013). Textile dyes: dyeing process and environmental impact. *Eco-Friendly Textile Dyeing and Finishing* 6 (6): 151–176.

Das, S. and Dash, H. (2019). *Microbial Diversity in the Genomic Era*. Elsevier.

Diphare, M.J., Pilusa, J., Muzenda, E., and Mollagee, M. (2013). A review of waste lubricating grease management. In: 2nd International Conference on Environment, Agriculture and Food Sciences, 131–134.

Dubber, D. and Gray, N.F. (2010). Replacement of chemical oxygen demand (COD) with total organic carbon (TOC) for monitoring wastewater treatment performance to minimize disposal of toxic analytical waste. *Journal of Environmental Science and Health Part A* 45 (12): 1595–1600. doi:10.1080/10934529.2010.506116.

Duffus, J.H. (2002). Heavy metals a meaningless term? (IUPAC Technical Report). *Pure and Applied Chemistry* 74 (5): 793–807.

Durak, J., Rokoszak, T., Skiba, A. et al. (2020). Environmental risk assessment of priority biocidal substances on Polish surface water sample. *Environmental Science and Pollution Research.* doi:10.1007/s11356-020-11581-7.

Duruibe, J.O., Ogwuegbu, M.O.C., and Egwurugwu, J.N. (2007). Heavy metal pollution and human biotoxic effects. *International Journal of Physical Sciences* 2 (5): 112–118.

ECHA (European Chemicals Agency). (2017). Guidance on the biocidal products regulation volume IV environment – assessment and evaluation (parts B + C): Version 2.0.

Eckenfelder, W.W., Ford, D.L., and Englande, A.J. (2008). *Industrial Water Quality*, 4e. McGraw-Hill.

Ekpu, A.O. (1995). Environmental impact of oil on water: a comparative overview of the law and policy in the United States and Nigeria. *Denv. J. Int'l L. & Pol'y* 24: 55.

Emsley, J. (2011). *Nature's Building Blocks: An A-Z Guide to the Elements.* Oxford University Press.

EPA (US Environmental Protection Agency). (2017). Technical overview of volatile organic compounds. www.epa.gov/indoor-air-quality-iaq/technical-overview-volatile-organic-compounds (accessed 10 April 2021).

EPA (US Environmental Protection Agency). (2020). Contaminants of emerging concern including pharmaceuticals and personal care products. www.epa.gov/wqc/contaminants-emerging-concern-including-pharmaceuticals-and-personal-care-products (accessed 11 April 2021).

Hallenbeck, W.H. (1986). Human health effects of exposure to cadmium. In: *Cadmium in the Environment* (ed. H. Mislin and O. Ravera), 131–137. Birkhäuser Basel. doi:10.1007/978-3-0348-7238-6_17.

Hamid, H. and Eskicioglu, C. (2012). Fate of estrogenic hormones in wastewater and sludge treatment: A review of properties and analytical detection techniques in sludge matrix. *Water Research* 46: 5813–5833.

Hargreaves, A.J., Vale, P., Whelan, J. et al. (2016). Mercury and antimony in wastewater: fate and treatment. *Water, Air, and Soil Pollution* 227 (3): 89.

Hashemi, M.S.H., Eslami, F., and Karimzadeh, R. (2018). Organic contaminants removal from industrial wastewater by CTAB treated synthetic zeolite Y. *Journal of Environmental Management* 233: 785–792, doi:10.1016/j.jenvman.2018.10.003.

Hassaan, M.A. and El Nemr, A. (2017). Health and environmental impacts of dyes: mini review. *Am J Environ Sci Eng* 1 (3): 64–67.

Katheresan, V., Kansedo, J., and Lau, S.Y. (2018). Efficiency of various recent wastewater dye removal methods: a review. *Journal of Environmental Chemical Engineering* 6 (4): 4676–4697.

Knapp, J.S. and Bromley-Challoner, K.C.A. (2003). Recalcitrant organic compounds. In: *Handbook of Water and Wastewater Microbiology*. doi:10.1016/b978-012470100-7/50035-2.

Kurniawan, T.A., Chan, G.Y., Lo, W.H., and Babel, S. (2006). Comparisons of low-cost adsorbents for treating wastewaters laden with heavy metals. *Science of the Total Environment* 366 (2–3): 409–426.

Lellis, B., Fávaro-Polonio, C.Z., Pamphile, J.A. et al. (2019). Effects of textile dyes on health and the environment and bioremediation potential of living organisms. *Biotechnology Research and Innovation* 3 (2): 297–290.

Li, X., Chen, W., Ma, L. et al. (2019). Characteristics and mechanisms of catalytic ozonation with Fe shaving-based catalyst in industrial wastewater advanced treatment. *Journal of Cleaner Production* 222: 174–181.

Mohr, S., Schröder, H., Feibicke, M. et al. (2008). Long-term effects of the antifouling booster biocide Irgarol 1051 on periphyton, plankton and ecosystem function in freshwater pond mesocosms. *Aquatic Toxicology* 90: 109–120. doi:10.1016/j.aquatox.2008.08.004.

Nawaz, T. and Sengupta, S. (2019). *Advances in Water Purification Techniques.* Elsevier. doi:10.1016/B978-0-12-814790-0.00004-1.

Odeigah, P.G.C., Nurudeen, O., and Amund, O.O. (1997). Genotoxicity of oil field wastewater in Nigeria. *Hereditas* 126 (2): 161–167.

Pirilä, M., Saouabe, M., Ojala, S. et al. (2015). Photocatalytic degradation of organic pollutants in wastewater. *Top Catal* 58: 1085–1099, doi:10.1007/s11244-015-0477-7.

Prada-Vásquez, M.A., Estrada-Flórez, S.E., Serna-Galvis, E.A., and Torres-Palma, R.A. (2020). Developments in the intensification of photo-Fenton and ozonation-based processes for the removal of contaminants of emerging concern in Ibero-American countries. *Science of the Total Environment* 765. doi:10.1016/j.scitotenv.2020.142699.

Salimi, M., Esrafili, A., Gholami, M. et al. (2017). Contaminants of emerging concern: a review of new approach in AOP technologies. *Environ Monit Assess* 189 (414): 1–22.

Sharma, H., Rawal, N., and Mathew, B.B. (2015). The characteristics, toxicity and effects of cadmium. *International Journal of Nanotechnology and Nanoscience* 3: 1–9.

Silva, V., Silva, C., Soares, P. et al. (2020). Isothiazolinone biocides: chemistry, biological, and toxicity profiles. *Molecules* 25 (4): 991. doi:10.3390/molecules25040991.

Sönnichsen, N. (2021). Lubricants – statistics and facts. www.statista.com/topics/5263/lubricants-industry (accessed 20 April 2021).

Stockholm Convention on Persistent Organic Pollutants (2001). http://chm.pops.int/TheConvention/ThePOPs/AllPOPs/tabid/2509/Default.aspx.

Suez. (n.d.). Oil industry. www.suezwaterhandbook.com/water-and-generalities/what-water-should-we-treat-and-why/industrial-effluent/oil-industry (accessed 13 April 2021).

Trojanowicz, M. (2020). Removal of persistent organic pollutants (POPs) from waters and wastewaters by the use of ionizing radiation. *Science of the Total Environment* 718: 134425.

Wainwright, M. (2008). Dyes in the development of drugs and pharmaceuticals. *Dyes and Pigments* 76 (3): 582–589.

Vanraes, P., Ghodbane, H., Davister, D. et al. (2017). Removal of several pesticides in a falling water film DBD reactor with activated carbon textile: Energy efficiency. *Water Research* 116: 1–12.

Vanraes, P., Wardenier, N., Surmont, P. et al. (2018). Removal of alachlor, diuron and isoproturon in water in a falling film dielectric barrier discharge (DBD) reactor combined with adsorption on activated carbon textile: Reaction mechanisms and oxidation by-products. *Journal of Hazardous Materials* 354: 180–190. doi:10.1016/j.jhazmat.2018.05.007.

WHO (World Health Organization). (1989). Indoor air uality: organic pollutants. Report on a WHO Meeting, Berlin, 23–27 August 1987. EURO Reports and Studies 111. Copenhagen, World Health Organization Regional Office for Europe.

WHO (World Health Organization). (2019). Preventing disease through healthy environments: exposure to arsenic: a major public health concern (No. WHO/CED/PHE/EPE/19.4.1).

WHO (World Health Organization). (2020). 10 chemicals of public health concern. https://www.who.int/news-room/photo-story/photo-story-detail/10-chemicals-of-public-health-concern (accessed 6 April 2021).

Yadav, M., Gupta, R., and Sharma, R.K. (2019). Green and sustainable pathways for wastewater purification. In: *Advances in Water Purification Techniques* (ed. S. Ahuja), 355–383. doi:10.1016/B978-0-12-814790-0.00014-4.

Yeh, S.H., Lai, C.H., Lin, C.H. et al. (2011). Estimating cancer risk increment from air pollutant exposure for sewer workers working in an industrial city. *Aerosol Air Qual. Res.* 11: 120–127.

Yu, L., Han, M., and He, F. (2017). A review of treating oily wastewater. *Arabian Journal of Chemistry* 10: S1913–S1922.

Zhang, Y., Wei, C., and Yan, B. (2019). Emission characteristics and associated health risk assessment of volatile organic compounds from a typical coking wastewater treatment plant. *Science of the Total Environment* 693: 133417.

2

Biosorption and Different Native Sources for Preparation of Biosorbents

A.B. Sathya[1], R. Sivashankar[2], J. Kanimozhi[3], R. Devika[1], and R. Balaji[1]

[1]*Department of Biotechnology, Aarupadai Veedu Institute of Technology, Chennai, Tamilnadu, India*
[2]*Department of Chemical Engineering, Hindustan Institute of Technology and Science, Chennai, Tamilnadu, India*
[3]*Department of Biotechnology, Kalasalingam Academy of Research and Education, Krishnankoil, Tamilnadu, India*

Introduction

Water plays a significant role in the global economy. Water covers most (71%) of the Earth's surface, but water from fresh sources like rivers, lakes, and streams contributes only 3% to total water resources. Freshwater bodies provide water suitable for human consumption. About 70% of freshwater is expended on irrigation. In many countries, these natural resources are limited, and their unavailability poses significant social and economic concerns (Baroni et al., 2007). As the quantity of water decreases, water quality is also seriously declining due to industrial, municipal, and agricultural wastes. Contamination can arise from wastewater leaking from damaged pipes and the flow of industrial waste into water. This contaminated water poses a constant danger to public health in the form of water-borne diseases (Arshad and Imran, 2020).

Heavy metal exposure is considered a major environmental concern and a hazard to human health among numerous contaminations in drinking water. Even in minute amounts, toxic metals can cause toxicity when they enter the body due to their non-degradable nature. Mercury, nickel, arsenic, chromium, manganese, copper, zinc, and cadmium are the most widely intoxicating heavy metals (Jaishankar et al., 2014). The wastewater streams from plating, casting, painting dyes, batteries, mining, and farming are key contributors to heavy metals entering the ecosystem (Ibrahim and Mutawie, 2013). Because of their toxic effects, most heavy metals produce various issues, including kidney problems, brain activity difficulties, and nervous weakness. Sleeplessness, impatience, anemic, dizziness, and muscle pain are some of their hazardous symptoms (Fu and Wang, 2011).

Several techniques for the biosorption of toxic metals from effluent have been developed. They include ion exchange, evaporation, chemical precipitation, and membrane filtration. Activated carbon is by far the most common and widely utilized material for treating

Biosorption for Wastewater Contaminants, First Edition.
Edited by Rangabhashiyam Selvasembian and Pardeep Singh.

wastewater pollution. Materials that are used to prepare higher-quality activated carbon cost more (Babel, 2003). In recent years, awareness has increased of the development of adsorbents to substitute for expensive activated carbon. The emphasis is on adsorbents that can bind to metal and remove undesirable heavy metals from polluted water at a low cost.

Biosorption and bioaccumulation are environmentally friendly solutions with benefits compared to traditional methods. Owing to their metal-binding functional groups, abundant natural materials have been considered as viable biosorbents for eliminating heavy metals: for example, microbial biomass, agro waste, and industrial byproducts. Parameters such as pH, temperature, metal ion concentration, biosorbent dose, and agitation rate influence biosorption. After the removal of heavy metals, biosorbents can be regenerated and reused, thus making the process economical (Kanamarlapudi et al., 2018). This chapter provides a summary of some low-cost biosorbents discussed in recent publications.

Biosorption and Its Mechanism

Biosorption is a simple physical and passive mechanism involving the attachment of biosorbates (metal ions) to the biosorbent surface (of biological origin) (Mrvčić et al., 2012). Simple operation, no demand for supplementary nutrients, low sludge production, lower operational cost, good performance, biosorbent rejuvenation, and no increase in chemical oxygen demand (COD) in water are all advantages of this technology that are major drawbacks of older technologies. Pollutants with parts per billion (ppb) toxicity can be eliminated by biosorption even at diluted concentrations. This process is especially important for removing heavy metals.

The first step in biosorption is to suspend the biosorbent in the biosorbate-containing solution (metal ions). Equilibrium is achieved after incubation for a given period. The metal-enriched biomass is separated at this point (Chojnacka, 2010). Biosorption is a dynamic mechanism in which sorbate binds to a biosorbent. Physical attachment of metal ions (Van der Waals interaction or electrostatic forces) or chemical binding (replacement of ions), chelation, precipitation, reduction, and complexation are all possible with a wide range of natural materials as biosorbents. The significant factors that control the biosorption process are (Park et al., 2010) as follows:

- Nature of the biosorbent
- Category of the biological ligand
- Optimum parameters of sorbate and sorbent (temperature, pH, concentration)
- Accessibility of binding sites

Biosorbents

Biosorbents are biological elements such as bacteria, agricultural wastes, and municipal byproducts employed in the passive removal of toxins from a system. They have been encouraged since they are made from renewable resources, they are cheap and biologically degradable, and they do not produce toxic compounds after being used. Since they are biological material, they include several functional groups that act as the driving force for hydrophobic interactions, which are mostly pH-dependent during the sorption process. Biosorbent contains a range of functional sites, including carboxyl, imidazole, sulfuryl, amino, phosphate, sulfate, thioether, phenol, carbonyl, amide, and hydroxylate, in contrast with mono-functional ion-exchange resins. A large number of natural materials have been suggested as promising biological agents for removing pollutants from water. The sources can be inactive or dead microbial biomass, as well as living microorganisms. It is well understood how some biosorbents function, whereas others need more research. The toxicity of some of these biomaterials also requires in-depth analysis (Adewuyi, 2020).

Biosorbents, unlike mono-functional ion-exchange resins, possess functional groups that can bind and sequester metal ions. The functional groups may be amide, carbonyl, imidazole, sulfhydryl, thioether, amine, sulfonate, carboxyl, imine, and phosphodiester. The ability of a biosorbent to bind metal ions is characterized as its *biosorption capability* and defined as the number of metal ions biosorbed per unit weight of the biosorbent. This can be stated by using the mass balance equation as shown here:

$$q_e = \frac{(C_i - C_e)V}{m} \tag{2.1}$$

The metal's biosorption effectiveness (R %) is calculated using the following equation

$$R\% = \frac{C_i - C_e}{C_i} X100 \tag{2.2}$$

where q_e = total metal ions adsorbed by the adsorbent (mg/g); C_i = original metal ion concentration in the solution (mg/L); C_e = metal ion concentration in the solution (mg/L); V = medium volume (L); and m = quantity of biomass used during the process of adsorption (g) (Kanamarlapudi et al., 2018).

Types of Biosorbents

A wide range of materials available in nature can be used to remove metals from contaminated water resources as biosorbents. Biosorbents include microbial biomass (live and dead), plant and animal-derived materials, industrial and agriculture byproducts, biopolymers, and so on. Biosorbents are a less expensive and more efficient way to remove metallic elements from aqueous solutions, especially heavy metals. Choosing the most effective biosorbent type from a wide range of promising and cheap biomaterials is a significant challenge. Biosorbents with the ability to bind metal ions with larger affinities should be developed. The ideal characteristics of biosorbent are (Macek and Mackova, 2011):

- Biosorption ability
- Economical
- Ease of availability
- Effortless removal of adsorbed metal ions and promising reuse of the biosorbents.

In the case of biosorbents, proteins, several polyphenols (e.g. chitosan and pectic acid), and polyphenols (e.g. tannins, catechins) are chemical ingredients that bind metal ions efficiently. By applying some easy chemical treatments to biosorbents with poor adsorption properties, the biosorbents can be transformed into adsorbents with outstanding adsorption qualities (Inoue et al., 2017).

Microbial Biomass as Biosorbents

Microorganisms that can tolerate harsh conditions have been used as biosorbents to remove metal ions from wastewater. Most of the microbial groups consist of functional groups that exhibit their ability as biosorbents. A significant number of materials with microbial originshave been studied extensively as biosorbents to remove metal ions. Based on their good performance, low cost, and large available quantities, microorganisms (bacteria, algae, yeasts, and fungi) are receiving increased attention for heavy metal retrieval. Experiments concentrating on the usage of dead and living microorganisms have been carried out to provide alternativetypes of remediation (Hlihor et al., 2014).

When the biosorption process first began to use microorganisms, researchers revealed that inactive/dead microbial biomass could passively bind metal ions. Dead biomass offers various advantages over live microbial biomass: cost-effectiveness, toxicity limitation, ease of regeneration, displaying an exchange of ions, and a wide range of pH and temperatures (Adewuyi, 2020). Employing dead microbial biomass for metal ion binding is preferred over using alive biomass due to the lack of nutrient requirements and the management of biochemical oxygen demand (BOD) and COD in effluents. Therefore, the use of dead biomass is cost-effective (Rezaei, 2016).

It became obvious that microbial biomass biosorption is based not only on the microbial biomass's chemical composition but also on external physicochemical factors and the chemical matrix. These biosorbents are capable of effectively retrieving metal ions from solution and efficiently decreasing parts per million (ppm) to ppb levels. As a result, they are thought to be good candidates for treating high-volume, low-metal-ion-concentration complicated wastewaters (Chen and Wang, 2006).

Bacterial Biomass

Of the microorganisms, bacteria are the world's most abundant, versatile, and diverse creatures. Bacteria can be classified as gram-negative or gram-positive based on the cell wall and gram staining. In biosorption, the cell surface structure plays an important role.

The first aspect of biosorption is the bacterial cell wall that encounters the metal ion. The bacterial cell wall consists mainly of peptidoglycan, which defines its form and rigidity. Gram-positive bacteria consist of a thick peptidoglycan layer (90%) connected by amino acids. Teichoic acids bind to cytoplasmic membrane lipids in the cell wall, generating lipoteichoic acids that are strongly linked to the membrane. The phosphodiester linkages between the monomers of teichoic acid have an overall negative effect and are thus engaged in the biosorption of divalent cations (metal ions).Gram-negative bacteria have thinner walls that contain less peptidoglycan (20%). However, there is an additional external layer made up of phospholipids and lipopolysaccharides that has a negative charge, making metal binding easier (Shamim, 2018). For their survival, the majority of bacteria have a variety of resistance mechanisms as well as effective strategies for extracting metal ions. On the cell wall surface, certain bacteria produce slime or a capsule-like coating. This consists mainly of polysaccharides that lead to the detoxification of metal ions from wastewaters (Moat et al., 2002).

Bacteria are excellent biosorbents for sequestering metal ions because of their high surface-to-volume ratios and significant numbers of possible active sorption sites (Kanamarlapudi et al., 2018). The bacterial strains *Aeromonas sp*, *Pseudomonas sp*, *Acinetobacter sp*, *Enterobacter sp*, *Klebsiella sp*, and *Streptomonas sp*are commonly utilized for wastewater cleanup. The biosorption of toxic metals by several bacterial species is summarized in Table 2.1.

The functional groups of the cell wall, such as hydroxyl, amine, carboxyl, sulfate, and phosphate, play an essential part in the biosorption process. The metal ions become attached to the functional groups on the cell wall. The general process of metal absorption includes the attachment of metal ions to reactive groups located on bacterial cell walls and the internalization of metal ions within cells (Abbas et al., 2014). The presence of glycoproteins allows more metal to be taken by gram-positive bacteria. Less metal absorption is observed in gram-negative bacteria due to phospholipids and lipopolysaccharides (Das et al., 2008). The action of Van der Waals forces can lead to weak physical sorption, while relatively strong physical sorption may be due to electrostatic attractions between metal ions and functional groups in the surface (Aryal, 2020). The bacteria's biosorption mechanisms and characterization have been investigated by various methods such as X-ray diffraction, electron scanners, transmission electron microscopy, Fourier infrared spectroscopy transformation, potentiometric titrations, and energy-dispersive X-ray microanalysis. These methods have revealed the existence, type, and number of binding sites on the bacterial surface (Kazy et al., 2006).

Studies have identified numerous means of improving the bacterial capacity to function as biosorbents, including chemical and genetic modification. Some concentrated on improving the active sites to enhance biosorption, while the inhibition sites on the surface of the bacteria were less often taken into consideration (Adewuyi, 2020). For large-scale domestic and industrial wastewater treatment, several membrane bioreactors have been developed. The application of the membrane reactor has shown many advantages such as operational stability, prolonged preservation of the microorganism, maximum removal of suspended solids, low sludge production rate, ability to treat toxic organic and inorganic contaminants, compact plant size, and high degradation rate (Rana et al., 2017).

Algae as Biosorbents

Algae are aquatic plants that have no real roots or stems and that range in size from microalgae to macroalgae. Algae are considered autotrophic, meaning they can produce enormous amounts of biomass even when little nutrition is available. They are regarded as good biosorbents because they have low nutrient requirements, are readily available, have significant sorption capacity, have a high surface-to-volume ratio, produce less sludge, do not produce harmful compounds, and have the potential for regeneration and recovery of metals (Trinelli et al., 2013). Microalgae (mostly green), macroalgae (brown algae), and red algae are the three types. Of the three varieties, brown algae have the greatest potential to absorb metals (Shamim, 2018). Biosorption of heavy metals by different algal species are summarized in Table 2.2.

Table 2.1 Biosorption of heavy metals by different bacteria.

Bacterial biomass(biosorbent)	Metal ions(biosorbate)	Functional groups	Reference
B.subtilis	Chromium(III)	Carboxylic group	(Aravindhan et al., 2012)
E. coli	Chromium (VI)	Amine, phosphodiester groups	(Quiton et al., 2018)
B. thuringiensisOSM29	Nickel	Amine, carbonyl, carboxylic, and hydroxylgroups	(Oves et al., 2013)
Arthrobacter sp.	Copper	Amine, carboxylic, and phosphate groups	(Aryal et al., 2012)
C. mutjisii	Zinc	Amine, hydroxyl, ketone, and phosphategroups	(Saranya et al., 2018)
Rhodococcus sp.	Arsenic	Amine, carboxylic, and hydroxyl groups	(Prasad et al., 2011)
Pseudomonas sp.	Cadmium	Amine, carboxylic, hydroxyl, and phosphategroups	(Huang and Liu, 2013)
B. pumilus sp.	Lead	Carboxylic and hydroxyl groups	(Sayyadi et al., 2017)
Cupriavidusmetallidurans	Silver	Amino, Carboxylic, and hydroxyl groups	(Shamim and Rehman, 2014)
Cupriavidusmetallidurans	Selenium	Amino, Carboxylic, and hydroxyl groups	(Shamim et al., 2014)
Enterobacter cloacae	Mercury	Amino, hydroxyl groups	(Rani et al., 2010)
Cupriavidusmetallidurans	Gold	Amino, Carboxylic, and hydroxyl groups	(Monsieurs et al., 2015)
Thiobacillusthiooxidans	Copper	Amine, hydroxyl, ketone, and phosphategroups	(Nagashetti et al., 2013)
Rhodopseudomonaspalustris	Cobalt	Amine, carboxylic, and phosphate groups	(Gao et al., 2017)
Bacillus safensis	Aluminum	Carboxylic and hydroxyl groups	(Dhanarani et al., 2016)

Table 2.2 Biosorption of heavy metals by different algae.

Algal biomass(biosorbent)	Metal ions(biosorbate)	Functional groups	Reference
Palmaria palmate	Chromium	Amino, carboxyl, sulfur groups	(Murphy et al., 2008)
Ulva lactuca	Cadmium	Amino, hydroxyl, carboxyl groups	(Lupea et al., 2012)
Spirulina platensis	Copper	Amino, carboxylic groups	(Çelekli et al., 2010)
Oedogonium hatei	Nickel	Amino, amide, hydroxyl and carboxyl groups	(Gupta et al., 2010)
Maugeotia genuflexa	Arsenic	Carboxyl, hydroxyl, amide groups	(Sarı et al., 2011)
Cladophorasp	Lead	Carboxyl, amino, amide, hydroxyl groups	(Lee and Chang, 2011)
Laminaria japonica	Zinc	Carboxyl, hydroxyl groups	(Davis et al., 2003)
Spirogyra hyalina	Cobalt	Amino, hydroxyl groups	(Viraraghavan and Srinivasan, 2011)
Sargassum sp.	Mercury	Amino, sulfur groups	(Subhashini et al., 2011)

The cell wall components differ between algal species. Brown algae have three primary components in their cell walls: cellulose that provides structural support, alginic acid (mixture of polymers with their respective salts), and sulfated polysaccharide with large carboxylic groups implicated in metal biosorption. Red algae have garnered attention for biosorption due to the presence of sulfated polysaccharides formed from galactans (high concentrations of hydroxyl and carboxyl groups). Green algae have polysaccharide-bound cellulose with a large amount of protein that includes amino, sulfate, hydroxyl, and carboxyl functional groups (Flouty and Estephane, 2012).

The algal cell wall consists of polysaccharides that include chemical groups (sulfate, amino, hydroxyl, phosphate, imidazole, mannan) known to function as binding sites for metal (Romera et al., 2007). Electrostatic interaction attracts cations due to the negative charge of the cell wall. In addition to the charge, which plays an essential role in biosorption, hydrophobicity, species and structure, ionic strength of the metal, and chemical makeup of the ionic solution of metals are also important factors to consider when choosing algae for biosorption. (Adewuyi, 2020).

The mechanism of metal absorption by algae is similar to bacteria: metal ions are bound to the outer membrane, followed by internalization. Algal biosorption involves one of two mechanisms:

- Ion exchange process: ions such as calcium, magnesium, sodium, and potassium are replaced by metal ions on the algal surface.
- Functional groups and metal ions combine to form complexes.

Ionic strength and covalent interactions are hypothesized in the case of the metal uptake process. Carboxyl and sulfate groups form ionic bonds, but amino and carboxyl groups require covalent interaction between the metallic ligand and the chemical group. Phytochelatins are formed within the algal body in response to metal ions (Abbas et al., 2014).

Multiple researchers worked on exploiting algal biomass to remove heavy metals from polluted water sources. Algal biosorption capacity is 15–84% more than other microorganisms, according to research (Mustapha and Halimoon, 2015). As a result, algal biomass is seen as a cost-effective and ecologically beneficial wastewater treatment option.

Fungi as Biosorbents

Fungi are eukaryotic species that include yeasts, mushrooms, molds, and so on. They are used as biosorbents due to their distinguishing features: i.e. easy to grow, greater biomass yield, and ease of alteration either genetically or chemically (Mulligan et al., 2001). Both dead and living forms of fungi can be used as biosorbent material (Wang and Chen, 2006). The cell wall of fungal organisms possesses outstanding binding characteristics because of the presence of chitin, mannans, and glucans in addition to lipids, polyphosphates, and proteins (Javaid et al., 2011). The fungal cell wall is rich in polysaccharides (90%) and glycoproteins that contain different metal-binding groups, such as amines, phosphates, carboxylate, and hydroxyls (Remacle, 1990). Active and passive metal absorption by fungi have been reported:

- Active or intracellular absorption or bioaccumulation depends on the metabolism of the cell.
- Passive absorption, or biosorption, involves metallic ions binding to the exterior of the cell membrane and is unrelated to cell metabolism.

Uptake of metals in active mode occurs only with living cells. In this circumstance, metal ions may bind with cell surface functional groups via ion exchange complex formation or simple physical binding. The metal absorption process, which is independent of energy, may be influenced by temperature, metabolic inhibitors, etc. (Shamim, 2018). Physical and chemical treatments such as thermal treatment, dimethyl sulfoxide, detergents, orthophosphoric acid, glutaraldehyde, formaldehyde, and alkali can alter the biosorption potential of fungal populations (Das et al., 2008). At the industrial level, fungi can be easily generated to adsorb metal ions from huge amounts of polluted supplies. Moreover, biomass can be created using inexpensive growth media or byproducts from a variety of fermentation processes. Furthermore, fungi are somewhat sensitive to nutritional variations as well as other process factors, including temperature, pH, and aeration. They are easily separated by simple techniques such as filtration due to their filamentous existence. As a result, they are also regarded as cost-effective and environmentally beneficial biosorbents (Leitão, 2009). Table 2.3 lists some fungal species that have been utilized as biosorbents.

Yeasts as Biosorbents

Yeasts are single-celled organisms in which the bulk of the biomass either biosorbs a wide range of metals or is selective for a single metal ion. Yeasts also possess a high potential to accumulate and, as a result, can be utilized as biosorbents to absorb heavy metals.

Table 2.3 Biosorption of heavy metals by different fungi.

Fungal biomass (biosorbent)	Metal ions (biosorbate)	Functional groups	References
Pleurotus ostreatus	Chromium	Carboxyl, amine groups	(Arbanah et al., 2013)
Hydrilla verticillata	Cadmium	Carboxyl, hydroxyl, amine groups	(Acosta Rodríguez et al., 2013)
Aspergillus terreus	Copper	Carboxyl groups	(Gulati et al., 2002)
Trametes versicolor	Nickel	Carboxyl, hydroxyl, amine groups	(Subbaiah and Yun, 2013)
Penicillium chrysogenum	Arsenic	Carboxyl, amino groups	(Mamisahebei et al., 2007)
Phanerochaete chrysosporium	Lead	Hydroxyl, amino groups	(Haluk Ceribasi and Yetis, 2004)
Pencillium simpliccium	Zinc	Carboxyl, amino groups	(Fan et al., 2008)
Aspergillus fumigatus	Mercury	Amino, hydroxyl groups	(Mamisahebei et al., 2007)

Saccharomyces cerevisiae is a model organism for biosorption research. They remain non-pathogenic and easy to cultivate, and they produce a large amount of biomass using a basic growing media (Gaensly et al., 2014). Biosorption properties of yeast were studied using various forms, including free vs. immobilized cells, dead vs. live cell, engineered vs. non-engineered cell, etc. (Park et al., 2003). The yeast cells in free form are not suitable for biosorption as there is a problem in solid-fluid phase separation. In flocculating cells, this problem tends less effective (Veglio and Beolchini, 1997). Yeast cells must be pretreated to increase the surface-to-volume ratio of metal binding sites (Mapolelo, 2004).

Different parameters of metal ions such as valency, diameter, cell age, culture conditions, concentration of metals, temperature, and pH influence the process of biosorption (Wang and Chen, 2006). Biosorption involves amino, carboxyl, and hydroxyl functional groups. The isothermal model of Langmuir was better adapted to a biosorption ion exchange mechanism (Luo et al., 2010). Yeasts' huge size, on the other hand, makes them potential candidates for metal bioremediation. Table 2.4 lists the many yeast strains utilized in biosorption.

Biosorbents Derived from Plant and Animal Waste

Solid wastes derived from flora and fauna are plentiful, low-cost, renewable resources. They're made in vast quantities every year, and disposing of them is usually a problem. An important area of research is to find meaningful uses for these materials. They can be used to minimize waste and create cost-effective products (Kulkarni, 2014).

Plants disposed of as agricultural waste and food industry waste can be used as biosorbents. This is a method of repurposing and recycling discarded materials, so using plant materials has no significant cost (Ali Redha, 2020). Plant-derived wastes are predominantly made up of cellulose, with structural components such as lignin, proteins, hemicellulose,

Table 2.4 Biosorption of heavy metals by different yeasts.

Yeast biomass(biosorbent)	Metal ions(biosorbate)	References
Candida utilis	Chromium	(Anaemene, 2012)
Saccharomyces cerevisiae	Cadmium	(Das et al., 2008)
Saccharomyces cerevisiae	Cobalt	(Arakaki et al., 2011)
Candida pelliculosa	Copper	(Apinthanapong and Phensaijai, 2009)
Mucor rouxii	Lead	(Muraleedharan et al., 1991)
Saccharomyces cerevisiae	Mercury	(Anaemene, 2012)
Saccharomyces cerevisiae	Nickel	(Siñeriz et al., 2009)
Thiobacillusthiooxidans	Zinc	(Nagashetti et al., 2013)

Table 2.5 Biosorption of heavy metals by different plant materials.

Plant waste	Metal	Adsorption capacity	Reference
Wheat bran	Mercury	82%	(Farajzadeh and Monji, 2004)
Black gram husk	Lead	93%	(Saeed et al., 2005)
Rice bran	Cadmium	80%	(Montanher et al., 2005)
Baggase	Zinc	90–95%	(Mohan and Singh, 2002)
Activated carbon of peanut shells	Nickel	75%	(Wilson et al., 2006)
Barley straw	Copper	80%	(Pehlivan et al., 2012)
Coconut shell fibers	Chromium	80%	(Mohan et al., 2006)

carbohydrates, lipids, and starch (Rajapaksha et al., 2015). Plant biosorbents can absorb water because of the presence of carboxylic and phenolic functional groups in the cellulosic matrix and components linked with cellulose, such as hemicellulose and lignin (Abdi and Kazemi, 2015). Based on cation exchange between binding sites and metal ions, the metal ions bind with functional groups, resulting in biosorption and, therefore, the removal of the metal ions from media (Abdi and Kazemi, 2015). Table 2.5 lists the various plants employed as biosorbents.

Animal waste products are dumped as solid waste in the environment without being processed or composted or simply washed into water canals, posing a health risk to humans and other living species. The chance to use animal waste for a useful purpose is lost if they are discarded without preparation. Some research has looked into using these animal waste products as a viable adsorbent to adsorb heavy metals from effluent because they are inexpensive and conveniently accessible. Heavy metal adsorption has been researched using animal manure as an adsorbent. Animal bones, pretreated fish bones, crab shells, pretreated arca shells, pretreated crab and arca shells, eggshells, Muscadomestica, and so on are a few other examples (Srivastava et al., 2016). Table 2.6 summarizes the results of the adsorption analysis performed on animal byproducts.

Table 2.6 Biosorption of heavy metals by different animal wastes.

Animal waste	Metal	Adsorption capacity	Reference
Pretreated fish bones	Copper	150.7 mg/g.	(Kizilkaya et al., 2010)
Dried animal bones	Zinc	0.1764 mmol/g.	(Banat et al., 2002)
Crab shell	Cobalt	322.6 mg/g	(Vijayaraghavan et al., 2006)
Pretreated arca shell biomass	Lead	18.33 mg/g	(Dahiya et al., 2008)
Animal bone	Nickel	7.22 mg/g	(Al-Asheh et al., 1999)

These adsorbents could offer significant advantages compared to currently available economically priced activated carbons and help with waste reduction. Furthermore, research is required to adapt the simulation methodology to larger manufacturing facilities rather than small experimental applications.

Biocomposites

Biocomposites consist of composite materials up of multiple ingredients that are mixed to make a new product that outperforms the individual constituent materials. They constitute biomass-based products that are biodegradable, high-performing, and environmentally friendly and can be utilized for wastewater treatment. Biopolymers such as cellulose, chitosan, starch, chitin, alginate, and others continue to be the most important part of biocomposites. Biopolymers' advantages include their non-toxicity, availability, economics, and environmentally friendly nature (Zhang et al., 2013).

Alteration of Biosorbents

The amount and accessibility of binding sites on the surface of an adsorbent determine the biosorption technique. Usage of biosorbents in their natural state has shown a number of drawbacks due to their poor biosorption potential and unpredictable physical stability. Modifying the surface features of biosorbents can have a huge impact on the biosorbents' ability to remove metal particles (Gupta et al., 2002). Several researchers concentrated on altering the biomass chemically such that structural stability and effective heavy metal ion biosorption capability can be achieved.

Hydrophobicity, water absorbency, thermal resistance, cation/anion exchange capacity, and ability to resist microbial attack have been improved significantly by modifying the surface of biomass. Physical pretreatment methods such as heating, chilling, drying, lyophilization, and autoclaving are used to alter the cell surface. Several researchers have reported using chemical pretreatments for surface alteration, including washing with mineral and organic acid solutions, detergents, alkaline solutions, and organic compounds (Vieira and Volesky, 2000). These pretreatments assisted in modifying surface binding sites by eliminating or hiding the groups or exposing further sites to bind metals. Biosorptive performance has been boosted by carboxylation, phosphorylation, and amination of amine, carboxyl, and hydroxyl groups, saponification of ester groups, sulfonation, halogenation, oxidation, and so on (Wan Ngah and Hanafiah, 2008). Various biomasses, irrespective of their source, including pine bark, rose petals, spirogyra, walnut shell, rubber leaves, and sawdust, showed promising biosorption capabilities after alkaline pretreatment (Argun et al., 2009).

Desorption and Regeneration

The ability of biosorbents to recover after use is one of their most important achievements. Adsorbate is cleared off the biosorbent surface after use, and the biosorbent reverts to its original structure and efficiency (Adewuyi, 2020).Biosorbents' economic value and sustainability are primarily determined by the number of cycles they can be reused. It is essential to develop an effective desorption method. High performance is not enough for a biosorbent; it also needs to be reusable. As a result, when choosing biosorbents, desorption and regeneration are the important processes to consider. Still, some of them are difficult to regenerate, making their long-term usage doubtful since they would need to be discarded after a few cycles. Disposing of such materials can result in contamination of the environment. The separation of spent biosorbents and the regeneration and recycling of the medium after the sorption process are very significant.

Eluent utilization is now the most frequently used desorption mechanism. Choosing the right eluent is essential and depends on the form of biosorbent, adsorbent, and biosorption mechanism. A proper eluent should not harm or modify the biosorbent structure, it should be environmentally friendly and affordable, it should have a high level of adsorbing ability, it should not alter adsorbing or biosorbing substances, and it must be easily disconnected. To dislodge metal ions while concurrently replenishing the filled biosorbent, chemicals like hydrochloric acid and sodium hydroxide and chelators like EDTA have been utilized (Ahemad and Kibret, 2013).

Cost Evaluation

Approximating the cost of biosorption and biosorbents is a challenging process that is rarely published. Cost assessment depends on many factors that make generalization difficult. Variables such as process handling, storage, energy use, repair, optimization of process, rejuvenation, discharge, and desorption are considered when evaluating cost. The type of water or effluent to be treated, as well as the amount, will affect the cost. Capital costs and operating costs, on the other hand, are determined by the type and size of the treatment plant.

It is preferred to employ wastes as a substrate for processing biosorbents in order to lower process expenses. Agricultural waste, home trash, and industrial waste, such as bacterial drainage from fermentation, fungal waste from food manufacturing, and waste from other industrial processes, can all be used to make biosorbents. Waste management is a significant environmental issue, and using these wastes addresses the concern while also lowering the cost of manufacturing biosorbents for treating water.

Pretreatment of feedstock is one aspect that increases production costs. Certain products need to be pretreated before manufacturing, and additional pretreatment increases the cost of production. It is important to note that feedstock treatment should be efficient but also inexpensive and accessible. It is critical that the treatment plant be placed near the effluent source in order to reduce expenses.

The cost of disposal must also be taken into account in biosorption, as once the biomass is fully used in repeated cycles it must be substituted. However, it may be applied in other areas, such as particulate board processing, cement, biogas, etc.

Conclusion

Biosorption is a low-cost treatment for complicated commercial wastewater with high volumes and low heavy metal concentrations. Metal ions in an aqueous solution are replaced for a counter ion bound to biomass in this process, classified as ion exchange. The efficacy of the various biosorbents is unquestionably essential to the biosorption technology's promising potential. Several types of biomass that are abundant and exhibit strong metal binding characteristics have been recognized. Many natural biomass biosorbents have been established from cellulosic and microbial origins, with effective biosorption features. But several of these biosorbents have shown poor efficiency. Surface changes made to biosorbents have helped enhance their metal-binding characteristics, as is apparent from extensive research; however, the adjustments increase the average operational costs to match the cost of commercialized ion-exchange resins. The existence of multifunctional groups on the biomass surface causes it to be non-selective to a particular metal ion. The biosorbents' non-selective and unspecific character, as well as their lesser resilience, pose a significant marketing challenge. The creation of biosorption models and the identification of biosorption mechanisms, as well as the identification of better and more selective biosorbents, are all key future research topics for biosorption technology.

References

Abbas, S.H., Ismail, I.M., Mostafa, T.M., and Sulaymon, A.H. (2014). Biosorption of heavy metals: A review. *Journal of Chemical Science and Technology* 3 (4): 74–102.

Abdi, O. and Kazemi, M. (2015). A review study of biosorption of heavy metals and comparison between different biosorbents. *Journal of Materials and Environmental Science* 6 (5): 1386–1399.

Acosta Rodríguez, I., Martínez-Juárez, V.M., Cárdenas-González, J.F. et al. (2013). Biosorption of Arsenic(III) from Aqueous Solutions by Modified Fungal Biomass of *Paecilomyces* sp. *Bioinorganic Chemistry and Applications* 2013: 1–5. doi:10.1155/2013/376780.

Adewuyi, A. (2020). Chemically modified biosorbents and their role in the removal of emerging pharmaceutical waste in the water system. *Water* 12 (6): 1551. doi:10.3390/w12061551.

Ahemad, M. and Kibret, M. (2013). Recent trends in microbial biosorption of heavy metals: A review. *Biochemistry and Molecular Biology* 1 (1): 19–26.

Al-Asheh, S., Banat, F., and Mohai, F. (1999). Sorption of copper and nickel by spent animal bones. *Chemosphere* 39: 2087–2096.

Ali Redha, A. (2020). Removal of heavy metals from aqueous media by biosorption. *Arab Journal of Basic and Applied Sciences* 27 (1): 183–193. doi:10.1080/25765299.2020.1756177.

Anaemene, I.A. (2012). The use of Candida sp. in the biosorption of heavy metals from industrial effluent. *European Journal of Experimental Biology* 2 (3): 484–488.

Apinthanapong, M. and Phensaijai, M. (2009). Biosorption of copper by spent yeast immobilized in sodium alginate beads. *Kasetsart Journal (Natural Science)* 43: 326–332.

Arakaki, A.H., Vandenberghe, S., de Soccol, L.P. et al. (2011). Optimization of biomass production with copper bioaccumulation by yeasts in submerged fermentation. *Brazilian Archives of Biology and Technology* 54 (5): 1027–1034. doi:10.1590/S1516-89132011000500021.

Aravindhan, R., Fathima, A., Selvamurugan, M. et al. (2012). Adsorption, desorption, and kinetic study on Cr(III) removal from aqueous solution using Bacillus subtilis biomass. *Clean Technologies and Environmental Policy* 14 (4): 727–735. doi:10.1007/s10098-011-0440-7.

Arbanah, M., Najwa, M., and Ku Halim, K. (2013). Utilization of *Pleurotusostreatus* in the removal of Cr (VI) from chemical laboratory waste. *International Refereed Journal of Engineering Science* 2: 29–39.

Argun, M.E., Dursun, S., and Karatas, M. (2009). Removal of Cd (II), Pb (II), Cu (II) and Ni (II) from water using modified pine bark. *Desalination* 249: 519–527.

Arshad, N. and Imran, S. (2020). Indigenous waste plant materials: An easy and cost-effective approach for the removal of heavy metals from water. *Current Research in Green and Sustainable Chemistry* 3: 100040. doi:10.1016/j.crgsc.2020.100040.

Aryal, M. (2020). A comprehensive study on the bacterial biosorption of heavy metals: materials, performances, mechanisms, and mathematical modellings. *Reviews in Chemical Engineering* 20190016. doi:10.1515/revce-2019-0016.

Aryal, M., Ziagova, M.G., and Liakopoulou-Kyriakides, M. (2012). Cu(II). Biosorption and competitive studies in multi-ions aqueous systems by Arthrobacter sp. Sphe3 and Bacillus sphaericus cells: equilibrium and thermodynamic studies. *Water, Air, and Soil Pollution* 223 (8): 5119–5130. doi:10.1007/s11270-012-1263-9.

Babel, S. (2003). Low-cost adsorbents for heavy metals uptake from contaminated water: a review. *Journal of Hazardous Materials* 97 (1–3): 219–243. doi:10.1016/S0304-3894(02)00263-7.

Banat, F., Al-Asheh, S., and Mohai, F. (2002). Multi-metal sorption by spent animal bones. *Separation Science and Technology* 37: 311–327.

Baroni, L., Cenci, L., Tettamanti, M. et al. (2007). Evaluating the environmental impact of various dietary patterns combined with different food production systems. *European Journal of Clinical Nutrition* 61 (2): 279–286. doi:10.1038/sj.ejcn.1602522.

Çelekli, A., Yavuzatmaca, M., and Bozkurt, H. (2010). An eco-friendly process: Predictive modelling of copper adsorption from aqueous solution on Spirulina platensis. *Journal of Hazardous Materials* 173 (1–3): 123–129. doi:10.1016/j.jhazmat.2009.08.057.

Chen, C. and Wang, J.-L. (2006). [Cation (K^+, Mg^{2+}, Na^+, Ca^{2+}) release in Zn(II) biosorption by Saccharomyces cerevisiae]. *Huan Jing Ke Xue= Huanjing Kexue* 27 (11): 2261–2267.

Chojnacka, K. (2010). Biosorption and bioaccumulation – the prospects for practical applications. *Environment International* 36 (3): 299–307. doi:10.1016/j.envint.2009.12.001.

Dahiya, S., Tripathi, R.M., and Hegde, A.G. (2008). Biosorption of lead and copper from aqueous solutions by pretreated crab and arca shell biomass. *Bioresource Technology* 99: 179–187.

Das, N., Vimala, R., and Karthika,: (2008). Biosorption of heavy metals – An overview. *Indian Journal of Biotechnology* 7: 159–169.

Davis, T.A., Volesky, B., and Mucci, A. (2003). A review of the biochemistry of heavy metal biosorption by brown algae. *Water Research* 37 (18): 4311–4330. doi:10.1016/S0043-1354(03)00293-8.

Dhanarani, S., Viswanathan, E., Piruthiviraj,: et al. (2016). Comparative study on the biosorption of aluminum by free and immobilized cells of Bacillus safensis KTSMBNL 26 isolated from explosive contaminated soil. *Journal of the Taiwan Institute of Chemical Engineers* 69: 61–67. doi:10.1016/j.jtice.2016.09.032.

Fan, T., Liu, Y., Feng, B. et al. (2008). Biosorption of cadmium(II), zinc(II) and lead(II) by Penicillium simplicissimum: Isotherms, kinetics and thermodynamics. *Journal of Hazardous Materials* 160 (2–3): 655–661. doi:10.1016/j.jhazmat.2008.03.038.

Farajzadeh, M.A. and Monji, A.B. (2004). Adsorption characteristics of wheat bran towards heavy metal cations. *Sep Purif Technol* 38: 197–207.

Flouty, R. and Estephane, G. (2012). Bioaccumulation and biosorption of copper and lead by a unicellular algae Chlamydomonas reinhardtii in single and binary metal systems: A comparative study. *Journal of Environmental Management* 111: 106–114. doi:10.1016/j.jenvman.2012.06.042.

Fu, F. and Wang, Q. (2011). Removal of heavy metal ions from wastewaters: A review. *Journal of Environmental Management* 92 (3): 407–418. doi:10.1016/j.jenvman.2010.11.011.

Gaensly, F., Picheth, G., Brand, D. et al. (2014). The uptake of different iron salts by the yeast Saccharomyces cerevisiae. *Brazilian Journal of Microbiology* 45 (2): 491–494. doi:10.1590/S1517-83822014000200016.

Gao, R., Wang, Y., Zhang, Y. et al. (2017). Cobalt(II) bioaccumulation and distribution in *Rhodopseudomonas palustris*. *Biotechnology and Biotechnological Equipment* 31 (3): 527–534. doi:10.1080/13102818.2017.1292148.

Gulati, R., Saxena, R.K., and Gupta, R. (2002). Fermentation waste of Aspergillus terreus: a potential copper biosorbent. *World Journal of Microbiology and Biotechnology* 18 (5): 397–401. doi:10.1023/A:1015540921432.

Gupta, R., Saxena, R., Mohapatra, H., and Ahuja, P. (2002). Microbial variables for bioremediation of heavy metals from industrial effluents. In: *Progress in Industrial Microbiology* (ed. V.P. Singh and R.D. Stapleton), vol. 36: 189–229. Elsevier.

Gupta, V.K., Rastogi, A., and Nayak, A. (2010). Biosorption of nickel onto treated alga (Oedogonium hatei): Application of isotherm and kinetic models. *Journal of Colloid and Interface Science* 342 (2): 533–539. doi:10.1016/j.jcis.2009.10.074.

Haluk Ceribasi, I. and Yetis, U. (2004). Biosorption of Ni(ii) and Pb(ii) by *Phanerochaete chrysosporium* from a binary metal system – kinetics. *Water SA* 27 (1): 15–20. doi:10.4314/wsa.v27i1.5004.

Hlihor, R.M., Bulgariu, L., Sobariu, D.L. et al. (2014). Recent advances in biosorption of heavy metals: Support tools for biosorption equilibrium, kinetics and mechanism. *Revue Roumaine de Chimie* 59: 527–538.

Huang, W. and Liu, Z.-M. (2013). Biosorption of Cd (II)/Pb(II) from aqueous solution by biosurfactant-producing bacteria: isotherm kinetic characteristic and mechanism studies. *Colloids Surf B* 105: 113–119.

Ibrahim, W.M. and Mutawie, H.H. (2013). Bioremoval of heavy metals from industrial effluent by fixed-bed column of red macroalgae. *Toxicology and Industrial Health* 29 (1): 38–42. doi:10.1177/0748233712445044.

Inoue, K., Parajuli, D., Ghimire, K. et al. (2017). Biosorbents for removing hazardous metals and metalloids. *Materials* 10 (8): 857. doi:10.3390/ma10080857.

Jaishankar, M., Tseten, T., Anbalagan, N. et al. (2014). Toxicity, mechanism and health effects of some heavy metals. *Interdisciplinary Toxicology* 7 (2): 60–72. doi:10.2478/intox-2014-0009.

Javaid, A., Bajwa, R., Shafique, U. et al. (2011). Removal of heavy metals by adsorption on Pleurotus ostreatus. *Biomass and Bioenergy* 35 (5): 1675–1682. doi:10.1016/j.biombioe.2010.12.035.

Kanamarlapudi, S.L.R.K., Chintalpudi, V.K., and Muddada, S. (2018). Application of biosorption for removal of heavy metals from wastewater. In: *Biosorption* (ed. J. Derco and B. Vrana). InTech. doi: 10.5772/intechopen.77315.

Kazy, S.K., Das, S.K., and Sar, P. (2006). Lanthanum biosorption by a Pseudomonas sp.: equilibrium studies and chemical characterization. *Journal of Industrial Microbiology and Biotechnology* 33 (9): 773–783. doi:10.1007/s10295-006-0108-1.

Kizilkaya, B., Tekinay, A.A., and Dilgin, Y. (2010). Adsorption and removal of Cu (II) ions from aqueous solution using pretreated fish bones. *Desalination* 264: 37–47.

Kulkarni, S.J. (2014). Use of biotechnology for synthesis of various products from different feedstocks – a review. *Int. J. Adv. Res. Bio-Technol.* 2: 1–3.

Lee, Y.-C. and Chang, S.-P. (2011). The biosorption of heavy metals from aqueous solution by Spirogyra and Cladophora filamentous macroalgae. *Bioresource Technology* 102 (9): 5297–5304. doi:10.1016/j.biortech.2010.12.103.

Leitão, A.L. (2009). Potential of Penicillium Species in the Bioremediation Field. *International Journal of Environmental Research and Public Health* 6 (4): 1393–1417. doi:10.3390/ijerph6041393.

Luo, J., Xiao, X., and Luo, S.-L. (2010). Biosorption of cadmium(II) from aqueous solutions by industrial fungus Rhizopus cohnii. *Transactions of Nonferrous Metals Society of China* 20 (6): 1104–1111. doi:10.1016/S1003-6326(09)60264-8.

Lupea, M., Bulgariu, L., and Macoveanu, M. (2012). Biosorption of Cd(II) from aqueous solution on marine green algae biomass. *Environmental Engineering and Management Journal* 11 (3): 607–615. doi:10.30638/eemj.2012.076.

Macek, T. and Mackova, M. (2011). Potential of biosorption technology. In: Microbial Biosorption of Metals (ed.: Kotrba, M. Mackova, and T. Macek). Springer. doi:10.1007/978-94-007-0443-5_2.

Mamisahebei, S., Khaniki, G.R.J., Torabian, A. et al. (2007). Removal of arsenic from an aqueous solution by pretreated waste tea fungal biomass. *Journal of Environmental Health Science and Engineering* 4 (2): 85–92.

Mapolelo, M. (2004). Trace enrichment of metal ions in aquatic environments by Saccharomyces cerevisiae. *Talanta* 64 (1): 39–47. doi:10.1016/j.talanta.2003.10.058.

Moat, A.G., Foster, J.W., and Spector, M.P. (2002). *Microbial Physiology*. New York: Wiley-Liss.

Mohan, D. and Singh, K.P. (2002). Single- and multi-component adsorption of cadmium and zinc using activated carbon derived from bagasse-an agricultural waste. *Water Res* 36: 2304–2318.

Mohan, D., Singh, K.P., and Singh, V.K. (2006). Chromium (III) removal from wastewater using low cost activated carbon derived from agriculture waste material and activated carbon fabric filter. *J Hazard Mater* 135: 280–295.

Monsieurs P., Hobman J., Vandenbussche G. et al. (2015). Response of *Cupriavidus metallidurans* CH34 to Metals. In: *Metal Response in Cupriavidus metallidurans* (ed. M. Mergeay and R. Van Houdt), 45–89. Springer Briefs in Molecular Science. Springer. doi:10.1007/978-3-319-20594-6_3.

Montanher, S.F., Oliveira, E.A., and Rollemberg, M.C. (2005). Removal of metal ions from aqueous solutions by sorption onto rice bran. *J Hazard Mater* 117: 207–211.

Mrvčić, J., Stanzer, D., Šolić, E. et al. (2012). Interaction of lactic acid bacteria with metal ions: opportunities for improving food safety and quality. *World Journal of Microbiology and Biotechnology* 28 (9): 2771–2782. doi:10.1007/s11274-012-1094-2.

Mulligan, C., Yong, R., and Gibbs, B. (2001). Remediation alternative treatment option for heavy metal bearing wastewaters: A review. *Bioresource Technology* 53: 195–206.

Muraleedharan, T.R., Iyengar, L., and Venkobachar, C. (1991). Biosorption: An attractive alternative for metal removal and recovery. *Current Science* 61 (6): 379–385.

Murphy, V., Hughes, H., and McLoughlin, P. (2008). Comparative study of chromium biosorption by red, green and brown seaweed biomass. *Chemosphere* 70 (6): 1128–1134. doi:10.1016/j.chemosphere.2007.08.015.

Mustapha, M.U. and Halimoon, N. (2015). Microorganisms and biosorption of heavy metals in the environment: a review paper. *Journal of Microbial and Biochemical Technology* 07 (05). doi:10.4172/1948-5948.1000219.

Nagashetti, V., Mahadevaraju, G.K., Muralidhar, T.S. et al. (2013). Biosorption of heavy metals from soil by Pseudomonas aeruginosa. *International Journal of Innovative Technology and Exploring Engineering (IJITEE)* 2 (6): 9–17.

Oves, M., Khan, M.S., and Zaidi, A. (2013). Biosorption of heavy metals by Bacillus thuringiensis strain OSM29 originating from industrial effluent contaminated north Indian soil. *Saudi Journal of Biological Sciences* 20 (2): 121–129. doi:10.1016/j.sjbs.2012.11.006.

Oyedepo, T.A. (2011). Biosorption of lead (II) and copper (II) metal ions on *Calotropisprocera* (Ait.). *Science Journal of Pure and Applied Chemistry* 1: 1–7.

Park, D., Yun, Y.-S., and Park, J.M. (2010). The past, present, and future trends of biosorption. *Biotechnology and Bioprocess Engineering* 15 (1): 86–102. doi:10.1007/s12257-009-0199-4.

Park, J.K., Lee, J.W., and Jung, J.Y. (2003). Cadmium uptake capacity of two strains of Saccharomyces cerevisiae cells. *Enzyme and Microbial Technology* 33 (4): 371–378. doi:10.1016/S0141-0229(03)00133-9.

Pehlivan, E., Altun, T., and Parlayici, S. (2012). Modified barley straw as a potential biosorbent for removal of copper ions from aqueous solution. *Food Chem* 135: 2229–2234.

Prasad, K.S., Srivastava, P., Subramanian, V., and Paul, J. (2011). Biosorption of As(III) ion on Rhodococcus sp. WB-12: biomass characterization and kinetic studies. *Separ Sci Technol* 46: 2517–2525.

Quiton, K.G., Doma, B., Futalan, C.M. et al. (2018). Removal of chromium(VI) and zinc(II) from aqueous solution using kaolin-supported bacterial biofilms of Gram-negative E. coli and Gram-positive Staphylococcus epidermidis. *Sustainable Environment Research* 28 (5): 206–213. doi:10.1016/j.serj.2018.04.002.

Rajapaksha, A.U., Vithanage, M., Ahmad, M. et al. (2015). Enhanced sulfamethazine removal by steam-activated invasive plant-derived biochar. *Journal of Hazardous Materials* 290: 43–50. doi:10.1016/j.jhazmat.2015.02.046.

Rana, R.S., Singh, P., Kandari, V. et al. (2017). A review on characterization and bioremediation of pharmaceutical industries' wastewater: an Indian perspective. *Applied Water Science* 7 (1): 1–12. doi:10.1007/s13201-014-0225-3.

Rani, M.J., Hemambika, B., Hemapriya, J., and Kannan, V.R. (2010). Comparative assessment of heavy metal removal by immobilized and dead bacterial cells: A biosorption approach. *African Journal of Environmental Science and Technology* 4 (2): 077–083.

Remacle, J. (1990). The cell wall and metal binding. In: *Biosorption of Heavy Metals* (ed. B. Volesky), 83–92. Boca Raton, Florida: CRC Press.

Rezaei, H. (2016). Biosorption of chromium by using Spirulina sp. *Arabian Journal of Chemistry* 9 (6): 846–853. doi:10.1016/j.arabjc.2013.11.008.

Romera, E., González, F., Ballester, A. et al. (2007). Comparative study of biosorption of heavy metals using different types of algae. *Bioresource Technology* 98 (17): 3344–3353. doi:10.1016/j.biortech.2006.09.026.

Saeed, A., Iqbal, M., and Akhtar, M.W. (2005). Removal and recovery of lead (II) from single and multiple (Cd, Ni, Cu, Zn) solutions by crop milling waste (black gram husk). *J Hazard Mater* 117: 65–73.

Saranya, K., Sundaramanickam, A., Shekhar, S. et al. (2018). Biosorption of multi-heavy metals by coral associated phosphate solubilising bacteria Cronobacter muytjensii KSCAS2. *Journal of Environmental Management* 222: 396–401. doi:10.1016/j.jenvman.2018.05.083.

Sarı, A., Uluozlü, Ö.D., and Tüzen, M. (2011). Equilibrium, thermodynamic and kinetic investigations on biosorption of arsenic from aqueous solution by algae (Maugeotia genuflexa) biomass. *Chemical Engineering Journal* 167 (1): 155–161. doi:10.1016/j.cej.2010.12.014.

Sayyadi, S., Ahmady-Asbchin, S., Kamali, K. et al. (2017). Thermodynamic, equilibrium and kinetic studies on biosorption of Pb +2 from aqueous solution by Bacillus pumilus sp. AS1 isolated from soil at abandoned lead mine. *Journal of the Taiwan Institute of Chemical Engineers* 80: 701–708. doi:10.1016/j.jtice.2017.09.005.

Shamim, S. (2018). Biosorption of heavy metals. In: *Biosorption* (ed. J. Derco and B. Vrana), 21–49. InTech.

Shamim, S. and Rehman, A. (2014). Physicochemical surface properties of *Cupriavidus metallidurans* CH34 and *Pseudomonas putida* mt2 under cadmium stress: Physicochemical surface properties under Cd stress. *Journal of Basic Microbiology* 54 (4): 306–314. doi:10.1002/jobm.201200434.

Shamim, S., Rehman, A., and Qazi, M.H. (2014). Cadmium-Resistance Mechanism in the Bacteria Cupriavidus metallidurans CH34 and Pseudomonas putida mt2. *Archives of Environmental Contamination and Toxicology* 67 (2): 149–157. doi:10.1007/s00244-014-0009-7.

Siñeriz, M.L., Kothe, E., and Abate, C.M. (2009). Cadmium biosorption by Streptomyces sp. F4 isolated from former uranium mine. *Journal of Basic Microbiology* 49 (S1): S55–S62. doi:10.1002/jobm.200700376.

Srivastava, S., Agrawal, S.B., and Mondal, M.K. (2016). Animal wastes: an alternative adsorbent for removal of toxic heavy metals from industrial wastewater. *Journal of Scientific Research* 60: 65–72.

Subbaiah, M.V. and Yun, Y.S. (2013). Biosorption of Nickel(II) from aqueous solution by the fungal mat of Trametes versicolor (rainbow) biomass: equilibrium, kinetics, and thermodynamic studies. *Biotechnology and Bioprocess Engineering* 18 (2): 280–288. doi:10.1007/s12257-012-0401-y.

Subhashini, S., Kaliappan, S., and Velan, M. (2011). Removal of heavy metal from aqueous solution using *Schizosaccharomycespombe* in free and alginate immobilized cells. In: *2nd International Conference on Environmental Science and Technology IPCBEE*, vol. 6: V2107–V2111. IACSIT Press.

Trinelli, M.A., Areco, M.M., and dos Santos Afonso, M. (2013). Co-biosorption of copper and glyphosate by Ulva lactuca. *Colloids and Surfaces B: Biointerfaces* 105: 251–258. doi:10.1016/j.colsurfb.2012.12.047.

Veglio, F. and Beolchini, F. (1997). Removal of metals by biosorption: A review. *Hydrometallurgy* 44: 301–316.

Vieira, R.H. and Volesky, B. (2000). Biosorption: A solution to pollution? *International Microbiology* 3: 17–24.

Vijayaraghavan, K., Palanivelu, K., and Velan, M. (2006). Biosorption of copper (II) and cobalt (II) from aqueous solutions by crab shell particles. *Bioresource Technology* 97: 1411–1419.

Viraraghavan, T. and Srinivasan, A. (2011). Fungal biosorption and biosorbents. In: *Microbial Biosorption of Metals* (ed. Kotrba, M. Mackova, and T. Macek), 143–158. Dordrecht: Springer Netherlands. doi:10.1007/978-94-007-0443-5_6.

Wan Ngah, W.S. and Hanafiah M.A.K.M. (2008). Removal of heavy metal ions from wastewater by chemically modified plant wastes as adsorbents: a review. *Bioresource Technol* 99: 3935–3948.

Wang, J. and Chen, C. (2006). Biosorption of heavy metals by Saccharomyces cerevisiae: A review. *Biotechnology Advances* 24 (5): 427–451. doi:10.1016/j.biotechadv.2006.03.001.

Wilson, K., Yang, H., Seo, C.W., and Marshall, W.E. (2006). Select metal adsorption by activated carbon made from peanut shells. *Biores Technol* 97: 2266–2270.

Zhang, L., Xia, W., Teng, B. et al. (2013). Zirconium cross-linked chitosan composite: Preparation, characterization and application in adsorption of Cr (VI). *Chem. Eng. J.* 229: 1–8.

3

Biosorption for Eliminating Inorganic Contaminants (IOCs) from Wastewater

Rahul Sharma[1], *Pinki Rani Agrawal*[2,3], *Ravi Kumar*[2,3], *Ittishree*[4], *and Gaurav Gupta*[5]

[1] *Department of Chemistry, Sabarmati University, Ahmedabad, Gujarat, India*
[2] *CSIR National Physical Laboratory, New Delhi, India*
[3] *Academy of Scientific and Innovative Research (AcSIR), Ghaziabad, Uttar Pradesh, India*
[4] *G.V.M College of Pharmacy, Pt. B.D. Sharma University of Health Sciences, Sonipat, Haryana, India*
[5] *Department of Physics and Astrophysics, University of Delhi, New Delhi, India*

Introduction: Water Pollution by Inorganic Contaminants (IOCs)

Although the Industrial Revolution played a vital role in the world's economic development, it also led to a wide spectrum of environmental problems (Rhind, 2009; Yap et al., 2020; Bragina et al., 2012). Consequently, environmental conservation is a significant concern, and numerous initiatives have been undertaken to handle air, water, soil, noise, and electromagnetic pollution (Bragina et al., 2012; Shepherd et al., 2018, Ciu et al., 2020; Yun et al., 2020; Sordello et al., 2019). In particular, the scarcity of drinking water and the fast depletion of freshwater resources due to anthropogenic and geogenic activities demand urgent attention.

The primary source of water pollution is the discharge of unchecked industrial, domestic, and agricultural wastes into freshwater bodies and groundwater reservoirs. However, water pollution from industrial wastes and wastewater streams is the most severe. Importantly, the lack of adequate treatment technologies is mainly responsible for discharging insufficiently treated industrial wastewater streams into bodies of water. It has exacerbated water scarcity, leading to critical situations for water quality and quantity in many parts of the globe (Boretti and Rosa, 2019; Kummu and Guillaume, 2016; Maddan and Smith, 2020).

Today, both developed and developing economies are facing the catastrophic effects of water pollution. Heavy metal ions like lead, arsenic, mercury; anionic species like fluoride, cyanides, chlorides, sulfates, mineral acids, and nitrates; synthetic organic dyes such as methylene blue and methyl orange; and other organic compounds like crude oil and organic solvents have been recognized as a few of the major categories of water pollutants.

Biosorption for Wastewater Contaminants, First Edition.
Edited by Rangabhashiyam Selvasembian and Pardeep Singh.

Broadly, they can be classified into two main categories: organic contaminants (OCs), such as organic dyes, pharmaceutical solvents, crude oil, and related components; and inorganic contaminants (IOCs), such as heavy metal salts, mineral acids, cyanides, etc. (Zolkefli et al., 2020; Sharma and Bhattacharya, 2017; Wen et al., 2017; Amoatey and Baawain, 2019). Their presence in bodies of water imposes serious threats to the sustainability of the ecosystem due to adverse effects on human health, aquatic life, and plants. Water pollution is considered an offshoot of scientific and technological growth, the Industrial Revolution, and the unavailability of efficient waste-management strategies. The discharge of domestic and industrial effluents/wastewater into bodies of water not only makes water sources unfit for drinking but also adversely affects their utility for pharmaceutical, agricultural, chemical, and industrial applications (Bukola et al., 2015; Feng et al., 2020; Ali et al., 2019; Tabrez et al., 2021).

Water pollution due to IOCs originating from either anthropogenic or geogenic activities is a major global issue, especially in developing and populous countries. Apart from their adverse effects on human beings, because they are non-biodegradable and have complex chemical structures and natures, IOCs inhibit self-purification of water systems, deplete dissolved oxygen, and destroy aquatic life. Moreover, due to their bioaccumulation and biomagnification, IOCs tend to remain in the environment and are known to adversely affect plant life, soil fertility, and animal health via food chains and direct consumption (Tabrez et al., 2021; Ahmed et al., 2019; Xie et al., 2016; Chibuike and Obiora, 2014). Unfortunately, in the past few decades, it has been recognized that most IOCs exist naturally and are dispersed in water sources through volcanic eruptions, sea-salt spray, forest fires, rock weathering, biogenic sources, and wind-borne soil particles (Mullenix, 2014; Shankar and Shankar, 2014; Greenstone and Hanna, 2014; Adejumoke, 2018). Natural weathering processes generally lead to the release of IOCs, particularly heavy metal ions and their sulfate, fluoride, and cyanide salts from their endemic spheres. However, humans have caused IOCs to increase multifold in water systems through various unchecked practices, particularly electronic waste, mining, processing, manufacturing, and metallurgical operations. This has led to stringent environmental legislation and regulations for controlling and reducing IOCs in bodies of water (Greenstone and Hanna, 2014; Adejumoke, 2018; Moore et al., 2018).

In addition to legislation, global demand for sustainable wastewater treatment technology to eliminate IOCs is always evident. Of several wastewater treatment technologies, the adsorption technique is considered better due to its lower cost, simple operation, and versatile nature and the availability of different kinds of adsorbents ranging from activated carbon powders to new-generation nano-adsorbents such as metal oxide nanoparticles, graphene, carbon nanotubes (CNTs) and their composites, etc. (Mustapha et al., 2020; Ali et al., 2019; Shen et al., 2015; Anjum et al., 2016). Recently, new-generation nano-adsorbents have shown tremendous potential to separate IOCs from wastewater due to their large surface area, low density, and tunable physicochemical properties (Hu et al., 2016). However, the requirement of special synthesis methods, low yield, high cost, and poor environmental benignity are few challenging issues associated with such systems and restrict their commercial viability.

In this context, biosorption has emerged as one of the most promising alternatives to conventional methods due to its cost-effectiveness, scalability, environmental friendliness, contaminant removal efficiency, tunable surface functionalities, recyclability, and excellent post-life utility of the active biosorbents (Kanamarlapudi et al., 2018; Saini et al., 2020). Like the adsorption mechanism, biosorption involves binding adsorbate species (contaminants present in water) on the surface of an adsorbent of biological origin through physicochemical interactions such as H-bonding, Van der Waals interaction, chemical binding such as chelation, complexation, and ion-exchange between the adsorbate and adsorbent.

Biosorbents are primarily composed of bio-wastes isolated or extracted from agricultural and industrial by-products, plant-derived substances, microorganisms, synthetic or natural bio-polymers, and many similar materials (Reinhardt, 1960-61; Mo et al., 2018; Kikuchi and Tanaka, 2012; Ramrakhiani et al., 2017). Amine (-NH_2), amide (-$CONH_2$), imidazole, thioether (-S-), sulfonate (-SO_3H), carbonyl (-CO-), sulfhydryl, carboxyl (-COOH), phosphodiester, phenolic (C_6H_5OH), imine (=NH), and phosphate groups (PO_4^{3-}) are major surface functionalities of the biosorbents that attract and sequester IOCs (Adewuyi, 2020; Mallampati and Valiyaveettil, 2014). The concentration; size; nature and magnitude of charge; hydration extent; molecular mass; chemical, stereochemical, and coordination behavior of the IOCs; nature and number of sorption sites on the biosorbent; sorption process pH; temperature, etc. are the key factors that affect the efficiency of biosorption processes.

Based on these parameters, a number of biosorbents have been developed over the years to eliminate IOCs from wastewater streams using cellulose, lignin, and starch-based agricultural wastes; industrial wastes of tea, sugarcane, and food industries; and plant-derived materials like neem bark, papaya wood, date pits, guava pilings, wheat shells, green coconut powder, etc. (Alalwan et al., 2020; Şen et al., 2015; Ali Redha, 2020). Generally, batch adsorption studies are employed for assessing the sorption efficiency of biosorbents and understanding the biosorption mechanism. Freundlich and Langmuir adsorption isotherm models and pseudo first-order and pseudo second-order kinetic models are among those universally adopted for in-depth exploration of biosorption mechanisms.

To a great extent, engineered biosorbents have revealed promising potential in the sequestration of IOCs from wastewater at laboratory scale. However, a continuous need for more efficient, technologically feasible, and commercially viable biosorbent systems exists for the sustainability of fixed sorbent bed biosorption systems (biofilters) that can be used for continuous treatment of wastewater streams containing large amounts of IOCs. Although the challenges are numerous in developing highly efficient biofilters, a few have been proposed recently. These biofilters can fix inorganic/organic contaminants from a wastewater stream (Farnane et al., 2018; Chaudhary et al., 2003; Reungoat et al., 2011).

This chapter presents a comprehensive review of this field. A number of case studies are discussed from an intensive assessment of literature reports showing the adverse impacts on local habitats of IOCs in water systems. The first section of the chapter highlights water pollution by IOCs and related issues. The chapter then discusses sources of IOCs in water systems and their permissible limits in drinking and surface water. The next section throws light on environmental and health hazards of IOC pollution in water, followed by a description of recent treatment technologies for eliminating IOCs from wastewater streams. The potential of biosorption methods and recent advances in their designing and development are addressed. Finally, the chapter concludes with comments on the current situation, significant challenges, and future scope.

Permissible Limits and Sources of IOCs in Water Systems

Standard Permissible Limits of Some IOCs in Water

IOCs generally end up in water systems from both anthropogenic and natural processes, and they cause devastating and catastrophic effects on all spheres of the environment. Owing to their grave impacts, various international and national agencies and foundations including the World Health Organization (WHO), Bureau of Indian Standards (BIS), United States Environmental Protection Agency (EPA), European Commission on the Environment (ECE), Australian Drinking Water Guidelines (ADWG), and Norma Official Mexicana (NOM)-127-SSA1-1994) have imposed stringent concentration limits for IOCs in water (Puri and Kumar, 2012; WHO, 1993). The primary motive behind this legislation is to prohibit, reduce, and regulate harmful practices that are responsible for severe water pollution. However, in the era of globalization, environmental legislation and regulations are continually evolving. The standard permissible concentration limits of IOCs in water systems as per BIS guidelines are given in Table 3.1.

Table 3.1 The acceptable limits of some IOCs in drinking water as per BIS guidelines.

	Permissible limits (mg/L)				
Heavy metals	**Drinking water**	**Surface water**	**Public sewers**	**Irrigation water**	**Coastal water**
Arsenic	0.05	0.2	0.2	0.2	0.2
Mercury	0.001	0.01	0.01	0.01	0.01
Cadmium	0.01	2	1	—	2
Chromium	0.05	0.1	2	—	1
Selenium	0.01	0.05	0.05	—	0.05
Lead	0.1	0.1	1	—	1
Nickel	0.1	3	3	—	5
Zinc	15	5	—	15	15
Sulfate	250–600	—	—	—	—
Fluoride	1.5	2	—	—	15
Nitrate	10	10	—	—	20
Cyanide	0.5	0.2	2	2	2

Sources of IOCs in Water Systems

Natural Sources

The major sources of inorganic pollutant contamination in water systems can be classified into two broad categories: natural and anthropogenic. Heavy metals are considered the most toxic contaminants among IOCs and mainly enter bodies of water through natural activities like volcanic eruptions, sea-salt spray, forest fires, rock weathering, biogenic sources, and wind-borne soil particles. Moreover, weathering phenomena often lead to the emission of heavy metal ions from their endemic spheres. In addition, heavy metals exist as hydroxides, oxides, sulfides, sulfates, phosphates, silicates, and organic compounds in natural environments and can end up in water via these deposits (Briffa et al., 2020; Masindi and Muedi, 2018; Albert et al., 1999).

Fluoride contamination, specifically in groundwater, is another issue of significant concern. It is known to cause chronic dental and skeletal fluorosis. Fluoride enters groundwater via weathering of igneous and sedimentary rocks containing fluoride-based minerals like apatite, fluorite, biotite, and hornblende. Rainfall-mediated infiltration through these rocks is also responsible for the enhancement of fluoride concentration in groundwater. Moreover, volcanic ash enriched with huge amounts of highly water-soluble fluoride constitutes another natural source of fluoride contamination (Raj and Shaji, 2017).

Cyanides are highly powerful and rapid-acting poisonous IOCs present in wastewater streams. They are produced naturally by many species of bacteria, fungi, and algae. Cyanides are found in many natural foods and plants, such as almonds, millet sprouts, lima beans, soy, spinach, bamboo shoots, and cassava roots. However, the contribution of such systems in cyanide contamination in water sources is negligible.

Nitrates and sulfates compose a large portion of IOCs in water sources, and their contamination of water is among the most prevalent environmental problems. Due to their adverse effects, these species are included in environmental legislation and regulations. As no specific sulfate contamination hazards are recognized, the maximum permissible limit of sulfate in drinking water is a bit high: about 250 mg/L. The main sources of nitrate and sulfate contamination in water systems include natural fertilization, bacterial production, and atmospheric deposition (Jaszczak et al., 2017; Lovasoa et al., 2017; Jezierski et al., 2006; Villareal and Carpenter, 2003; Xue et al., 2016).

Anthropogenic Sources

From the perspective of the catastrophic consequences of IOC pollution in water systems, anthropogenic sources (human-made practices) are solely responsible. These practices include mining activities, manufacturing and finishing industries, coal combustion, fertilizers, pharmaceutical waste, and recycling units. To a great extent, unchecked discharge of industrial wastewater streams is responsible for contamination by IOCs in surface and groundwater. These streams contain huge amounts of heavy metal ions (arsenic, mercury, chromium, etc.), fluoride, cyanides, nitrates, and sulfates. For instance, arsenic (As) is a well-known poison that is deadly for aquatic environments and human health, and it is is dispersed in both ground and surface water from coal-fired plants, smelting, alkali processing, and other industrial actions. The effluents from manufacturing industries like batteries (NiCd and lead-acid), pigments, plastic, and electroplating are increasing the cadmium, mercury, and lead contamination in water. These effluents also disturb the pH conditions of aquatic systems due to the large amount of acidic impurities they contain (Masindi and Muedi 2018; Reza and Singh, 2010; Tchounwou et al., 2012).

Furthermore, agricultural practices like phosphate fertilizers, pesticides, and herbicides are primary reasons for IOC pollution in both surface and groundwater. In many reports, it has been shown that the fluoride concentration in irrigation water can reach up to 0.34 mg/l, and successive irrigation can further enhance the fluoride level in groundwater. Datta et al. (2014) have revealed that in many parts of Bengal, soil contains up to 10 mg/kg of water-soluble fluoride where phosphate-based fertilizers are used, and hence these fertilizers are considered potential threats as far as fluoride contamination is concerned. Industrial activities like aluminum smelting, ceramic firing, and cement production are also among the anthropogenic sources which are continuously causing fluoride contamination in water.

Wastewater streams from metal mining, steel plants, and organic chemical industries release extremely toxic cyanides in water sources in large amounts, along with vehicle exhaust, burning of municipal wastes, and use of cyanide-containing pesticides in agricultural practices (Liu et al., 2020; Adimalla et al., 2019). Thiocyanates are present in water from coal processing, gold and silver extraction, and mining industries. In agricultural lands, thiocyanates include pesticides, insecticides, rodenticides, and weed killers, from which they find their way to groundwater and surface water (Jaszczak et al., 2017; Spurr et al., 2019). The discharge of untreated domestic sewage is another primary anthropogenic source from which IOCs enter natural bodies of water.

In a nutshell, the cocktail of industrial, agricultural, and domestic sewage creates a complex chemical composition of IOCs in water systems where they adversely affect aquatic habitats, land fertility, groundwater composition, plant life, and human health through food chains (Edokpayi et al., 2017; Dwivedi, 2017). Water pollution from IOCs via anthropogenic pathways has reached a concerning level and is on the verge of becoming a major global threat. The ill effects of these IOCs on the environment and animal and plant health are discussed in the next section, along with some case studies.

IOCs in Water: Environmental and Health Hazards

Assessment studies on water pollution show that IOCs are the most toxic and hazardous water pollutants and not only affect the local aquatic systems but also have a very strong tendency to persist due to non-biodegradability, bioaccumulation, and biomagnification (Camargo, 2003; Maurya et al., 2019). The presence of IOCs, even in very low concentrations, is also known to affect environmental sustainability. Global communities have made many attempts to restrict and regulate contamination of IOCs through environmental legislation after realizing their catastrophic and devastating effects. However, most of the developing countries have not been able to do so. The rapid growth of industrialization in these countries is making the situation even worse. This leads to chronic exposure of biospheres to IOCs, as such pollutants have the potential to survive in various weather and climate conditions and to be transported long distances (Singh et al., 2011; Thompson and Darwish, 2019; Lu et al., 2015).

Specifically, heavy metals are strongly bioaccumulated in the biota, and consequently, all living organisms encounter heavy metal ions. They destroy the population of aquatic species, reduce soil fertility, and damage plant life (Kumari et al., 2017; Rahman et al., 2012; Arantes et al., 2016). For instance, the presence of arsenic in shallow aquifers is extremely hazardous as these are the primary source of drinking water for large populations, particularly rural populations in many parts of the world. The direct consumption of such water for drinking and irrigation processes gives arsenic multiple routes to enter the human body. After being ingested, arsenic flows with the bloodstream, conjugates with hemoglobin, and finally reaches the kidneys, liver, lungs, and other parts of the body, where it accumulates. This leads to different types of cancers, including skin, bladder, lung, kidney, liver, and prostate, among others (Hong et al., 2014).

Surprisingly, there is no specific concentration limit of lead in the human body, although this metal is fatal for humans. In the human body, lead is mainly absorbed by the gastrointestinal and respiratory systems and ultimately binds with red blood cells. In the liver, lead is highly resistant to metabolic degradation or detoxification, and it substitutes for calcium in bones. Minute exposure to lead is also known to cause cardiovascular diseases like stroke and coronary heart diseases, cancer, lack of intelligence in children, and hypertension (Assi et al., 2016).

Owing to the inertness of elemental mercury in the human body, it penetrates the brain and kidney tissues. Further, mercuric ions are generally indistinguishable from other inorganic ions, which leads to their higher degree of accumulation in kidneys. Thereafter, mercury acts as a potent neurotoxin, and exposure causes fetus damage, nephrotoxicity, lung damage, and even sudden death (Jyothi and Farook, 2020).

Since the daily excretion of cadmium cannot exceed 0.01% of its intake, it is extremely toxic even when consumed in very low concentrations. It adversely affects the kidneys, bones, heart, blood vessels, lungs, reproductive system, intestinal functioning, and immune system and is also known to cause cancer. Women have a higher body burden of Cd than men due to greater intestinal absorption of Cd. Itai-itai disease, mainly found in women, is caused by exposure to cadmium. According to a report by Roy and McDonald, carrots grown in cadmium-contaminated soil are found to cause toxicological problems in men, women, and young children (Roy and McDonald, 2015). Cd uptake by carrot roots was about 5 times more than the regulatory limits for men, 8 times the limit for women, and 12 times the limit for children.

IOCs reach agricultural lands through irrigation processes and severely affect the physicochemical properties of soil. This makes soil unfit for crop production. Further, IOCs find their way to groundwater through the contaminated soil. Additionally, IOCs develop multiple exposure routes to surrounding animals, plants, and humans from contaminated soils. The most common routes are the well-established food chains.

Mineral acids also impose a serious threat to aquatic life. They are produced via geogenic processes: for instance, minerals containing sulfur (pyrites of iron and copper) interact with air and water to form sulfuric acid, which enters water resources. Charged impurities like chlorides contaminate groundwater and surface water, particularly in countries where sodium and calcium salts of chlorides are used to melt snow on roadsides. Fluoride acts as a nutrient for human health in very low concentrations; however, exposure to high concentrations is responsible for skeletal and dental fluorosis and causes teeth mottling and deformation of bones in humans (Peckham and Awofeso, 2014; Rostern, 2017; Mohamed et al., 2020).

Regular exposure to nitrate-contaminated water may cause specific cancers and birth defects as it initiates the formation of highly toxic nitroso compounds. Intake of nitrates from drinking water also causes adverse impacts on reproductive outcomes and other health effects (Ward et al., 2018; Jain et al., 2020).

Overall, water pollution by IOCs is a serious threat to environmental sustainability and the health of living species, including humans. As all domains of the environment are bonded with each other in a synergic fashion, damage to one sphere can critically affect the sustainability of others – and unfortunately, this damage is irreversible. Therefore, water contamination due to inorganic pollutants needs to be addressed immediately. In recent decades, a number of water pollution management strategies and remediation techniques have been globally developed and adopted for the sequestration of inorganic water pollutants. Some of the recent strategies and treatment technologies for eliminating IOCs from wastewater streams are illustrated in the subsequent section.

Elimination of IOCs from Wastewater: Recent Strategies and Remediation Techniques

Today, the shortage of fresh water has become a global reality due to the rapid contamination of water resources. In some cases, this has become a national security issue. Many diseases and other ill effects are emerging each day due to the consumption of contaminated water. Apart from imposing hazardous impacts on local aquatic environments, these water contaminants move rapidly through food chains via biomagnification and bioaccumulation. Therefore, immediate action plans must be implemented for water remediation. In addition to awareness programs and legislation, advancement in treatment technologies is also necessary for the sequestration of water contaminants, particularly IOCs. Specifically, prior to the discharge of industrial effluents to natural water streams, extensive treatment of wastewater streams is highly desirable. If these waste streams are not adequately treated, they increase the pollutant load in natural water resources. This further aggravates the situation, as it is very difficult to perform remediation actions on large bodies of water.

By extensively probing the nature of IOCs, several treatment technologies have been developed to eliminate them from wastewater. However, no single treatment method or technique can resolve the issue; therefore, integrated systems are used that combine two or more treatment techniques. In a very general approach, three major steps are followed for water decontamination. The primary step includes the removal of organic matter and suspended solids from wastewater. For this physical separation, methods such as screening separation, gravity separation, filtration, centrifugation, separation, and sedimentation are usually employed. The second step involves eliminating insoluble inorganic and organic contaminants. The third step involves removing soluble contaminants from wastewater and is the most crucial step in terms of both treatment cost and efficiency (Bolisetty et al., 2019; Barakat, 2011; Selvi et al., 2019).

The various treatment techniques and methods included in the third step of wastewater treatment are oxidation/precipitation, ion exchange, coagulation, electrolysis, membrane filtration, and sorption methods. The most common treatment techniques used for eliminating IOCs are discussed below (Gunatilake, 2015; Ince and Ince, 2019; Sharma and Bhattacharya, 2017).

Oxidation/Precipitation

In oxidation methods, powerful oxidizing agents like hydroxyl and sulfate radicals are used to eliminate IOCs. This method is very common in treating industrial wastewater streams (Bolisetty et al., 2019; Sharma and Bhattacharya, 2017; Chen et al., 2013). The oxidizing agents cause the oxidation of contaminants and transform them into their precipitated forms, which can be easily removed via filtration (Bora and Dutta, 2019; Dutta et al., 2004; Miller et al., 2011). The use of chlorine in the domestic water supply is a common example of such a method. In a very detailed report, Deng and Zhao elaborated the role of various oxidizing agents. Further, the report throws light on recent advancements in the field (Deng and Zhao, 2015).

Ion Exchange

In 1850, ion exchange was developed by H.S. Thompson. This method has a high capability of producing potable water by capturing ionic impurities such as metal ions, nitrates, chlorides, fluorides, etc. In a typical chemical process, the ionic impurities of water are bound with ion-exchange resin, and similar charge species are released into the aqueous medium. Water hardness is generally removed by this method, in which divalent calcium and magnesium ions are sequestrated from water. Many synthetic ion-exchange resins are now commercially available; however, their performance heavily depends on the nature of the ionic impurity (Kurniawan et al., 2006; Hamdan et al., 2014).

Electrokinetics (EK)

EK processes are the best fit for IOC decontamination, especially from soil and water. EK primarily involves three steps: electro-osmosis, electromigration, and electrophoresis. In this method, a low-voltage direct current is used as a cleaning agent. In addition to the electro-osmosis, electromigration, and electrophoresis transport mechanisms, phenomena such as electrolysis and electrodeposition occur. This method has the potential to remove other contaminants including polychlorinated biphenyls, pesticides, and heavy metals, as revealed in many research reports (Selvi et al., 2019; Hamdan et al., 2014).

Membrane Filtration / Reverse Osmosis

This is the most widely employed method for obtaining drinking water. With the increase in membrane casting technologies, this method has gained significant attention recently. It can remove any kind of water contaminant, including fungus, bacteria, and even viruses. It works based on the size exclusion principle. Impurities larger than the pores of the membrane are retained on the membrane.

Different kinds of filtration such as ultrafiltration, nano-filtration, and reverse-osmosis methods have been developed. In these methods, a semi-permeable membrane is generally used that allows water molecules through and discards impurities depending on their size. For instance, permeable membranes with a pore size of 5–20 nm are used in ultrafiltration. It only allows separation of molecules with 1000–100,000 Da molecular weights (Bolisetty et al., 2019; Selvi et al., 2019; Azimi et al., 2017; Shahalam et al., 2002).

To further enhance the efficiency of the process, researchers have developed nano-filtration and reverse-osmosis methods. Among several wastewater treatment techniques and membrane filtration techniques, reverse osmosis has been recognized as the most effective for water purification (Singh et al., 2018).

Sorption Methods

Owing to their simplicity, versatile nature, cost-effectiveness, and high efficiency, sorption methods are universally considered best for treating wastewater at bulk scale, particularly IOCs (Lasat, 2002). In particular, carbon nanomaterials – graphene, carbon nanotubes (CNTs), and their hybrids and composites – are the most promising sorbents of modern times due to their large specific surface area, tunable surface properties, low density, and hierarchical porous 3D architectures. However, due to high-cost synthesis processes, their realization in commercial technologies has not yet been achieved. Activated carbon-based sorbents are still used most frequently.

Interestingly, biosorbents have emerged as potential alternatives to conventional sorbents because they can be prepared from inexpensive industrial, agricultural, and domestic wastes. Further, they have better environmental and biological compatibility than traditional adsorbents (Cherian and Oliveira, 2005; Behera, 2014; Dickinson et al., 2009; Kanamarlapudi et al., 2018). Moreover, they strengthen waste-to-wealth approaches and strategies. In the past decade, many biosorbents have been developed that have better sorption capacity than other classes of adsorbent materials. The role of biosorption methods in the sequestration of IOCs is discussed in the subsequent section.

Biosorption Methods for Eliminating IOCs from Wastewater

In the past few decades, particularly due to the growing interest of the global community in renewable resources, biosorption-mediated water decontamination has gained massive attention from the scientific and industrial communities. In this context, various types of biosorbents based on bacteria, fungi, algae, plants, and domestic, agricultural, and industrial wastes have been identified, modified, and utilized to capture IOCs from wastewater streams.

Wastewater remediation using bacteria-based biosorbents is a relatively new but fast-growing field. Interestingly, numerous functional groups like amine, carboxyl, hydroxyl, and sulfhydryl groups and phosphate on the surface of bacterial cell walls have been recognized for playing a crucial role in the biosorptive elimination of IOCs. The important chemical processes generally involved in the biosorption of IOCs by bacterial biomass include physical interactions, coordination complexations, chelation, ion exchange, H-bonding, and precipitation or combination of such processes. The physicochemical characteristics of the cell wall material of such bacterial species are the key parameters that govern the efficiency of biosorptive elimination of contaminants.

Fungal-based biosorbents in this context are highly advantageous due to their large cell-wall content, which contains more and diverse functional groups and thus provides tunable surface functionalities. This makes fungal biosorbents extremely efficient in the sequestration of IOCs. Proteins, carbohydrates, chitin, chitosan, polyphosphates, and lipids are the major constituents of fungal cell walls. However, the composition of such materials varies from one type of fungal taxonomic group to another.

Another important class of biosorbents is constituted by algae, which are known for their oxygenic photosynthesis. The term *algae* refers to a broad spectrum of chlorophyll-containing eukaryotic organisms. These organisms are vast in number and have diverse natures. Different types of carbohydrates are present in the cell walls of algae, particularly polysaccharides such as cellulose, chitin, alginate, glycan and glycoproteins, etc. These materials enrich the cell walls with oxygen-rich functionalities such as carbonyl groups, hydroxyl groups, and phenolic groups. These functionalities interact with IOCs and bind them to the surface. Consequently, algae-based biosorbents are highly promising in IOC decontamination (Adewuyi, 2020; Anastopoulos et al., 2019; Kanamarlapudi et al., 2018).

Currently, the sustainability of natural resources like water, air, and soil has become a significant challenge. To tackle and avoid the worst situations, a waste-to-wealth approach is imperative and in high demand, as such strategies and protocols to restrict the extent of catastrophic forms of pollution and decrease the load of natural resources that are continuously undergoing unchecked exploitation to meet global demand. Plant waste materials are an excellent source of biosorbent materials. They are present everywhere in abundance, and they offer a unique opportunity to transform waste into value-added products. As they are enriched with cellulose and lignin components, they provide a strong platform for developing engineered biosorbents.

Moreover, agricultural wastes are also recognized as a potential source for the production of biosorbents. They also include cellulose, hemicellulose, proteins, and many electron-density-rich functionalities such as alcoholic, ketonic, carboxylic, ester, ether, and phenolic groups (Gao et al., 2018; Jin et al., 2020; Xu and Zhu, 2017). These functionalities are the binding sites in the biosorptive mechanism of IOCs. Moreover, due to such tunable functionalities, the sorption characteristics of waste-derived biosorbents can be further improved. Recently, many biosorbents have been designed using the above-mentioned bioactive materials for the decontamination of wastewater streams. The role of such biosorbents in eliminating IOCs from waste streams is illustrated here.

A report by Kamde et al. (Kamde et al., 2018) reveals the role of *Acidothiobacillus ferrooxidans* bacteria in the removal and transformation of highly toxic arsenite [As(III)] into arsenate [As(V)] from simulated water samples and arsenic-contaminated groundwater. First, *A .ferrooxidans* is immobilized on granulated activated carbon. The active Fe(II) center of the bacterium is converted to Fe(III) upon interaction with activated carbon. Second, the arsenic-contaminated water samples are treated in a bioreactor. The arsenic removal efficiency of the proposed biosorbent based filter was more than 95%. Moreover, the effects of the initial Fe(II) concentration, flow rate of the feed water, and arsenic concentration on biosorption efficiency are also investigated. In the treated water sample, the concentration of arsenic was found to be less than 10 μg/L, which is under the prescribed limits as per WHO guidelines.

Arsenic contamination in groundwater is a very serious problem in many developing countries. Biosorption methods, with their simplicity, can be employed at the point source. For instance, two rod-shaped gram-positive bacteria were first isolated and then utilized for arsenic removal by Dey et al. (2016). The bacteria were isolated from arsenic-contaminated groundwater of the Purbasthali block of Burdwan, West Bengal, India. The report suggests that the bacteria have an arsenic tolerance limit as high as 4500 ppm for arsenate and 550 ppm for arsenite. Through intensive biochemical analysis and 16S rRNA sequencing, the researchers identified *Bacillus* sp. (SW2) and *Aneurinibacillus aneurinilyticus* (SW4). The removal efficiency of SW2 for arsenite and arsenate was 51.45% and 53.29%, respectively. SW4 exhibited slightly less removal efficiency than SW2: around 51.99 and 50.37% for arsenite and arsenate, respectively. Sorption equilibrium was achieved within 72 hours of the sorption process. Moreover, both species can withstand a solution of up to 10% sodium chloride. Furthermore, both SW2 and SW4 species can convert arsenite into arsenate and hence cement the biosorption pathway to eliminate heavy metals.

Fluoride is another challenging toxic inorganic contaminant. It is very hazardous to consume water having a fluoride concentration of more than 1 mg/L. Mohan et al. (2012) reported low-cost pine and wood bark chars to eliminate fluoride from water. The report suggests that the sorption efficiency is observed to be maximum under highly acidic conditions, around a pH of 2. The kinetic studies show that the sorption equilibrium was reached in 48 hours with the adsorbent dose of 10 g/L. Sorption isotherm studies carried out by the group confirm the validity of the Langmuir adsorption isotherm model for pinewood and the Toth isotherm model for pine bark, respectively. Further, it is claimed that new internal pores are created that enhance the material's pore volume. This creation of pore volume can be attributed to the swelling effect in char in water media due to its high oxygen content. Noticeably, the proposed sorbents with only 1–3 m^2/g surface area have more fluoride removal efficiency than activated carbons having ~1000 m^2/g surface area.

Another report by Chen et al. (2016) reported a spent mushroom compost biochar (SMCB) based biosorbent with aluminum hydroxide coatings for eliminating fluoride from drinking water. The maximum adsorption capacity of SMCB is found to be 36.5 mg/g. The sorption of fluoride over the SMCB surface followed the Langmuir adsorption isotherm. Importantly, this biosorbent adsorbs fluoride in a pH range of 6.0 to 8.0. Within the equilibrium time, the sorbent reduces the fluoride concentration to 1 mg/L from 10 mg/L. This indicates the capability of mushroom cultivation waste biomass-derived biosorbents, and such adsorbents can be used in real-time applications.

Cyanides are among the most toxic class of inorganic contaminants. Even when present in very low concentrations, they impose serious threats to human and aquatic life. Therefore, their decontamination from water is crucial. Dwivedi et al. (2016) show the role of *Prunus amygdalus* (almond) shell (PAS) and *Tectona grandis* (sagwan) leaves (TGL) powder in removing cyanide from simulated water samples. The various sorption parameters such as solution pH, biosorbent dose, contact time, and initial cyanide concentration were optimized. The authors depicted the validity of the Langmuir isotherm model on experimental sorption data, and the corresponding sorption capacities of PAS and TGL powder for cyanide were found to be 32.05 mg/g and 18.45 mg/g, respectively. The kinetic studies show the applicability of a pseudo second-order kinetic model on the sorption process of both biosorbents. Such studies open a new pathway for developing low-cost biosorbents from waste materials and strengthening waste-to-wealth approaches.

A nearly 32- to 38-fold reduction in concentration of copper cyanide and nickel cyanide was achieved by Patil and Paknikar (1999) using a *Cladosporium cladosporioides* biomass derived biosorbent. The report reveals that it takes five to six hours for complete degradation of unrecoverable metal cyanide residuals. Further, it is claimed that the treated water is safe to be discharged in open environments and water resources.

Currently, the field of biosorptive elimination of IOCs is in its developing phase. Most research reports only involve lab-scale investigations. However, diverse spectrums of biosorbents have been synthesized to remove different types of IOCs, including heavy metals (arsenic, mercury, lead, etc.), fluorides, cyanides, nitrates, etc. The various reported biosorbents along with their performance in removal of IOCs are listed in Table 3.2.

These literature reports illustrate the synthesis protocols of biosorbents and their promising potential in the sequestration of IOCs from wastewater streams. The reported biosorbents exhibited a versatile nature and showed excellent removal efficiencies in eliminating every type of IOCs. However, it is also emphasized that their performance is significantly affected by various parameters such as the nature of the sorbent and adsorbate, surface functionalities of the sorbent, sorbent dose, solution pH, initial adsorbate concentration, temperature, etc.

As biosorbents can be easily fabricated from waste products and biomass, biosorption methods are promising alternatives for activated carbons and other such adsorbents. Although they are still limited to laboratory boundaries, they can be realized in real wastewater decontamination procedures with or without integration with other techniques. The next section describes various challenging aspects of the field along with future perspectives of biosorption methods.

Concluding Remarks and Future Perspectives

There is no doubt that water pollution due to IOCs has reached an alarming level and is intensifying. Water quality and quantity are deteriorating, and the sustainability of ecosystems is in extreme danger. Although many initiatives and steps have been taken, such as creating awareness of water quality problems, promoting advanced wastewater treatment techniques, and implementing stringent environmental legislation on effluent discharge, the issue is persistent and has different levels of impact in different places. The situation demands synergic and synchronized management and mitigation and sequestration techniques that not only eliminate IOCs but also include novel solid waste management, waste recycling, and waste upcycling themes.

Biosorption can play a central role in this regard and can be efficiently used in eliminating IOCs from industrial and environmental effluents. Although this is an emerging field, many biosorbents derived from waste substances have shown the capability to remove

Table 3.2 Some recently reported biosorbents for eliminating IOCs from wastewater.

Biosorbent	Inorganic contaminant	Sorption capacity	Ref.
Pomegranate peel derived AC (PPAC)	Ni (II)	83.02%	(Elsayed et al., 2020)
Tea factory waste (TFW)	Cr (VI)	54.65 mg/g	(Malkoc and Nuhoglu, 2007)
Peach and apricot stones	Pb (II)	97.64% and 93%	(Rashed, 2006)
Chemically modified *Ziziphus spina-christi* stones (ZSCs)	Pb (II) Cd (II) Zn (II)	9.06 mg/g 4.19 mg/g 5.38 mg/g	(Assirey et al., 2020)
Neem bark	Pb (II)	86.7%	(Naiya et al., 2008)
Maugeotia genuflexa (algae)	As (III)	57.48 mg/g	(Sarı et al., 2011)
Bacillus subtilis bacterium	Pb (II) Sb (III)	17.34 mg/g 2.32 mg/g	(Cai et al., 2018)
Spirulina platensis (cyanobacterium, microalgae)	Cu (II)	90.6%	(Al-Homaidan et al., 2014)
Oak shell waste derived magnetic biosorbent	Pb(II) Sr(II) Ba(II) Ag(I)	133.84 mg/g 40.39 mg/g 50.12 mg/g 53.12 mg/g	(Adibmehr and Faghihian, 2018)
Cymodocea rotundata seagrass beads	NO_3^- PO_4^{3-}	62% 71%	(Soumya et al., 2015)
Algal biosorbent (*Ulva fasciata*)	F^-	~88%	**(Rao et al., 2009)**
White-rot fungus *Pleurotus eryngii* ATCC 90888	F^-	66.6 mg/g	(Amin et al., 2015)
Grape pomace	F^-	7.54 mg/g	(Zhang et al., 2014)
Zostera marina plant	Mn (II)	120.6 mg/g	(Deniz and Tezel Ersanli, 2020)
Sugarcane bagasse	CN^-	79.30%	(Singh and Balomajumder, 2017)
Olive cake	Cr Cu Pb Zn	3.08 mg/g 5.17 mg/g 13.21 mg/g 5.51 mg/g	(Fernández-González et al., 2019)

more IOCs than conventional materials and techniques. More importantly, biosorption methods are extremely useful in recovering contaminants as resource material. This further reduces the load of natural resources. Uniquely, biosorbents have environmentally benign disposability and can have excellent post-life utility where they can be further transformed into other kinds of secondary materials with other applications. Compared to advanced nano-adsorbents, the precursors of biosorbents are inexpensive, and their fabrication is facile and less energy-intensive. Easily accessible biomass, agricultural and plant wastes, and industrial wastes are among potential sources of biosorbents.

Despite many fascinating advantages, enormous hurdles are also associated with developing and designing biosorbents, thus restricting their commercial viability. In particular, weathering stability and recyclability are critical issues. Biosorbents generally suffer from poor recyclability and often undergo degradation under mild environmental conditions. More in-depth studies are required to address these issues.

Further, synthesis protocols of biosorbents need to be scaled up. Since large volumes of wastewater streams need to be treated in real-time applications, at least pilot-scale biosorption methods need to be developed. Such systems can be further modified for continuously treating wastewater streams of industrial effluents, agricultural run-off, and domestic sewage. In this context, integrating biosorbents with new novel nano-adsorbents via the aid of nanoscience can serve the purpose. By designing biosorbents with extremely high sorption capacity, tunable surface functionalities, hierarchical porous architecture, and ultralow density with multi-functionality can be realized. The current advancements in the field are encouraging, and biosorption processes are likely to be common in the near future for the decontamination of wastewater streams.

References

Adejumoke, I. (2018). *Water Pollution: Effects, Prevention, and Climatic Impact*. IntechOpen.

Adewuyi, A. (2020). Chemically modified biosorbents and their role in the removal of emerging pharmaceutical waste in the water system. *Water* 12 (6): 1551. doi:10.3390/w12061551.

Adimalla, N., Venkatayogi, S., and Das, S.V.G. (2019). Assessment of fluoride contamination and distribution: a case study from a rural part of Andhra Pradesh, India. *Applied Water Science.* 9 (4). doi:10.1007/s13201-019-0968-y.

Ahmed, A.S.S., Sultana, S., Habib, A. et al. (2019). Bioaccumulation of heavy metals in some commercially important fishes from a tropical river estuary suggests higher potential health risk in children than adults. *PloS ONE*, 14 (10): e0219336.

Alalwan, H.A., Kadhom, M.A., and Alminshid, A.H. (2020). Removal of heavy metals from wastewater using agricultural byproducts. *Journal of Water Supply: Research and Technology-Aqua* 69 (2): 99–112. doi:10.2166/aqua.2020.133.

Albert, M.J., Faruque, A.S.G. et al. (1999). Case-control study of enteropathogens associated with childhood diarrhea in Dhaka, Bangladesh. *Journal of Clinical Microbiology* 37 (11): 3458–3464. doi:10.1128/jcm.37.11.3458-3464.1999.

Ali, H., Khan, E., and Ilahi, I. (2019). Environmental chemistry and ecotoxicology of hazardous heavy metals: environmental persistence, toxicity, and bioaccumulation. *Journal of Chemistry.* 2019: 1–14. doi:10.1155/2019/6730305.

Ali Redha, A. (2020). Removal of heavy metals from aqueous media by biosorption. *Arab Journal of Basic and Applied Sciences* 27 (1): 183–193. doi:10.1080/25765299.2020.1756177.

Amoatey, P. and Baawain, M.S. (2019). Effects of pollution on freshwater aquatic organisms. *Water Environment Research* 91 (10): 1272–1287. doi:10.1002/wer.1221.

Anastopoulos, I., Robalds, A., Tran, H.N. et al. (2019). Removal of heavy metals by leaves-derived biosorbents. *Environmental Chemistry Letters* 17 (2): 755–766. doi:10.1007/s10311-018-00829-x.

Anjum, M., Miandad, R., Waqas, M. et al. (2019). Remediation of wastewater using various nano-materials. *Arabian Journal of Chemistry* 12 (8): 4897–4919. doi:10.1016/j.arabjc.2016.10.004.

Arantes, F.P., Savassi, L.A., Santos, H.B. et al. (2016). Bioaccumulation of mercury, cadmium, zinc, chromium, and lead in muscle, liver, and spleen tissues of a large commercially valuable catfish species from Brazil. *Anais da Academia Brasileira de Ciencias* 88 (1): 137–147.

Assi, M.A., Hezmee, M.N.M., Haron, A.W. et al. (2016). The detrimental effects of lead on human and animal health. *Veterinary World* 9 (6): 660–671. doi:10.14202/vetworld.2016.660-671.

Azimi, A., Azari, A., Rezakazemi, M. et al. (2017). Removal of Heavy Metals from Industrial Wastewaters: A Review. *ChemBioEng Reviews* 4 (1): 37–59. doi:10.1002/cben.201600010.

Barakat, M.A. (2011). New trends in removing heavy metals from industrial wastewater. *Arabian Journal of Chemistry* 4 (4): 361–377. doi:10.1016/j.arabjc.2010.07.019.

Behera, K.K. (2014). Phytoremediation, transgenic plants and microbes. *Sustainable Agriculture Reviews*: 65–85. doi:10.1007/978-3-319-00915-5_4.

Bolisetty, S., Peydayesh, M., and Mezzenga, R. (2019). Sustainable technologies for water purification from heavy metals: review and analysis. *Chemical Society reviews* 48 (2): 463–487.

Bora, A.J. and Dutta, R.K. (2019). Removal of metals (Pb, Cd, Cu, Cr, Ni, and Co) from drinking water by oxidation-coagulation-absorption at optimized pH. *Journal of Water Process Engineering.* 31: 100839. doi:10.1016/j.jwpe.2019.100839.

Boretti, A. and Rosa, L. (2019). Reassessing the projections of the World Water Development Report. *npj Clean Water* 2 (1). doi:10.1038/s41545-019-0039-9.

Bragina, A., Berg, C., Cardinale, M. et al. (2012). Sphagnum mosses harbour highly specific bacterial diversity during their whole lifecycle. *The ISME Journal* 6 (4): 802–813. doi:10.1038/ismej.2011.151.

Briffa, J., Sinagra, E., and Blundell, R. (2020). Heavy metal pollution in the environment and their toxicological effects on humans. *Heliyon* 6 (9): e04691.

Bukola, D. and Zaid, A. (2015). Consequences of anthropogenic activities on fish and the aquatic environment. *Poultry, Fisheries and Wildlife Sciences* 3 (2). doi:10.4172/2375-446x.1000138.

Camargo, J.A. (2003). Fluoride toxicity to aquatic organisms: a review. *Chemosphere* 50 (3): 251–264. doi:10.1016/s0045-6535(02)00498-8.

Chaudhary, D.S., Vigneswaran, S., Ngo, H.-H. et al. (2003). Biofilter in water and wastewater treatment. *Korean Journal of Chemical Engineering* 20 (6): 1054–1065. doi:10.1007/bf02706936.

Chen, G.-J., Peng, C.-Y., Fang, J.-Y. et al. (2016). Biosorption of fluoride from drinking water using spent mushroom compost biochar coated with aluminum hydroxide. *Desalination and Water Treatment* 57 (26): 12385–12395. doi:10.1080/19443994.2015.1049959.

Chen, X., Huang, G., and Wang, J. (2013). Electrochemical reduction/oxidation in the treatment of heavy metal wastewater. *Journal of Metallurgical Engineering (ME)* 2 (4).

Cherian, S. and Margarida Oliveira, M. (2005). Transgenic plants in phytoremediation: recent advances and new possibilities. *Environmental Science and Technology* 39 (24): 9377–9390. doi:10.1021/es051134l.

Chibuike, G.U. and Obiora, S.C. (2014). Heavy metal polluted soils: effect on plants and bioremediation methods. *Applied and Environmental Soil Science* 2014: 1–12. doi:10.1155/2014/752708.

Cui, L., Zhou, J., Peng, X. et al. (2020). Analyses of air pollution control measures and co-benefits in the heavily air-polluted Jinan city of China, 2013–2017. *Scientific Reports* 10 (1). doi:10.1038/s41598-020-62475-0.

Datta, A.S., Chakrabortty, A., De Dalal, S.S., and Lahiri, S.C. (2014). Fluoride contamination of underground water in West Bengal, India. *Fluoride* 47: 241–248.

Deng, Y. and Zhao, R. (2015). Advanced oxidation processes (AOPs) in wastewater treatment. *Current Pollution Reports* 1 (3): 167–176. doi:10.1007/s40726-015-0015-z.

Dey, U., Chatterjee, S., and Mondal, N.K. (2016). Isolation and characterization of arsenic-resistant bacteria and possible application in bioremediation. *Biotechnology Reports* 10: 1–7. doi:10.1016/j.btre.2016.02.002.

Dickinson, N.M., Baker, A.J.M., Doronila, A. et al. (2009). Phytoremediation of inorganics: realism and synergies. *International Journal of Phytoremediation* 11 (2): 97–114. doi:10.1080/15226510802378368.

Dutta, P.K., Ray, A.K., Sharma, V.K. et al. (2004). Adsorption of arsenate and arsenite on titanium dioxide suspensions. *Journal of Colloid and Interface Science* 278 (2): 270–275.

Dwivedi, A.K. (2017). Researches in water pollution: A review. *International Research Journal of Natural and Applied Sciences* 4: 118–142.

Dwivedi, N., Balomajumder, C., and Mondal, P. (2016). Comparative investigation on the removal of cyanide from aqueous solution using two different bioadsorbents. *Water Resources and Industry* 15: 28–40. doi:10.1016/j.wri.2016.06.002.

Edokpayi, J.N., Odiyo, J.O., and Durowoju, O.S. (2017). Impact of wastewater on surface water quality in developing countries: a case study of South Africa. *Water Quality* doi:10.5772/66561.

Farnane, M., Machrouhi, A., Elhalil, A. et al. (2018). New sustainable biosorbent based on recycled deoiled carob seeds: optimization of heavy metals remediation. *Journal of Chemistry* 2018: 1–16. doi:10.1155/2018/5748493.

Feng, C., Yu, L., Xiao, Y. et al. (2020). Environmental behavior and effects of pollutants in water. *Journal of Chemistry* 2020: 1–2. doi:10.1155/2020/2639389.

Gao, X., Zhang, H., Chen, K. et al. (2018). Removal of heavy metal and sulfate ions by cellulose derivative-based biosorbents. *Cellulose* 25 (4): 2531–2545. doi:10.1007/s10570-018-1690-x.

Greenstone, M. and Hanna, R. (2014). Environmental regulations, air and water pollution, and infant mortality in India. *American Economic Review* 104: 3038–72.

Gunatilake, S. (2015). Methods of removing heavy metals from industrial wastewater. *Methods* 1: 14.

Hamdan, S.H., Molelekwa, G.F., and Van der Bruggen, B. (2014). Electrokinetic remediation technique: an integrated approach to finding new strategies for restoration of saline soil and to control seawater intrusion. *ChemElectroChem* 1 (7): 1104–1117. doi:10.1002/celc.201402071.

Hong, Y.-S., Song, K.-H., and Chung, J.-Y. (2014). Health effects of chronic arsenic exposure. *Journal of Preventive Medicine and Public Health* 47 (5): 245–252. doi:10.3961/jpmph.14.035.

Hu, P., Tan, B., and Long, M. (2016). Advanced nanoarchitectures of carbon aerogels for multifunctional environmental applications. *Nanotechnology Reviews* 5 (1). doi:10.1515/ntrev-2015-0050.

Ince, M. and Ince, O.K. (2020). Heavy metal removal techniques using response surface methodology: water/wastewater treatment. *Biochemical Toxicology – Heavy Metals and Nanomaterials* doi:10.5772/intechopen.88915.

Jain, D., Chaudhary, P., Varshney, N. et al. (2020). Carcinogenic effects of N-nitroso compounds in the environment. *Environment Conservation Journal*: 25–41. doi:10.36953/ecj.2020.21304.

Jaszczak, E., Polkowska, Ż., Narkowicz, S. et al. (2017). Cyanides in the environment—analysis—problems and challenges. *Environmental Science and Pollution Research.*\ 24 (19): 15929–15948. doi:10.1007/s11356-017-9081-7.

Jezierski, P., Szynkiewicz, A., and Jsunędrysek, M.-O. (2006). Natural and anthropogenic origin sulphate in an mountainous groundwater system: S and O isotope evidences. *Water, Air, and Soil Pollution* 173 (1-4): 81–101. doi:10.1007/s11270-005-9028-3.

Jin, C., Liu, G., Wu, G. et al. (2020). Facile fabrication of crown ether functionalized lignin-based biosorbent for the selective removal of Pb(II). *Industrial Crops and Products* 155: 112829. doi:10.1016/j.indcrop.2020.112829.

Jyothi, N.R. and Farook, N.A.M. (2020). Mercury toxicity in public health. *Heavy Metal Toxicity in Public Health* doi:10.5772/intechopen.90333.

Kamde, K., Pandey, R.A., Thul, S. et al. (2018). Removal of arsenic by Acidothiobacillus ferrooxidans bacteria in bench scale fixed-bed bioreactor system. *Chemistry and Ecology* 34 (9): 818–838. doi:10.1080/02757540.2018.1504927.

Kanamarlapudi, S.L.R.K., Chintalpudi, V.K., and Muddada, S. (2018). Application of biosorption for removal of heavy metals from wastewater. In: *Biosorption* (ed. J. Derco and B. Vrana). InTech. doi: 10.5772/intechopen.77315.

Kikuchi, T. and Tanaka, S. (2012). Biological removal and recovery of toxic heavy metals in water environment. *Critical Reviews in Environmental Science and Technology* 42 (10): 1007–1057. doi:10.1080/10643389.2011.651343.

Kumari, B., Kumar, V., Sinha, A.K. et al. (2017). Toxicology of arsenic in fish and aquatic systems. *Environmental Chemistry Letters* 15 (1): 43–64. doi:10.1007/s10311-016-0588-9.

Kummu, M., Guillaume, J.H.A., de Moel, H. et al. (2016). The world's road to water scarcity: shortage and stress in the 20th century and pathways towards sustainability. *Scientific Reports* 6 (1). doi:10.1038/srep38495.

Kurniawan, T.A., Chan, G.Y.S., Lo, W.-H. et al. (2006). Physico-chemical treatment techniques for wastewater laden with heavy metals. *Chemical Engineering Journal* 118 (1–2): 83–98. doi:10.1016/j.cej.2006.01.015.

Lasat, M.M. (2002). Phytoextraction of toxic metals. *Journal of Environmental Quality* 31 (1): 109–120. doi:10.2134/jeq2002.1090.

Liu, P., Zhan, X., Wu, X. et al. (2020). Effect of weathering on environmental behavior of microplastics: Properties, sorption and potential risks. *Chemosphere* 242: 125193. doi:10.1016/j.chemosphere.2019.125193.

Lovasoa, C.R., Hela, K., Harinaivo, A.A. et al. (2017). Bioremediation of soil and water polluted by cyanide: A review. *African Journal of Environmental Science and Technology* 11 (6): 272–291. doi:10.5897/ajest2016.2264.

Lu, Y., Song, S., Wang, R. et al. (2015). Impacts of soil and water pollution on food safety and health risks in China. *Environment international* 77: 5–15.

Maddan, H. and Smith, B. (2021). Water, sanitation and hygiene in refugee camps in Uganda: photo essay. *Research in Drama Education: The Journal of Applied Theatre and Performance* 26 (2): 240–246. doi:10.1080/13569783.2020.1792282.

Mallampati, R. and Valiyaveettil, S. (2014). Eggshell membrane-supported recyclable catalytic noble metal nanoparticles for organic reactions. *ACS Sustainable Chemistry and Engineering* 2 (4): 855–859. doi:10.1021/sc4004899.

Masindi, V. and Muedi, K.L. (2018). Environmental contamination by heavy metals. *Heavy Metals* doi:10.5772/intechopen.76082.

Maurya, P.K., Malik, D.S., Yadav, K.K. et al. (2019). Bioaccumulation and potential sources of heavy metal contamination in fish species in River Ganga basin: Possible human health risks evaluation. *Toxicology Reports* 6: 472–481. doi:10.1016/j.toxrep.2019.05.012.

Miller, S.M., Spaulding, M.L., and Zimmerman, J.B. (2011). Optimization of capacity and kinetics for a novel bio-based arsenic sorbent, TiO2-impregnated chitosan bead. *Water Research* 45 (17): 5745–5754. doi:10.1016/j.watres.2011.08.040.

Mo, J., Yang, Q., Zhang, N. et al. (2018). A review on agro-industrial waste (AIW) derived adsorbents for water and wastewater treatment. *Journal of Environmental Management* 227: 395–405. doi:10.1016/j.jenvman.2018.08.069.

Mohamed, K., Majid, M.I.A., Leong, Y.-H. et al. (2020). Dioxins in peat and its formation: An overview. *Cogent Environmental Science* 6 (1): 1864870. doi:10.1080/23311843.2020.1864870.

Mohan, D., Sharma, R., Singh, V.K. et al. (2012). Fluoride removal from water using bio-char, a green waste, low-cost adsorbent: equilibrium uptake and sorption dynamics modeling. *Industrial and Engineering Chemistry Research* 51 (2): 900–914. doi:10.1021/ie202189v.

Moore, J.W., Nowlan, L., Olszynski, M. et al. (2018). Towards linking environmental law and science. *FACETS* 3 (1): 375–391. doi:10.1139/facets-2017-0106.

Mullenix, P.J. (2014). A new perspective on metals and other contaminants in fluoridation chemicals. *International Journal of Occupational and Environmental Health* 20 (2): 157–166. doi:10.1179/2049396714y.0000000062.

Mustapha, S., Tijani, J.O., Ndamitso, M.M. et al. (2020). The role of kaolin and kaolin/ZnO nanoadsorbents in adsorption studies for tannery wastewater treatment. *Scientific Reports* 10 (1). doi:10.1038/s41598-020-69808-z.

Patil, Y. and Paknikar, K. (1999). Removal and recovery of metal cyanides using a combination of biosorption and biodegradation processes. *Biotechnology Letters* 21: 913–919.

Peckham, S. and Awofeso, N. (2014). Water fluoridation: a critical review of the physiological effects of ingested fluoride as a public health intervention. *The Scientific World Journal* 2014: 1–10. doi:10.1155/2014/293019.

Puri, A. and Kumar, M. (2012). A review of permissible limits of drinking water. *Indian Journal of Occupational and Environmental Medicine* 16 (1): 40. doi:10.4103/0019-5278.99696.

Rahman, M.A., Hasegawa, H., and Lim, R.P. (2012). Bioaccumulation, biotransformation and trophic transfer of arsenic in the aquatic food chain. *Environmental Research* 116: 118–135.

Raj, D. and Shaji, E. (2017). Fluoride contamination in groundwater resources of Alleppey, southern India. *Geoscience Frontiers* 8 (1): 117–124. doi:10.1016/j.gsf.2016.01.002.

Ramrakhiani, L., Halder, A., Majumder, A. et al. (2017). Industrial waste derived biosorbent for toxic metal remediation: Mechanism studies and spent biosorbent management. *Chemical Engineering Journal* 308: 1048–1064. doi:10.1016/j.cej.2016.09.145.

Reinhardt, A. (1960-61). *Abstract Painting*. New York: Museum of Modern Art.

Reungoat, J., Escher, B.I., Macova, M. et al. (2011). Biofiltration of wastewater treatment plant effluent: Effective removal of pharmaceuticals and personal care products and reduction of toxicity. *Water Research* 45 (9): 2751–2762. doi:10.1016/j.watres.2011.02.013.

Reza, R. and Singh, G. (2010). Heavy metal contamination and its indexing approach for river water. *International Journal of Environmental Science and Technology* 7 (4): 785–792. doi:10.1007/bf03326187.

Rhind, S.M. (2009). Anthropogenic pollutants: a threat to ecosystem sustainability? *Philosophical Transactions of the Royal Society B: Biological Sciences* 364 (1534): 3391–3401. doi:10.1098/rstb.2009.0122.

Rostern, N., T. (2017). The effects of some metals in acidified waters on aquatic organisms. *Oceanography and Fisheries Open access Journal* 4 (4). doi:10.19080/OFOAJ.2017.04.555645.

Roy, M. and McDonald, L.M. (2015). Metal uptake in plants and health risk assessments in metal-contaminated smelter soils. *Land Degradation and Development* 26 (8): 785–792. doi:10.1002/ldr.2237.

Saini, S., Gill, J.K., Kaur, J. et al. (2020). Biosorption as environmentally friendly technique for heavy metal removal from wastewater. *Fresh Water Pollution Dynamics and Remediation*: 167–181. doi:10.1007/978-981-13-8277-2_10.

Selvi, A., Rajasekar, A., Theerthagiri, J. et al. (2019). *Integrated remediation processes toward heavy metal removal/recovery from various environments – a review. Frontiers in Environmental Science.* 7. doi:10.3389/fenvs.2019.00066.

Şen, A., Pereira, H., Olivella, M.A. et al. (2015). Heavy metals removal in aqueous environments using bark as a biosorbent. *International Journal of Environmental Science and Technology* 12 (1): 391–404. doi:10.1007/s13762-014-0525-z.

Shahalam, A.M., Al-Harthy, A., and Al-Zawhry, A. (2002). Feed water pretreatment in RO systems: unit processes in the middle east. *Desalination* 150 (3): 235–245. doi:10.1016/s0011-9164(02)00979-7.

Shankar, S., Shanker, U., and Shikha (2014). Arsenic contamination of groundwater: a review of sources, prevalence, health risks, and strategies for mitigation. *The Scientific World Journal* 2014: 1–18. doi:10.1155/2014/304524.

Sharma, S. and Bhattacharya, A. (2017). Drinking water contamination and treatment techniques. *Applied Water Science* 7 (3): 1043–1067. doi:10.1007/s13201-016-0455-7.

Shen, Y., Fang, Q., and Chen, B. (2015). Environmental applications of three-dimensional graphene-based macrostructures: adsorption, transformation, and detection. *Environmental Science and Technology* 49 (1): 67–84. doi:10.1021/es504421y.

Shepherd, S., Lima, M.A.P., Oliveira, E.E. et al. (2018). Extremely low frequency electromagnetic fields impair the cognitive and motor abilities of honey bees. *Scientific Reports* 8 (1). doi:10.1038/s41598-018-26185-y.

Singh, N.B., Nagpal, G., Agrawal, S. et al. (2018). Water purification by using adsorbents: a review. *Environmental Technology and Innovation* 11: 187–240. doi:10.1016/j.eti.2018.05.006.

Singh, R., Gautam, N., Mishra, A. et al. (2011). Heavy metals and living systems: An overview. *Indian Journal of Pharmacology* 43 (3): 246. doi:10.4103/0253-7613.81505.

Sordello, R., De Lachapelle, F.F., Livoreil, B. et al. (2019). Evidence of the environmental impact of noise pollution on biodiversity: a systematic map protocol. *Environmental Evidence* 8 (1). doi:10.1186/s13750-019-0146-6.

Spurr, L.P., Watts, M.P., Gan, H.M. et al. (2019). Biodegradation of thiocyanate by a native groundwater microbial consortium. *Peer J* 7: e6498. doi:10.7717/peerj.6498.

Tabrez, S., Zughaibi, T.A., and Javed, M. (2021). Bioaccumulation of heavy metals and their toxicity assessment in Mystus species. *Saudi Journal of Biological Sciences* 28 (2): 1459–1464. doi:10.1016/j.sjbs.2020.11.085.

Tchounwou, P.B., Yedjou, C.G., Patlolla, A.K. et al. (2012). Heavy metal toxicity and the environment. In: *Molecular, Clinical and Environmental Toxicology* (ed. A. Luch), 133–164. Springer Basel.

Thompson, L.A. and Darwish, W.S. (2019). Environmental chemical contaminants in food: review of a global problem. *Journal of Toxicology* 2019: 1–14. doi:10.1155/2019/2345283.

Villareal, T.A. and Carpenter, E.J. (2003). Buoyancy regulation and the potential for vertical migration in the oceanic cyanobacterium trichodesmium. *Microbial Ecology* 45 (1): 1–10. doi:10.1007/s00248-002-1012-5.

Ward, M., Jones, R., Brender, J. et al. (2018). Drinking water nitrate and human health: an updated review. *International Journal of Environmental Research and Public Health* 15 (7): 1557. doi:10.3390/ijerph15071557.

Wen, Y., Schoups, G., and van de Giesen, N. (2017). Organic pollution of rivers: Combined threats of urbanization, livestock farming and global climate change. *Scientific Reports* 7 (1). doi:10.1038/srep43289.

World Health Organization (WHO). (1993). Guidelines for drinking-water quality.

Xie, Y., Fan, J., Zhu, W. et al. (2016). Effect of heavy metals pollution on soil microbial diversity and bermudagrass genetic variation. *Frontiers in Plant Science* 7. doi:10.3389/fpls.2016.00755.

Xu, F., Zhu, T.-T., Rao, Q.-Q. et al. (2017). Fabrication of mesoporous lignin-based biosorbent from rice straw and its application for heavy-metal-ion removal. *Journal of Environmental Sciences* 53: 132–140. doi:10.1016/j.jes.2016.03.026.

Xue, Y., Song, J., Zhang, Y. et al. (2016). Nitrate pollution and preliminary source identification of surface water in a semi-arid river basin, using isotopic and hydrochemical approaches. *Water* 8 (8): 328. doi:10.3390/w8080328.

Yap, P.L., Hassan, K., Auyoong, Y.L. et al. (2020). Removal of multiple water pollutants: all-in-one bioinspired multifunctional graphene biopolymer foam for simultaneous removal of multiple water pollutants. *Advanced Materials Interfaces* 7 (18): 2070103. doi:10.1002/admi.202070103.

Yun, T., Kim, H., Iqbal, A. et al. (2020). Electromagnetic interference shielding: electromagnetic shielding of monolayer MXene assemblies. *Advanced Materials* 32 (9): 2070064. doi:10.1002/adma.202070064.

Zolkefli, N., Sharuddin, S.S., Yusoff, M.Z.M. et al. (2020). A review of current and emerging approaches for water pollution monitoring. *Water* 12 (12): 3417. doi:10.3390/w12123417.

4

Biosorption for Eliminating Organic Contaminants from Wastewater

Pinki Rani Agrawal[1], *Rahul Sharma*[1], *and Abhishek Agrawal*[2]

[1] *AcSIR, CSIR-National Physical Laboratory, New Delhi, India*
[2] *School of Pharmaceutical Sciences, Pratap University, Jaipur, Rajasthan, India*

Introduction

The consumption of natural sources (air, water, etc.) is increasing globally as the world's population grows. Water is essential to all environments. As the world develops, industries (pharmaceuticals, oil, paint, etc.) are constructed near rural and urban cities situated on the bank of bodies of water (Chaukura et al., 2016). These industries discharge their waste directly into the water with or without treatment, leading to water pollution.

Generally, industrial waste and agricultural waste draw more attention for releasing organic pollution into bodies of water. Organic contaminants such as natural and synthetic dyes, pharmaceutical wastes, and agricultural waste create water pollution and draw the attention of researchers because they are hazardous to humans (Harrad, 2001). Moreover, organic contaminants in water are lethal to aquatic species because they are highly soluble in water, do not degrade even after a long time, penetrate sunlight, and cause oxygen deficiency in water (Feng et al., 2013; Ngo et al., 2015; Ali et al., 2012; Ali and Aboul-Enein, 2004).

Humans can readily identify the presence of organic contaminants such as dyes in water because of their colors. Natural and synthetic dyes are among the organic contaminants. According to investigation, about 1,000,000 dyes are produced commercially, totalling more than 7×10^5 million tons, and dye-contaminated waste is released directly into the water supply (Aksu, 2005; Clarke and Anliker, 1980; Zollinger, 2002). The presence of dyes in water affects human life as well as aquatic life. In addition, agricultural waste plays a significant role in developing organic water contamination. Vast amounts of pesticides and herbicides flow into the water supply when it rains.

Additionally, pharmaceutical wastes have drawn attention as organic pollutants. Pharmaceutical companies produce huge amounts of pharmaceutical waste and discharge it into water without treatment. Pharmaceutical waste contains organic and inorganic molecules that are harmful to humans and aquatic species. As a result, the World Health

Biosorption for Wastewater Contaminants, First Edition.
Edited by Rangabhashiyam Selvasembian and Pardeep Singh.

Organization (WHO) and Bureau of Indian Standards (BIS) have prescribed limits for them in drinking water, and researchers are developing various methods such as ion exchange, membrane filtration, reverse osmosis, biosorption, and adsorption (Rashed, 2006; Van der Bruggen and Vandecasteele, 2002).

Biosorption is the best method to decontaminate organic impurities from water. Recently, this process has received more attention from researchers because of its low waste sludge, easy accessibility at a large scale, reproducibility, and repeatability; it is also economically and environmentally friendly and does not create secondary pollution (Park et al., 2010). This process is used on a large scale in industries to treat wastewater. As discussed in this chapter, researchers have recently reported in many articles that they can eliminate organic pollutants from water using biosorbents that are biodegradable microorganisms. The chapter presents types of organic pollutants and their effects on human health, along with a detailed study of the mechanism of biosorption, and ends with concluding remarks and future prospects.

Types of Organic Pollutants and Their Effects on Human Health

Organic Dyes

Organic dyes are a type of organic pollutant in contaminated water and a significant cause of aquatic pollution. Their presence in water is very harmful to humans as well as to aquatic life. Dye in water creates an oxygen deficiency that can kill aquatic life.

Various kinds of dyes are consumed at an industrial scale, both natural and synthetic (Lellis et al., 2019). Natural dyes are synthesized from natural sources such as plants (leaves, roots, berries, flowers, fungi, lichens, etc.). The first synthetic dye (organic aniline dye) was prepared in 1856 by William Henry Perkin (Chavan, 2011; Walsh, 1984). Today, many synthetic dyes are prepared using a chemical route at laboratory scale and industrial scale. Synthetic dyes are further categorized as cationic, anionic, and neutral. The consumption of synthetic dye has drawn a great deal of attention to various fields such as photolytic sensors, supramolecular chemistry, and dyeing clothes (Bouguettoucha et al., 2016). Some natural and synthetic dyes are explained in Figure 4.1.

The discharge of some textile dyes affects water quality because they are carcinogenic and non-biodegradable. Aromatic dyes are more carcinogenic, and they are not easily degraded even after a long time. Some aromatic dyes such as Congo red (CR), triphenylmethane dyes, o-Phenylenediamine (OPD), cationic dye methylene blue, crystal violet dye, and malachite green (MG) are used in industry and chemical labs on a large scale (Glossman-Mitnik, 2013; Abdullah et al., 2012; Kushwaha et al., 2014; Ismail et al., 2019). The discharge of dye contamination into water is harmful to human and aquatic life because consuming dye causes carcinogenic and mutagenic disease. Moreover, triphenylmethane dyes and OPD act like xenobiotics and are recalcitrant, directly affecting those who work in the industry (Papillomaviruses, 2011; Barceló, 2005; Chavan, 2011). For these reasons, degradation and removal methods have been developed to eliminate dye contaminants from water.

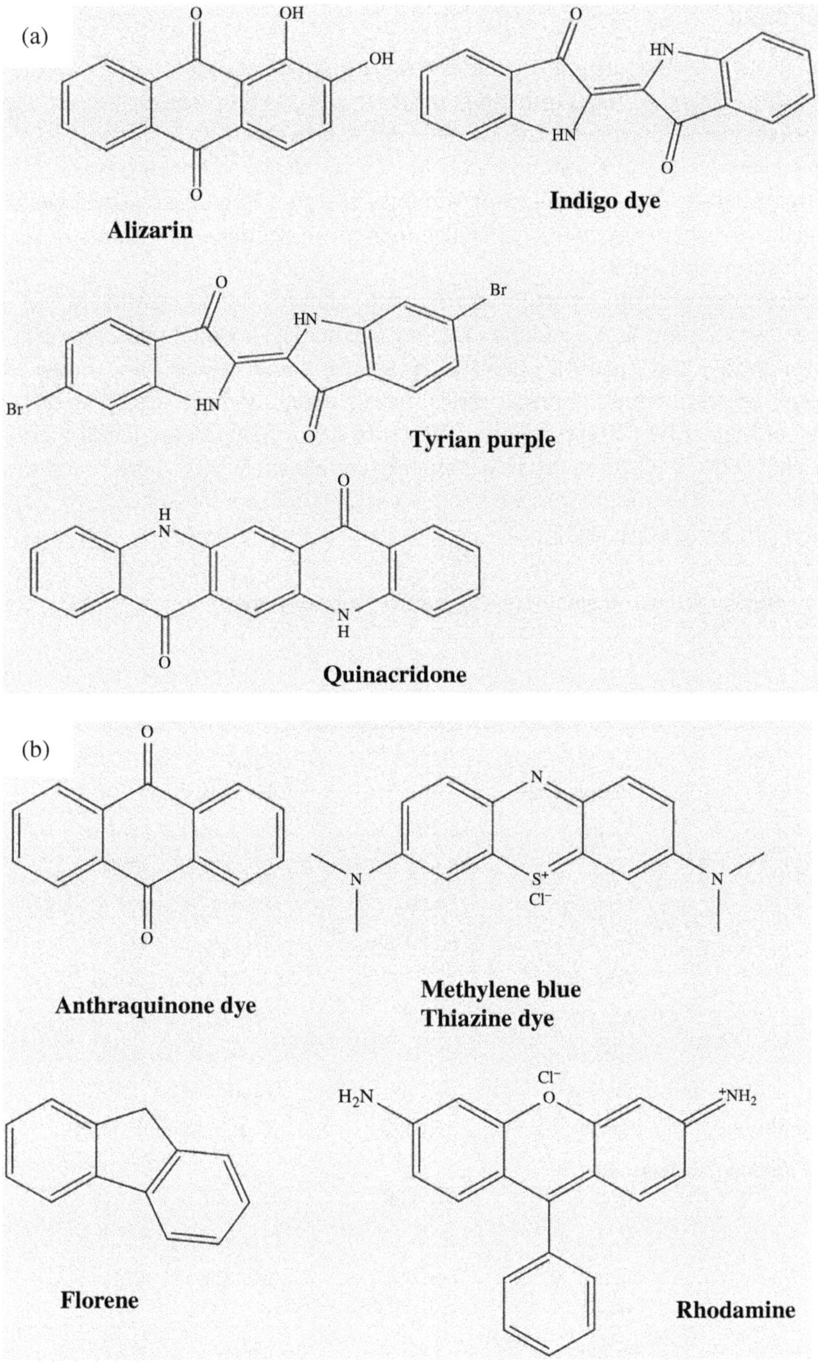

Figure 4.1 Structure of some synthetic dye molecules.

Pharmaceutical Waste

Pharmaceutical waste is also a form of organic waste. Human consumption of medicines has led to increasing pharmaceutical industries in urban areas. These companies are the reason for the emergence of pharmaceutical waste in drinking water, which creates organic water pollution (Adewuyi, 2020; Tran and Gin, 2017). The release of hospital waste into water without treatment leads to organic contaminants in water. The unrestrained use of antimicrobial agents in veterinary medicine is also increasing chemical pollutants in surface water as well as groundwater.

Even small amounts of pharmaceutical waste in water are very dangerous for the environment because they are non-biodegradable. Expired medicines, hospital waste, antibiotics, anti-inflammatories, and harmful chemical drugs are some of the wide range of chemical contaminants known as pharmaceutical active compounds (PACs) (Pal, 2018; Shannon and Woolridge, 2011). The presence of PACs in water has become lethal worldwide. Table 4.1 shows the PACs in water in developing countries. Due to these problems, the persistence of PACs is a concern for researchers working to decontaminate water supplies (Adewuyi, 2020; Aloui et al., 2020).

Table 4.1 Pharmaceutical active compounds (PACs) traced in surface water.

PAC	Country	Concentration (ng/L)	Reference
Antibiotics			
Erythromycin	Bangladesh	6.46	(Hossain et al., 2018)
Metronidazole	China	5.10	(Asghar et al., 2018)
Sulfamethoxazole	USA	14.73	(Bean et al., 2018)
Trimethoprim	Mexico	74.00	(Rivera-Jaimes et al., 2018)
Ciprofloxacin	India	6,500,000	(Fick et al., 2009)
Amoxicillin	Ghana	270	(Azanu et al., 2018)
Trimethoprim	Kenya	2650	(Madikizela et al., 2017)
Tetracycline	Ghana	30.00	(Azanu et al., 2018)
Anti-hyperglycemic Metformin	Sweden	8.40	(Lindim et al., 2016)
Lipid regular gemfibrozil	Spain	3735.00	(Carmona et al., 2014)
Analgesics and anti-inflammatories			
Acetaminophen	UK	9822.00	(Burns et al., 2018)
Diclofenac	Malaysia	15.49	(Praveena et al., 2018)
Ibuprofen	South Africa	846.00	(Matongo et al., 2015)
Ketoprofen	Portugal	86.90	(Pereira et al., 2017)
Naproxen	Poland	13.40	(Caban et al., 2015)
Antidepressives			
Diazepam	India	305.00	(Mutiyar et al., 2018)
Fluoxetine	Portugal	25.37	(Pereira et al., 2017)

Source: Adewuyi, 2020.

Agricultural Waste

All people depend on agriculture, and 70% of fresh water is used in agriculture farming. To increase the food supply, farmers use various types of fertilizers to produce crops in greater quantities. These fertilizers, which are sources of nitrogen and phosphorous nutrients, damage soil quality and water quality. Excess concentrations of nitrogen and phosphorous in water causes problems for aquatic plants due to interference in photosynthesis.

Agricultural contaminants such as pesticides are organic pollutants that create significant agrochemical waste in bodies of water and lead to organic water pollution. Pesticides are classified into subfamilies such as herbicides, insecticides, nematicides, and bactericides (Jurado et al., 2012). Pesticides are applied to grow crops quickly and protect crops from insects and weeds. However, agricultural contaminants are an enriched source of phosphorous and nitrogen nutrients.

Soil erosion and fertilizer runoff through rainwater into reservoirs leads to agricultural water pollution. Agricultural activities and their impact on surface and groundwater are described in Table 4.2. Agricultural irrigation is a significant problem worldwide. Long-term exposure to agrochemicals is carcinogenic to humans and also affects aquatic and animal life (Kumar et al., 2019a; Chaukura et al., 2016; Mahmoud et al., 2016; Jurado et al., 2012).

Increasing organic contaminants in reservoirs have become a challenging problem for researchers, who have developed conventional decontamination techniques to overcome the contamination of water bodies. At present, the most attractive method to eliminate organic water pollutants at a large scale is biosorption. This method has a low cost, and adsorbents are widely available, biodegradable, and recyclable (Nagendran, 2011). The next section discusses bioadsorbents and their mechanism.

Remediation Methods for Eliminating Organic Contaminants from Wastewater

Biosorption as a Remediation Method for Organic Pollutants

Over the last decade, biosorption has been discussed for eliminating water impurities (metals, dyes, pesticides, pharmaceutical waste). Generally, *biosorption* can be described as a physiochemical process that helps to eliminate natural industrial and agricultural contaminants from water (Park et al., 2010; Michalak et al., 2013; Wu et al., 2012). The different types of bioadsorbents have been evaluated for water treatment (Wang et al., 2014). Fomina and his team have described biosorption using various terms. For instance, biosorption is a phenomenon of using living and dead microorganisms to eliminate pollutants from water (Fomina and Gadd, 2014). Natural materials have been used as biosorbents since ancient times. Some biosorbents have been engineered from waste biomass to decontaminate organic pollutants from industrial water.

In biosorption, organic water pollutants are attracted to biosorbents by physiorption and chemisorption; the biosorbents can be regenerated, and the materials can be recycled many times (Figure 4.2). Biosorption is a more fruitful and attractive method for decontaminating organic water pollutants because of their low cost and because they are environmentally

Table 4.2 Sources and hazards of organic contaminants.

Agricultural activity	Impacts: Groundwater	Impacts: Surface water
Tillage/ plowing	Sediment/turbidity: sediments carry phosphorus and pesticides that are adsorbed into sediment particles; siltation of river beds and loss of habitat, spawning grounds, etc.	
Fertilizing	Runoff of nutrients, especially phosphorus, leading to eutrophication and causing tastes and odors in public water supplies; excess algae growth, leading to deoxygenation of water and fish kills.	Leaching of nitrates into groundwater; excessive levels are a threat to public health.
Manure spreading	Carried out as a fertilizer activity; spreading on frozen ground results in high levels of contamination in receiving waters in the form of pathogens, metals, phosphorus, and nitrogen, leading to eutrophication.	Contamination of groundwater, especially by nitrogen.
Pesticides	Runoff of pesticides leads to contamination of surface water and biota; dysfunction of the ecological system in surface waters due to the loss of top predators because of growth inhibition and reproductive failure; and public health impacts from eating contaminated fish. The wind carries pesticides as dust over very long distances, and they can contaminate aquatic systems thousands of miles away (e.g. tropical/ subtropical pesticides found in Arctic mammals).	Some pesticides may leach into groundwater, causing human health problems from contaminated wells.
Feedlots and corrals	Contamination of surface water with many pathogens (bacteria, viruses, etc.), leading to chronic public health problems. Also contamination by metals contained in urine and feces.	Potential leaching of nitrogen, metals, etc. into groundwater.
Irrigation	Runoff of salts leading to salinization of surface waters; runoff of fertilizers and pesticides to surface waters with ecological damage, bioaccumulation in edible fish species, etc. High levels of trace elements such as selenium can occur with serious ecological damage and potential human health impacts.	Enrichment of groundwater with salts, nutrients (especially nitrate)
Clear-cutting	Erosion of land, leading to high levels of turbidity in rivers, siltation of bottom habitats, etc. Disruption and change of the hydrologic regime, often with a loss of perennial streams; causes public health problems due to the loss of potable water.	Erosion of land, leading to high levels of turbidity in rivers, siltation of bottom habitats, etc. Disruption and change of hydrologic regimes, often with loss of perennial streams; causes public health problems due to the loss of potable water.

Source: Nagendran, 2011.

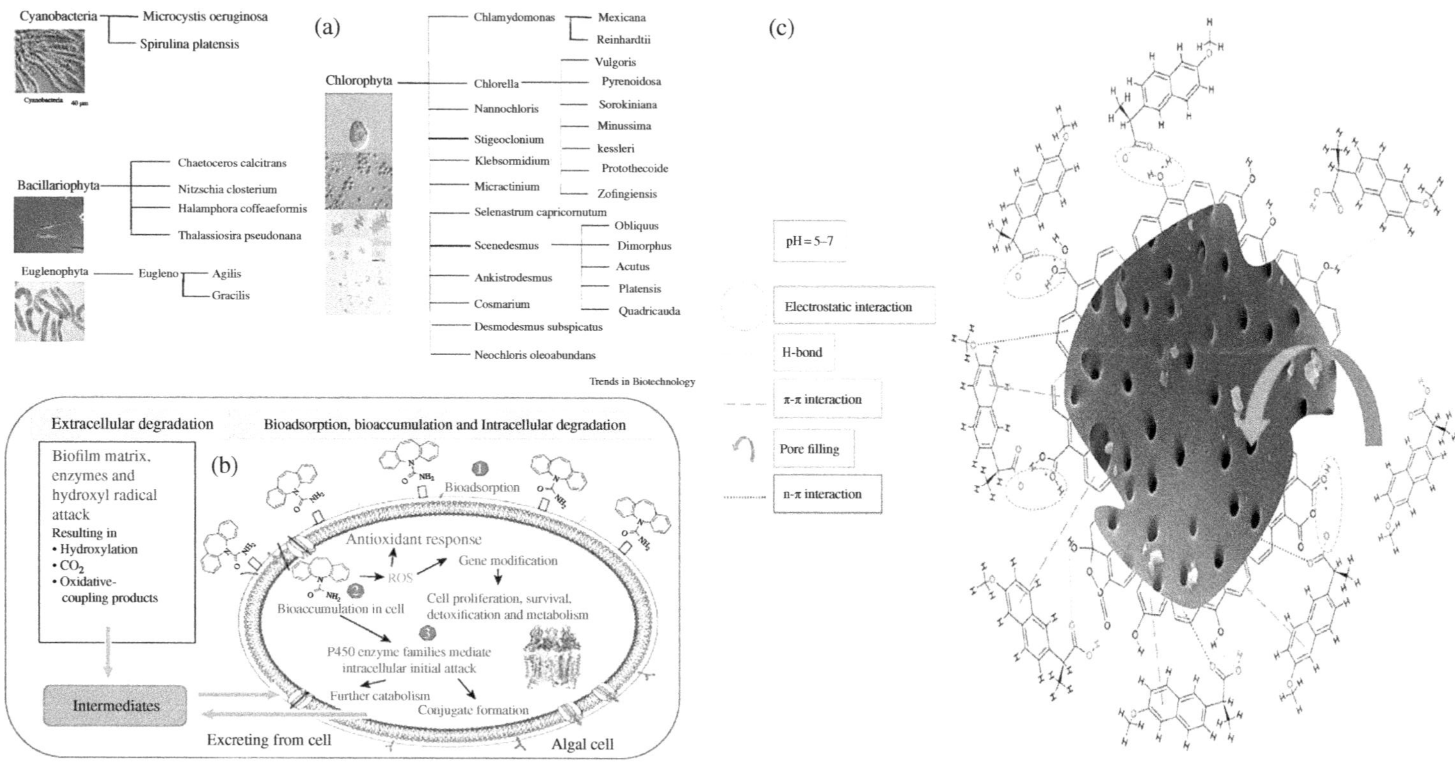

Figure 4.2 Proposed mechanism for bioadsorbents and bioadsorbates. *Source:* Xiong 2018, p. 30-44 / with permission of Elsevier.

friendly, cause low sludge formation, and are regenerative (Wang et al., 2015; Nharingo and Moyo, 2016). Many bioadsorbents are engineered by natural organisms and agricultural waste. Macroalgae, fungi, bacteria, and some aquatic plants have been introduced as biosorbents to remove dyes, pharmaceutical contaminants, and agricultural contaminants. Others biosorbents have been synthesized from agricultural waste and industrial waste.

Microalgae, fungi, and bacteria have been demonstrated to remove pharmaceutical waste, dyes, organic solvents, and pesticides from water (Adewuyi, 2020; Aravindhan et al., 2009; Wu et al., 2012; Aloui et al., 2020; Silva et al., 2019). These sorbents have been reported as green materials, and after adsorption, the pollutants can be remediated; the biosorbents are also regenerated to eliminate organic pollutants from water. Many review articles have described these adsorbents' vital role in the adsorption of organic pollutants from water. These articles have also explained the biosorption mechanism during the sorption process. Microalgae, fungi, and bacteria use their cell wall to evacuate organic pollutants from water. The cell wall is a source of nutrients, cellulose, hemicellulose, lignin, and protein, which work to adsorb organic pollutants from water by electrostatic attraction on the sorbents. These sorbents can be used at a large scale to remove organic pollutants from water. Microalgae are a primary source of carbon content, and they take part in photosynthesis and produce oxygen in large quantities and fixation of carbon dioxide gas (CO_2) (Rangabhashiyam et al., 2014).

Biopolymers are an excellent attracting material for water decontamination because of their functionality and low cost; they are also easy to fabricate and environmentally friendly. Biopolymers can be synthesized from natural sources such as plants that consist of biopolymers such as lignin, cellulose, and hemicellulose. Chitosan is a biopolymer that is also reported as a good adsorbent for water treatment. Some researchers have demonstrated lignin cellulosic biopolymer to remove organic water contaminants (Abdolali et al., 2014; Crini et al., 2019; Ngo et al., 2015).

Researchers are currently focusing on resolving societal problems such as agricultural and industrial waste (Dai et al., 2018). They have prepared porous activated carbon from agricultural and industrial waste (Liakos et al., 2021), and the activated carbon has been used for applications such as sodium ion batteries, removing heavy metal and organic water pollutants, gas purification, decolorization and liquid treatment, oil treatment, and air treatment (Kumar et al., 2019b; Reza et al., 2020; Mezohegyi et al., 2012; Jafarinejad, 2017; Ro et al., 2008). Most researchers developed activated carbon from rice husks (Shamsollahi and Partovinia, 2019), orange peels, banana peels (Temesgen et al., 2018; Mohammed and Chong, 2014), nuts, jute, coconut husks, sugarcane waste (Sarker et al., 2017), and grapefruit peels (Rosales et al., 2016). They have evaluated the activated carbon to remove organic pollutants such as dyes (Shan et al., 2016). Some biosorbents are listed in Table 4.3 with their performance.

Mechanism of Biosorption for Adsorption of Organic Pollutants

The mechanism of biosorption depends primarily on three factors: pH, temperature, and type of biosorbents (for instance, lignocellulose, microalgae, fungi, and agricultural waste materials) (Mavioglu Ayan et al., 2012). Some of the literature investigated the mechanism of biosorption, by no researchers gave a reason for the adsorption process. They described

Table 4.3 Biosorbents and their performance.

Adsorbent	Adsorbate	Adsorption capacity	Reference
Hydrogel composite	Methylene blue	1968	(Melo et al., 2018)
Hydrogel composite	Methylene blue	1952	(Vaz et al., 2017)
Hydrogel micro beads	Acid orange 7	1670	(Kuroiwa et al., 2017)
Urea diammonium tartrate modified chitosan	Congo red	1597	(Zahir et al., 2017)
Chitosan granules	Reactive black 5	1559	(Kuki et al., 2011)
Diammonium tartrate modified chitosan	Congo red	1447	(Zahir et al., 2017)
Hyper-cross-linked hydrogel	Indigo carmine	118	(de Luna et al., 2017)
Hyper-cross-linked hydrogel	Rhodamine 6G	78	(de Luna et al., 2017)
Hyper-cross-linked hydrogel	Sunset yellow	72	(de Luna et al., 2017)
N-maleyl chitosan cross-linker	Methylene blue	66.89	(Nakhjiri et al., 2018)
N-maleyl chitosan cross-linker	Crystal violet	64.56	(Nakhjiri et al., 2018)
Rice husks	Methylene blue	221.75	(Vadivelan and Kumar, 2005)
Rice husk ash	Direct red 23	—	(Abdelwahab et al., 2005)
Rice husk ash	Methylene blue	690	(Chandrasekhar and Pramada, 2006)
Rice husk ash	Crystal violet	44.87	(Chakraborty et al., 2011)
Rice husk ash	Malachite green	—	(Chowdhury et al., 2011)
Rice husk ash	Methylene blue	≥45	(Chen et al., 2012)
Rice husks	Acid yellow 17	12.98	(Patil et al., 2015)
Powdered rice husks	Direct blue 199	8.6	(Saroj et al., 2015)
Rice husks	Methylene blue	1450-1950	(Vaz et al., 2017)
Rice husks	Eosine Y Red X-GRL	17.18 18.45	(Shafeii Darabi et al., 2018)

and presented hypothetical reasons for the mechanism of the adsorption process. On a hypothetical basis, the researchers explained that the interaction between adsorbate and adsorbent is either chemisorption or physical adsorption by adsorption isotherm studies. Some studies favored chemisorption as well as physical adsorption. Biosorption exhibits ion exchange reactions, electrostatic interactions, complexation, covalent bonding, weak Van der Waals force of attraction, adsorption, and micro-precipitation (Ahmad et al., 2014; Inyang and Dickenson, 2015; Rosales et al., 2017).

Abdolali and his team reviewed biosorption for wastewater treatment with lignocellulose materials (2014). In the review article, they explained the biosorption mechanism for metal and organic contaminants on lignocellulose biosorbents. During biosorption, lignocellulose biosorbents attract by chelating and ion exchange and form complex ions with a functional group. Fomina and Gadd described the biosorption of dye waste on chitosan biosorbents (2014). They also discussed factors that affected biosorption. The pH of the solution plays an important role in the interaction between adsorbate and adsorbent.

Conclusion and Future Prospects

Natural water resources such as lakes, rivers, and streams are freshwater reservoirs that ensure the water supply for domestic and agricultural practices and are crucial to maintain a balanced equilibrium between various spheres of the environment. However, in recent years, unchecked discharge of effluents into these water resources has emerged as a significant threat to environmental stability. Organic contaminants in water are especially devastating due to their complex and diverse chemistry. Wastewater streams from industries, agricultural runoff, and domestics sewage discharge large amounts of organic pollutants such as dyes, organic solvents, pesticides, and insecticides into bodies of water. As a result, stringent environmental laws have been imposed globally at various levels to reduce water contamination. In addition, scientific communities are developing state-of-the-art, apex-level technologies to treat wastewater streams. Of these techniques, adsorption has gained universal recognition because of its simple operation, low energy consumption, cost-effectiveness, and versatility. Furthermore, the chances of secondary contamination are significantly less with adsorption.

Among several classes of adsorbents, biosorbents have proven to be the most promising and efficient due to their large surface area, easy and scalable synthesis, low cost, recyclability, and excellent post-life utility. Importantly, biosorption methods provide an opportunity for the recovery of contaminants as resource material from polluted waters. Also, biosorbents can be made from extracted and isolated components of industrial, agricultural, and domestic wastes.

Through engineered nanostructuring, novel properties have been introduced into biosorbents. Such strategies open new avenues for developing and designing advanced biosorbents with extremely high efficiency for the sequestration of organic contaminants from water. Interestingly, recent reports reveal that biosorbents, especially those derived from agricultural wastes, can play a game-changing role in wastewater decontamination applications. Furthermore, waste-to-wealth approaches that include solid waste management, waste recycling, and upcycling strategies and approaches can be strengthened by fabricating biosorbents from industrial, agricultural, and domestic biowastes. However, specific challenges are associated with the commercial viability and technological feasibility of biosorbents, such as poor long-term environmental stability and fast weathering. Therefore, dedicated efforts in this direction are required to realize the commercial viability of biosorption in real-time wastewater treatment techniques.

The problem is complex and multidimensional, and a collective social, political, and scientific approach is indispensable for mitigation. Replenishing potable water is necessary for all humans, and researchers as well as government officials should take immediate steps to avoid wiping away a requirement for life: water.

References

Abdelwahab, O., El Nemr, A., El Sikaily, A., and Khaled, A. (2005). Use of rice husk for adsorption of direct dyes from aqueous solution: a case study of Direct F. *Scarlet. Egyptian Journal of Aquatic Research* 31 (1): 1–11.

Abdolali, A., Guo, W.S., Ngo, H.H. et al. (2014). Typical lignocellulosic wastes and by-products for biosorption process in water and wastewater treatment: A critical review. *Bioresource Technology* 160: 57–66. doi:10.1016/j.biortech.2013.12.037.

Abdullah, N.A., Othaman, R., Abdullah, I. et al. (2012). Studies on the adsorption of phenol red dye using silica-filled ENR/PVC beads. *Journal of Emerging Trends in Engineering and Applied Sciences* 3 (5): 845–850.

Adewuyi, A. (2020). Chemically modified biosorbents and their role in the removal of emerging pharmaceutical waste in the water system. *Water* 12 (6): 1551. doi:10.3390/w12061551.

Ahmad, M., Rajapaksha, A.U., Lim, J.E. et al. (2014). Biochar as a sorbent for contaminant management in soil and water: A review. *Chemosphere* 99: 19–33. doi:10.1016/j.chemosphere.2013.10.071.

Aksu, Z. (2005). Application of biosorption for the removal of organic pollutants: a review. *Process Biochemistry* 40 (3-4): 997–1026. doi:10.1016/j.procbio.2004.04.008.

Ali, I. and Aboul-Enein, H.Y. (2004). *Chiral Pollutants: Distribution, Toxicity and Analysis by Chromatography and Capillary* Electrophoresis. John Wiley and Sons.

Ali, I., Asim, M., and Khan, T.A. (2012). Low cost adsorbents for the removal of organic pollutants from wastewater. *Journal of environmental management*, 113: 170–183.

Aloui, A., Bouziane, N., Bendjeffal, H. et al. (n.d.). Biosorption of pharmaceuticals from aqueous medium by luffa cylindrica fibres: application of the linear form of Redlich-Peterson isotherm equation. doi:10.21203/rs.3.rs-21496/v1.

Aravindhan, R., Rao, J.R., and Nair, B.U. (2009). Application of a chemically modified green macro alga as a biosorbent for phenol removal. *Journal of Environmental Management* 90 (5): 1877–1883. doi:10.1016/j.jenvman.2008.12.005.

Asghar, M.A., Zhu, Q., Sun, S. et al. (2018). Suspect screening and target quantification of human pharmaceutical residues in the surface water of Wuhan, China, using UHPLC-Q-Orbitrap HRMS. *Science of the Total Environment* 635: 828–837. doi:10.1016/j.scitotenv.2018.04.179.

Azanu, D., Styrishave, B., Darko, G. et al. (2018). Occurrence and risk assessment of antibiotics in water and lettuce in Ghana. *Science of the Total Environment* 622-623: 293–305. doi:10.1016/j.scitotenv.2017.11.287.

Barceló, D. (2010). *Emerging Organic Pollutants in Waste Waters and Sludge.* Springer Berlin Heidelberg.

Bean, T.G., Rattner, B.A., Lazarus, R.S. et al. (2018). Pharmaceuticals in water, fish and osprey nestlings in Delaware River and Bay. *Environmental Pollution* 232: 533–545. doi:10.1016/j.envpol.2017.09.083.

Bouguettoucha, A., Reffas, A., Chebli, D. et al. (2016). Novel activated carbon prepared from an agricultural waste, Stipa tenacissima, based on ZnCl2 activation—characterization and application to the removal of methylene blue. *Desalination and Water Treatment* 57 (50): 24056–24069. doi:10.1080/19443994.2015.1137231.

Burns, E.E., Carter, L.J., Kolpin, D.W. et al. (2018). Temporal and spatial variation in pharmaceutical concentrations in an urban river system. *Water Research* 137: 72–85. doi:10.1016/j.watres.2018.02.066.

Caban, M., Lis, E., Kumirska, J. et al. (2015). Determination of pharmaceutical residues in drinking water in Poland using a new SPE-GC-MS(SIM) method based on Speedisk extraction disks and DIMETRIS derivatization. *Science of the Total Environment* 538: 402–411. doi:10.1016/j.scitotenv.2015.08.076.

Carmona, E., Andreu, V., and Picó, Y. (2014). Occurrence of acidic pharmaceuticals and personal care products in Turia River Basin: From waste to drinking water. *Science of the Total Environment* 484: 53–63. doi:10.1016/j.scitotenv.2014.02.085.

Chakraborty, S., Chowdhury, S., and Saha, P.D. (2011). Adsorption of crystal violet from aqueous solution onto NaOH-modified rice husk. *Carbohydrate Polymers* 86 (4): 1533–1541. doi:10.1016/j.carbpol.2011.06.058.

Chandrasekhar, S. and Pramada, P.N. (2006). Rice husk ash as an adsorbent for methylene blue – effect of ashing temperature. *Adsorption* 12 (1): 27–43. doi:10.1007/s10450-006-0136-1.

Chaukura, N., Gwenzi, W., Tavengwa, N. et al. (2016). Biosorbents for the removal of synthetic organics and emerging pollutants: Opportunities and challenges for developing countries. *Environmental Development*19: 84–89. doi:10.1016/j.envdev.2016.05.002.

Chavan, R.B. (2011). Environmentally friendly dyes. *Handbook of Textile and Industrial Dyeing*: 515–561. doi:10.1533/9780857093974.2.515.

Chen, X.-G., Lv, S.-S., Liu, S.-T. et al. (2012). Adsorption of methylene blue by rice hull ash. *Separation Science and Technology* 47 (1): 147–156. doi:10.1080/01496395.2011.606865.

Chowdhury, S., Mishra, R., Saha, P. et al. (2011). Adsorption thermodynamics, kinetics and isosteric heat of adsorption of malachite green onto chemically modified rice husk. *Desalination* 265 (1-3): 159–168. doi:10.1016/j.desal.2010.07.047.

Clarke, E.A. and Anliker, R. (1980). Organic dyes and pigments. In: *Anthropogenic Compounds* (ed. R. Anliker, G.C. Butler, E.A. Clarke et al.), 181–215. Springer Berlin Heidelberg.

Crini, G., Torri, G., Lichtfouse, E. et al. (2019). Dye removal by biosorption using cross-linked chitosan-based hydrogels. *Environmental Chemistry Letters* 17 (4): 1645–1666. doi:10.1007/s10311-019-00903-y.

Dai, Y., Sun, Q., Wang, W. et al. (2018). Utilizations of agricultural waste as adsorbent for the removal of contaminants: A review. *Chemosphere* 211: 235–253. doi:10.1016/j.chemosphere.2018.06.179.

de Luna, M.S., Castaldo, R. et al. (2017). Chitosan hydrogels embedding hyper-crosslinked polymer particles as reusable broad-spectrum adsorbents for dye removal. *Carbohydrate Polymers* 177: 347–354. doi:10.1016/j.carbpol.2017.09.006.

Feng, Y., Dionysiou, D.D., Wu, Y. et al. (2013). Adsorption of dyestuff from aqueous solutions through oxalic acid-modified swede rape straw: Adsorption process and disposal methodology of depleted bioadsorbents. *Bioresource Technology* 138: 191–197. doi:10.1016/j.biortech.2013.03.146.

Fick, J., Söderström, H., Lindberg, R.H. et al. (2009). Contamination of surface, ground, and drinking water from pharmaceutical production. *Environmental Toxicology and Chemistry* 28 (12): 2522. doi:10.1897/09-073.1.

Fomina, M. and Gadd, G.M. (2014). Biosorption: current perspectives on concept, definition and application. *Bioresource Technology* 160: 3–14.

Glossman-Mitnik, D. (2013). Computational study of the chemical reactivity properties of the rhodamine b molecule. *Procedia Computer Science* 18: 816–825. doi:10.1016/j.procs.2013.05.246.

Harrad, S. (2012). *Persistent Organic Pollutants: Environmental Behaviour and Pathways of Human Exposure*. Springer Science and Business Media.

Hossain, A., Nakamichi, S., Habibullah-Al-Mamun, M. et al. (2018). Occurrence and ecological risk of pharmaceuticals in river surface water of Bangladesh. *Environmental Research* 165: 258–266. doi:10.1016/j.envres.2018.04.030.

International Agency for Research on Cancer (IARC). (2011). IARC monographs on the evaluation of carcinogenic risks to humans.

Inyang, M. and Dickenson, E. (2015). The potential role of biochar in the removal of organic and microbial contaminants from potable and reuse water: A review. *Chemosphere* 134: 232–240. doi:10.1016/j.chemosphere.2015.03.072.

Ismail, M., Akhtar, K., Khan, M.I. et al. (2019). Pollution toxicity and carcinogenicity of organic dyes and their catalytic bio-remediation. *Current Pharmaceutical Design* 25 (34): 3645–3663.

Jurado, A., Vàzquez-Suñé, E., Carrera, J. et al. (2012). Emerging organic contaminants in groundwater in Spain: a review of sources, recent occurrence and fate in a European context. *Science of the Total Environment* 440: 82–94.

Kuki, Á., Nagy, L., Zsuga, M. et al. (2011). Fast identification of phthalic acid esters in poly(vinyl chloride) samples by Direct Analysis In Real Time (DART) tandem mass spectrometry. *International Journal of Mass Spectrometry* 303 (2-3): 225–228. doi:10.1016/j.ijms.2011.02.011.

Kumar, D., Anand, S., Poonam et al. (2019a). Removal of inorganic and organic contaminants from terrestrial and aquatic ecosystems through phytoremediation and biosorption. *Environmental Biotechnology:* For Sustainable Future: 45–71. doi:10.1007/978-981-10-7284-0_3.

Kumar, U., Goonetilleke, D., Gaikwad, V. et al. (2019b). Activated carbon from e-waste plastics as a promising anode for sodium-ion batteries. *ACS Sustainable Chemistry and Engineering* 7 (12): 10310–10322. doi:10.1021/acssuschemeng.9b00135.

Kuroiwa, T., Takada, H., Shogen, A. et al. (2017). Cross-linkable chitosan-based hydrogel microbeads with pH-responsive adsorption properties for organic dyes prepared using size-tunable microchannel emulsification technique. *Colloids and Surfaces A: Physicochemical and Engineering Aspects* 514: 69–78. doi:10.1016/j.colsurfa.2016.11.046.

Kushwaha, A.K., Gupta, N., and Chattopadhyaya, M.C. (2014). Removal of cationic methylene blue and malachite green dyes from aqueous solution by waste materials of Daucus carota. *Journal of Saudi Chemical Society* 18 (3): 200–207. doi:10.1016/j.jscs.2011.06.011.

Lellis, B., Fávaro-Polonio, C.Z., Pamphile, J.A. et al. (2019). Effects of textile dyes on health and the environment and bioremediation potential of living organisms. *Biotechnology Research and Innovation* 3 (2): 275–290. doi:10.1016/j.biori.2019.09.001.

Liakos, E.V., Rekos, K., Giannakoudakis, D.A. et al. (2021). Activated porous carbon derived from tea and plane tree leaves biomass for the removal of pharmaceutical compounds from wastewaters. *Antibiotics* 10 (1): 65. doi:10.3390/antibiotics10010065.

Lindim, C., van Gils, J., Georgieva, D. et al. (2016). Evaluation of human pharmaceutical emissions and concentrations in Swedish river basins. *Science of the Total Environment* 572: 508–519. doi:10.1016/j.scitotenv.2016.08.074.

Madikizela, L.M., Tavengwa, N.T., and Chimuka, L. (2017). Status of pharmaceuticals in African water bodies: Occurrence, removal and analytical methods. *Journal of Environmental Management* 193: 211–220. doi:10.1016/j.jenvman.2017.02.022.

Mahmoud, A.F.A., Ikenaka, Y., Yohannes, Y.B. et al. (2016). Distribution and health risk assessment of organochlorine pesticides (OCPs) residue in edible cattle tissues from northeastern part of Egypt: High accumulation level of OCPs in tongue. *Chemosphere* 144: 1365–1371. doi:10.1016/j.chemosphere.2015.10.016.

Matongo, S., Birungi, G., Moodley, B. et al. (2015). Pharmaceutical residues in water and sediment of Msunduzi River, KwaZulu-Natal, South Africa. *Chemosphere* 134: 133–140. doi:10.1016/j.chemosphere.2015.03.093.

Mavioglu Ayan, E., Secim, P., Karakaya, S. et al. (2012). Oreganum Stalks as a New Biosorbent to Remove Textile Dyes from Aqueous Solutions. *CLEAN - Soil, Air, Water* 40 (8): 856–863. doi:10.1002/clen.201100153.

Melo, B.C., Paulino, F.A.A., Cardoso, V.A. et al. (2018). Cellulose nanowhiskers improve the methylene blue adsorption capacity of chitosan-g-poly(acrylic acid) hydrogel. *Carbohydrate Polymers* 181: 358–367. doi:10.1016/j.carbpol.2017.10.079.

Mezohegyi, G., van der Zee, F.P., Font, J. et al. (2012). Towards advanced aqueous dye removal processes: A short review on the versatile role of activated carbon. *Journal of Environmental Management* 102: 148–164. doi:10.1016/j.jenvman.2012.02.021.

Michalak, I., Chojnacka, K., and Witek-Krowiak, A. (2013). State of the art for the biosorption process—a review. *Applied Biochemistry and Biotechnology* 170 (6): 1389–1416.

Mohammed, R.R. and Chong, M.F. (2014). Treatment and decolorization of biologically treated palm oil mill effluent (POME) using banana peel as novel biosorbent. *Journal of Environmental Management* 132: 237–249. doi:10.1016/j.jenvman.2013.11.031.

Mutiyar, P.K., Gupta, S.K., and Mittal, A.K. (2018). Fate of pharmaceutical active compounds (PhACs) from River Yamuna, India: An ecotoxicological risk assessment approach. *Ecotoxicology and Environmental Safety* 150: 297–304. doi:10.1016/j.ecoenv.2017.12.041.

Nagendran, R. (2011). Agricultural Waste and Pollution. *Waste*: 341–355. doi:10.1016/b978-0-12-381475-3.10024-5.

Nakhjiri, M.T., Marandi, G.B., and Kurdtabar, M. (2018). Poly(AA-co-VPA) hydrogel cross-linked with N-maleyl chitosan as dye adsorbent: Isotherms, kinetics and thermodynamic investigation. *International Journal of Biological Macromolecules* 117: 152–166. doi:10.1016/j.ijbiomac.2018.05.140.

Ngo, H.H., Tran, V.S., Guo, W. et al. (2015). Typical low cost biosorbents for adsorptive removal of specific organic pollutants from water. *Bioresource Technology* 182: 353–363. doi:10.1016/j.biortech.2015.02.003.

Nharingo, T. and Moyo, M. (2016) Application of Opuntia ficus-indica in bioremediation of wastewaters. *A critical review. Journal of Environmental Management* 166: 55–72. doi:10.1016/j.jenvman.2015.10.005.

Pal, P. (2018). Treatment and disposal of pharmaceutical wastewater: toward the sustainable strategy. *Separation and Purification Reviews* 47 (3): 179–198. doi:10.1080/15422119.2017.1354888.

Park, D., Yun, Y.-S., and Park, J.M. (2010). The past, present, and future trends of biosorption. *Biotechnology and Bioprocess Engineering* 15 (1): 86–102. doi:10.1007/s12257-009-0199-4.

Patil, C., Ratnamala, G., and Channamallayya, S. (2015). Adsorption studies for removal of acid yellow 17 using activated rice husk. *Int Res J Eng Technol* 2: 769.

Pereira, A.M.P.T., Pereira, A.M.P., Silva, L.J.G. et al. (2017). Human pharmaceuticals in Portuguese rivers: The impact of water scarcity in the environmental risk. *Science of the Total Environment* 609: 1182–1191. doi:10.1016/j.scitotenv.2017.07.200.

Praveena, S.M., Shaifuddin, S.N.M., Sukiman, S. et al. (2018). Pharmaceuticals residues in selected tropical surface water bodies from Selangor (Malaysia): Occurrence and potential risk assessments. *Science of the Total Environment* 642: 230–240. doi:10.1016/j.scitotenv.2018.06.058.

Rangabhashiyam, S., Suganya, E., Selvaraju, N. et al. (2014). Significance of exploiting non-living biomaterials for the biosorption of wastewater pollutants. *World Journal of Microbiology and Biotechnology* 30 (6): 1669–1689. doi:10.1007/s11274-014-1599-y.

Rashed, M.N. (2006). Fruit *Stones from Industrial Waste for the Removal of Lead Ions from Polluted Water Environmental Monitoring and Assessment* 119 (1-3): 31–41. doi:10.1007/s10661-005-9002-y.

Reza, M.S., Yun, C.S., Afroze, S. et al. (2020). Preparation of activated carbon from biomass and its applications in water and gas purification, a review. *Arab Journal of Basic and Applied Sciences* 27 (1): 208–238. doi:10.1080/25765299.2020.1766799.

Rivera-Jaimes, J.A., Postigo, C., Melgoza-Alemán, R.M. et al. (2018). Study of pharmaceuticals in surface and wastewater from Cuernavaca, Morelos, Mexico: Occurrence and environmental risk assessment. *Science of the Total Environment* 613-614: 1263–1274.

Ro, K.S., McConnell, L.L., Johnson, M.H. et al. (2008). Livestock air treatment using PVA-coated powdered activated carbon biofilter. 2008 Providence, Rhode Island, June 29 - July 2, 2008. doi:10.13031/2013.24964.

Rosales, E., Meijide, J., Pazos, M. et al. (2017). Challenges and recent advances in biochar as low-cost biosorbent: From batch assays to continuous-flow systems. *Bioresource Technology* 246: 176–192. doi:10.1016/j.biortech.2017.06.084.

Rosales, E., Meijide, J., Tavares, T. et al. (2016). Grapefruit peelings as a promising biosorbent for the removal of leather dyes and hexavalent chromium. *Process Safety and Environmental Protection* 101: 61–71. doi:10.1016/j.psep.2016.03.006.

Sarker, T.C., Azam, S.M.G., Abd El-Gawad, A.M. et al. (2017). Sugarcane bagasse: a potential low-cost biosorbent for the removal of hazardous materials. *Clean Technologies and Environmental Policy* 19 (10): 2343–2362. doi:10.1007/s10098-017-1429-7.

Saroj, S., Singh, S.V., and Mohan, D. (2015). Removal of colour (direct blue 199) from carpet industry wastewater using different biosorbents (maize cob, citrus peel and rice husk). *Arabian Journal for Science and Engineering* 40 (6): 1553–1564. doi:10.1007/s13369-015-1630-0.

Shafeii Darabi, S.F., Bahramifar, N., and Khalilzadeh, M.A. (2018). Equilibrium, thermodynamic and kinetics studies on adsorption of eosin Y and red X-GRL from aqueous solution by treated rice husk. *Journal of Applied Research in Water and Wastewater* 5: 392–398.

Shamsollahi, Z. and Partovinia, A. (2019). Recent advances on pollutants removal by rice husk as a bio-based adsorbent: A critical review. *Journal of Environmental Management* 246: 314–323. doi:10.1016/j.jenvman.2019.05.145.

Shan, D., Deng, S., Zhao, T. et al. (2016). Preparation of ultrafine magnetic biochar and activated carbon for pharmaceutical adsorption and subsequent degradation by ball milling. *Journal of Hazardous Materials* 305: 156–163. doi:10.1016/j.jhazmat.2015.11.047.

Shannon, A.L. and Woolridge, A. (2011). Medical Waste. *Waste*: 329–339. doi:10.1016/b978-0-12-381475-3.10023-3.

Silva, A., Delerue-Matos, C., Figueiredo, S. et al. (2019). The use of algae and fungi for removal of pharmaceuticals by bioremediation and biosorption processes: a review. *Water* 11 (8): 1555. doi:10.3390/w11081555.

Temesgen, F., Gabbiye, N., and Sahu, O. (2018). Biosorption of reactive red dye (RRD) on activated surface of banana and orange peels: Economical alternative for textile effluent. *Surfaces and Interfaces* 12: 151–159. doi:10.1016/j.surfin.2018.04.007.

Tran, N.H. and Gin, K.Y.-H. (2017). Occurrence and removal of pharmaceuticals, hormones, personal care products, and endocrine disrupters in a full-scale water reclamation plant. *Science of the Total Environment* 599-600: 1503–1516. doi:10.1016/j.scitotenv.2017.05.097.

Vadivelan, V. and Vasanth Kumar, K. (2005). Equilibrium, kinetics, mechanism, and process design for the sorption of methylene blue onto rice husk. *Journal of Colloid and Interface Science* 286 (1): 90–100. doi:10.1016/j.jcis.2005.01.007.

Van der Bruggen, B., and Vandecasteele, C. (2002). Distillation vs. membrane filtration: overview of process evolutions in seawater desalination. *Desalination* 143 (3): 207–218. doi:10.1016/s0011-9164(02)00259-x.

Vaz, M.G., Pereira, A.G.B., Fajardo, A.R. et al. (2017). Methylene blue adsorption on chitosan-g-poly(acrylic acid)/rice husk ash superabsorbent composite: kinetics, equilibrium, and thermodynamics. *Water, Air, and Soil Pollution* 228 (1). doi:10.1007/s11270-016-3185-4.

Walsh, V. 1984. Invention and innovation in the chemical industry: demand-pull or discovery-push? *Research Policy* 13: 211–234.

Walsh, V. (1993). Invention and innovation in the chemical industry: demand-pull or discovery-push? *Research Policy* 22 (2): 115–116. doi:10.1016/0048-7333(93)90067-r.

Wang, H., Yuan, X., Zeng, G. et al. (2014). Removal of malachite green dye from wastewater by different organic acid-modified natural adsorbent: kinetics, equilibriums, mechanisms, practical application, and disposal of dye-loaded adsorbent. *Environmental Science and Pollution Research* 21 (19): 11552–11564. doi:10.1007/s11356-014-3025-2.

Wang, M.-X., Zhang, Q.-L., and Yao, S.-J. (2015). A novel biosorbent formed of marine-derived Penicillium janthinellum mycelial pellets for removing dyes from dye-containing wastewater. *Chemical Engineering Journal* 259: 837–844. doi:10.1016/j.cej.2014.08.003.

Wu, Y., Li, T., and Yang, L. (2012). Mechanisms of removing pollutants from aqueous solutions by microorganisms and their aggregates: A review. *Bioresource Technology* 107: 10–18. doi:10.1016/j.biortech.2011.12.088.

Xiong, J.-Q., Kurade, M.B., and Jeon, B.-H. (2018). Can microalgae remove pharmaceutical contaminants from water? *Trends in Biotechnology* 36 (1): 30–44. doi:10.1016/j.tibtech.2017.09.003.

Zahir, A., Aslam, Z., Kamal, M.S. et al. (2017). Development of novel cross-linked chitosan for the removal of anionic Congo red dye. *Journal of Molecular Liquids* 244: 211–218. doi:10.1016/j.molliq.2017.09.006.

Zollinger, H. (1989). Color chemistry: synthesis, properties and applications of organic dyes and pigments. *Leonardo* 22 (3/4): 456. doi:10.2307/1575449.

5

Recent Approaches in the Preparation of Various Biosorbents

Rajarathinam Nithya and Arunachalam Thirunavukkarasu

Department of Industrial Biotechnology, Government College of Technology, Coimbatore, Tamilnadu, India

Introduction

In this ultra-modern world, many industries have developed to meet our needs. They not only provide essentials but also cause significant problems that lead to environmental pollution. One type of pollution caused by industrial activity is water pollution. Exploitation of available water resources may affect future generations. Despite World Health Organization (WHO) recommendations of maximum acceptable concentration ranges that should be present in bodies of water, industries pollute fresh water by releasing toxic compounds like dyes, heavy metals, and other organics in effluents; the results are severe harmful effects on humans and aquatic life.

Among the toxic compounds in effluents, dyes are very complex and extremely challenging to degrade biologically (Forgacs et al., 2004; Gupta et al., 2009; Rafatullah et al., 2010). Dyes are released in large quantities as effluents from many industries such as textiles, paper, leather, food, cosmetics, etc.; worldwide, nearly 700,000 tons of dyes are released by industries each year (Kyzas et al., 2013). Dyes are further classified based on their charge as anionic, cationic, reactive, and non-ionic (Vijayaraghavan et al., 2013).

Other toxic compounds in effluents, such as phenols and phenolic compounds, can also cause harmful effects to human health even at very low concentrations (0.005 mg/L) (Aksu 2005). Phenol is produced by industries such as iron and steel, paper and pulp, petroleum, paint, and pharmaceuticals. Phenolic compounds such as nitro-phenols and chloro-phenols, along with heavy metals, are considered the most hazardous pollutants. Heavy metals are also discharged in effluent. These elements are toxic and carcinogenic; copper, chromium, lead, mercury, cadmium, and arsenic are some examples of heavy metals. Due to their non-biodegradable nature, these metal ions easily accumulate in the food chain in large amounts (Wang et al., 2009; Zhou and Haynes, 2010). The primary sources

Biosorption for Wastewater Contaminants, First Edition.
Edited by Rangabhashiyam Selvasembian and Pardeep Singh.

of these metals are industries such as automotive, semiconductor manufacturing, metal finishing and plating, textiles, and steel.

Removing toxic pollutants from the environment is essential and can be achieved using physical, chemical, or biological treatment processes. Physical treatment includes membrane filtration, coagulation, adsorption, and chelation of ion exchange. Chemical treatment includes oxidation and electrochemical treatment; and biological treatment includes anaerobic, aerobic, and enzymatic approaches (Wang et al., 2009; Gadd, 2009; Zhou and Haynes, 2010). These are considered conventional treatment methods and have drawbacks such as high energy requirements, low efficiency, and high cost (Gogate and Pandit, 2004; Bolong et al., 2009). In recent studies, researchers have proposed several approaches for effluent treatment by developing unconventional, low-cost adsorbents (Nithya et al., 2018; Thirunavukkarasu et al., 2018; Sivashankar et al., 2020; Rajarathinam et al., 2020; Thirunavukkarasu and Nithya, 2020). One of these unconventional methods is biosorption, which is an attractive technique because it is highly efficient and cost-effective and can be easily implemented (Chojnacka, 2010).

Biosorption is defined as a process of adsorption on the surface of a compound of biological origin. The process of biosorption is not bound to a particular mechanism or constrained to a specific type of pollutant, and hence it has a wide variety of applications including pollution control, element recovery, and recycling (Gadd 2009). The choice of low-cost biosorbent should satisfy the following conditions (Gadd 2009):

- A waste material or byproduct from other industries
- Easily available and present in abundance
- Does not cause any harm to the environment during the treatment process and requires little or no pretreatment.

The best suitable biosorbent has the potential to remove a pollutant efficiently. The choice of biosorbent includes, but is not restricted to, the mechanism used for the biosorption treatment, nature of the pollutant/element to be adsorbed, effectiveness of the biosorbent, possible regeneration or recycling of the biosorbent, and cost of the process.

Several reviews reported the use of biomass from agricultural waste (dead/living biomass) and other natural adsorbents for removing dyes (Pearce et al., 2003; Crini, 2006; Crini and Badot, 2008; Kaushik and Malik, 2009; Demirbas, 2009; Karthika et al., 2010; Ngah et al., 2011; Salleh et al., 2011) and toxic heavy metals (Ahluwalia and Goyal, 2007; Sud et al., 2008; Demirbas et al., 2008; Yadanaparthi et al., 2009; Lesmana et al., 2009; Farooq et al., 2010; Fosso-Kankeu and Mulaba-Bafubiandi, 2014). Recent research has focused on removing organic pollutants, pesticides, phenols, and phenolic compounds by using different natural and low-cost biosorbents (Jain et al., 2004; Yan et al., 2010; Mathialagan and Viraraghavan, 2005; Hai et al., 2012; Maciel et al., 2013; Nguyen et al., 2014). The most challenging problem in many industries is the cost associated with water treatment or disposal, accumulation of the waste, and transportation. This problem is commonly associated with food industries, which produce a huge amount of waste. The discharge of industrial wastes and their byproducts also causes environmental problems, and hence the utilization of industrial and agricultural wastes as effective, low-cost biosorbents provides a dual solution. Treating effluents with these low- or zero-cost waste materials adds more value and provides the best solution for the world's environmental issues.

Biosorbents

Biosorbents are materials with a biological origin that have versatile, functional groups on their surface, facilitating the passive translocation of pollutants or contaminants across the phase boundary. These biosorbents include microbial biomass and biomass/wastes from plants such as rice husks (Manique et al., 2012), coconut shells (Acheampong et al., 2013), plant bark (Reddy et al., 2011), leaves (Karthika et al., 2010; Reddy et al., 2010; Reddy et al., 2012), sawdust (Witek-Krowiak, 2013), sugarcane bagasse (Khoramzadeh et al., 2013), and peat moss (Ma and Tobin, 2003); animal wastes such as bones, crab shells, chicken feathers, animal horns, human hair, etc. (Srivastava et al., 2016); and bio-based industries (Ramrakhiani et al., 2017).

The use of such materials in pollution research has increased many-fold with remarkable removal efficiencies in recent decades as these are renewable, biodegradable, inexpensive, and readily available. The prime advantage of using biological materials is their diverse surface functional groups that facilitate hydrophobic interactions with solute molecules during the biosorption process (Tadkaew et al., 2011; Adewuyi et al., 2017). As this solute transfer process is mediated on the surface functional moieties, the acidic or alkaline nature of the liquid microenvironment is highly significant.

Thousands of works have been found in the literature, using a wide range of biosorbents for the potential removal of pollutants from wastewater. The majority of these scholarly works utilized plant-based biomass and living or dead microbial biomass and reported the biosorptive capacity of a diverse range of pollutants or contaminants. Some of the reports elucidated the mechanistic aspects of biosorption at the solid-liquid interface, along with their regeneration potential in batch and/or continuous flow studies.

As most of the possible research areas in conventional biosorption have been scientifically addressed, researchers are now actively searching for new-generation biosorbents with tailor-made functional groups to remove targeted pollutants or contaminants. Physical treatments such as milling, sizing, heating, steaming, freeze-drying, etc. and chemical-based treatments are effective strategies in the search for a new generation of bio-based sorbents. The literature also reported the comparative biosorptive analysis of unmodified biosorbents with modified materials. These reports showed that naive biosorbents could release certain organic compounds into the liquid, which augments the pollutant's concentration in wastewater and decreases the oxygen content (Ngah et al., 2008). Hence, pretreatment (physical) techniques are essential in preventing increased biochemical oxygen demand (BOD), chemical oxygen demand (COD), and total organic carbon in wastewater. Besides the physical treatment of biosorbents, the use of chemicals to modify the active binding pockets on the biosorbents' surface might provide the additional advantage of making them usable against a wide range of biosorbate molecules and improving the biosorption capacity.

Physical Treatment of Biosorbents

Sterilization

Sterilization is the process of destroying viable microbial cells on the surface of a material. In general, applying heat to the cell can destroy the cell wall integrity and cause the thermal denaturation of vital proteins and enzymes, resulting in cell death. This can be accomplished either by moist heat method or by dry heat sterilization.

Moist heat sterilization is the most efficient approach, as steam under high positive pressure acts as a microbicidal agent. Most processes employ steam with a temperature range of 121–134 °C. High positive pressure increases the boiling point of water and also facilitates the transfer of heat into the cellular organelles, which causes protein and/or enzyme coagulation and thus reduces microbial action. Autoclaving and boiling are the two methods of choice for the moist heat method, and applying steam under pressure reduces the time required to destroy the cell compared to dry heat sterilization.

Heat conduction is the principle of the dry heat method: it acts on cells and causes the release of oxidative free radicals, denaturing the proteins, drying the cells, and burning them to ashes. The lethal action of the dry heat method is primarily due to oxidative actions, which are comparatively less effective than the hydrolytic actions of moist heat sterilization. Red heat burning using Bunsen burners, flaming, incineration, hot-air ovens, and infra-red radiation are the choices for the dry heat method.

The choice of dry or moist heat depends on the type of biomaterials used. For instance, if dead microbes are used as a biosorbent, the only effective approach is autoclaving; and if plant-derived wastes are used, dry heat methods would be the choice.

Comminution

Comminution, or size reduction, is the process of reducing the average particle size of solid materials by crushing, grinding/milling, cutting, or vibrating. It is an energy-intensive process and can be mechanically operated or achieved with ultrasonic sounds. The energy to comminute can be driven by Rittinger's law, Kick's law, Bond's law, and Holmes's law.

The primary mechanism of mechanical size reduction is based on three forces: impact, shear, and compression. Using plant-based biomasses such as leaves, barks, seeds, roots, etc. as biosorbents, a mechanical sizing process is inevitable. After moisture removal, these biomasses are subjected to cutting, grinding, and milling to achieve the required average particle size. As the process of biosorption is highly dependent on the surface area of the biosorbents, the average particle size must be smaller to achieve the effective removal of contaminants or pollutants from wastewater. In general, roll crushing followed by ball milling is the preferable method for comminuting biosorbents. Ultrasonic waves are the least preferred approach as they require solid biosorbents to be suspended in the aqueous medium (Arunachalam et al., 2017). In roll crushing and ball milling, the impact and attrition created between the rollers and balls and the biosorbent material result in the sizing action.

Following comminution, milled biosorbents must be subjected to sieve analysis to assess the particle size distribution by allowing them to pass through a set of sieves with decreasing pore sizes. The degree of heterogeneity in the particle sizes of the crushed materials can be analyzed, and the required sizes of the materials can be easily screened out.

Cryodessication

Cryodessication, alternatively known as lyophilization or freeze-drying, is the process of low-temperature dehydration in which the desired product can be frozen by lowering its pressure; the ice that forms is sublimated. As this process operates at low temperature and pressure, the physicochemical properties of the materials can be restored. Free and bound moisture is completely removed from the biomaterials, which can then be effectively stored longer without the possibility of microbial contamination. Also, when using extracts of plant-derived products as biosorbents, freeze-drying is crucial as it converts liquid-phase constituents into solid biomasses.

Microwave Drying

Microwave drying is an alternative to conventional drying techniques to remove moisture from materials of interest. It works on the principle of heating mediated by exposing the material to electromagnetic radiation in the microwave frequency range of 1 to 1000 GHz (Ohring et al., 2015). Unlike conventional heating techniques such as hot-air ovens, spray dryers, dry baths, etc., which are driven by conduction and convection, the microwave technique uses magnetrons to produce microwaves that turn moisture into vapor.

The time required for drying using microwaves is 60 times better than the conventional oven-drying process. Rapid, violent microwaves can disrupt the biomass surface and release more volatile compounds from them; the dried biomass has low crystallinity. This lowers the net heating value, energy yield, and total elemental carbon compared to oven drying (Amer et al., 2019).

Chemical Treatment of Biosorbents

Acid Treatment

Concentrated and diluted acids, such as sulphuric acid, hydrochloric acid, nitric acid, etc., are used in acid treatment to break the rigid structure of biomasses. Dilute sulphuric acid is the most commonly used acid to treat a wide variety of biosorbents such as switchgrass (Digman et al., 2010; Li et al., 2010), corn stover (Du et al., 2010; Xu et al., 2010), spruce (softwood) (Shuai et al., 2010), poplar (Wyman et al., 2009; Kumar et al., 2009), etc. This acid treatment usually involves the addition of 0.2 to 2.5% w/v of either concentrated or dilute acids, followed by constant stirring at high temperatures in the range of 130 to 210 °C (Esteghlalian et al., 1997; Nguyen et al., 2000; Saha et al., 2005; Sun and Cheng, 2005; Silverstein et al., 2007; Sassner et al., 2008). This process improves the hydrolysis of cellulose, which increases the surface area and average pore volume of the biomaterial

(Martín-Lara et al., 2013). The type of acid, normality, pH of the solution, reaction temperature, and contact time are the prime influencing factors in the pretreatment process.

Besides sulphuric and nitric acids, the use of citric acid is also reported in the literature. For instance, 2M of citric acid treatment of walnut shells improved the adsorption capacity of Zn ions by 2.5 times with the incorporation of carboxylic groups into the biomaterial (Segovia-Sandoval et al., 2018). Despite having the potential to improve the physicochemical properties of the biomass, the corrosive nature, toxicity, and selection of material to withstand the required experimental conditions and the need for extensive washing and/or detoxification are potential challenges of acid treatment.

Alkali Treatment

The hydroxide forms of alkali compounds such as sodium, potassium, calcium, and ammonium are the key components in the alkali treatment of biosorbents. These basic compounds can alter the structural integrity of lignocellulosic biomass by degrading the ester and glycosidic side chains and also causes swelling and/or partial decrystallization of cellulose (Cheng et al., 2010; McIntosh and Vancov, 2010; Ibrahim et al., 2011) and partial solvation of hemicellulose (Sills and Gossett 2011). Of these compounds, sodium and calcium hydroxide are the most extensively studied as these alkali materials can disrupt the lignin structure of biologically derived materials and hence increase their surface-related properties (MacDonald et al., 1983; Soto et al., 1984; Zhao et al., 2008; Zhu et al., 2010).

Alkali treatment usually yields two streams consisting of wet solid fractions of cellulosic compounds and liquid fractions of dissolved hemicellulose, degraded lignin, and other unreacted chemicals. The wet solid fraction can be separated and neutralized before use in wastewater treatment processes. This technique is usually less severe than the other approaches as it can be operated at ambient temperature and pressure. The alkali loading, solid-to-liquid ratio, and contact or retention time are the other influencing factors affecting pretreatment performance.

Pyrolysis

Pyrolysis, or gasification, is an energy-intensive process in which biochar, bio-oil, and biogas are generated from biologically derived materials. Such products have distinct physicochemical properties and differ significantly from their parent biomasses. In general, gasification technology can be classified based on retention time, pyrolytic temperature, pressure, biosorbent size distribution, rate of heating, and fuel used for the heat supply.

After the recovery of bio-oil and bio-gas, the end product or residual matter is biochar that can be modified into activated carbon residues (Morni et al., 2018). Activated carbon can be produced using the different activation approaches and operating conditions (Yusuff, 2019). For instance, steam, carbon dioxide, or nitrogen can be used in the physical activation process, and acids, bases, or metallic oxides can be used in the chemical method of activation. Likewise, applying heat through microwaves and chemical reactions can synthesize activated carbon in physicochemical approaches (Ao et al., 2018). The nature of the precursor molecules, residual size, retention time, impregnation ratio, and chemical agents are the vital factors influencing the effectiveness of the activation process.

Numerous research has been carried out using activated carbon to remove contaminants and pollutants from wastewater. To further improvise the sorptive capacity of these materials, nanoparticles are loaded on their surface and employed in removing organic and inorganic compounds from wastewater (Lakshmi et al., 2018; Thirunavukkarasu et al., 2018; Thirunavukkarasu et al., 2020; Rajarathinam et al., 2020; Thirunavukkarasu and Nithya, 2021). Besides the advantages of synthesizing different forms of activated carbon, the generation and release of volatile organic compounds during the process and the energy requirements are the key concerns for this technology.

Solid-Liquid Extraction

Solid-liquid extraction or *leaching* is the process of separating desired components from solid materials using solvents. The relative affinity of the targeted compounds to the solvent over the solid material drives this separation process, and the degree of affinity determines the efficiency of the mass-transfer process. The prime driving force is the difference in the concentration gradient between the solid material and solvents used in the process. Besides the partitioning behavior of the solute, the solubility, solid-liquid ratio, contact time, solvent nature and temperature, etc. are the influencing factors of leaching.

Plant-derived biomass like seeds is collected and subjected to the leaching process before being used in biosorption studies to extract essential oils. Thus, the oil-free leached-out waste biomasses can be considered potential biosorbents to remove contaminants from aqueous media. For instance, Nemeş et al. (2016) reported the use of waste biomass from white mustard seeds to remove Pb(II), Zn(II), and Cd(II) from aqueous solution. This technique offers the advantage of recovering an essential product with the least complex operational conditions. However, the use of additional chemical solvents such as ethanol, methanol, hexane, or chloroform might increase the pollutant level in wastewater. Also, the residual biomasses showed considerable traces of such chemicals, and hence it is essential to search for alternatives to conventional chemical solvents.

Immobilization

Immobilization is a technique in which whole living or dead microbial cells are entrapped into inert and insoluble gels such as calcium alginate or strontium alginate. These immobilized cell systems provide increased resistance to fluctuation in the solution microenvironment in terms of pH or temperature. Besides the batch mode of operation, fixed-bed continuous flow studies are increasingly reported in the literature to simulate real-time industrial-scale analysis. The use of diverse classes of biosorbents as the packing materials in such fixed beds has increased tremendously in the past decade. Before column loading, microbial biomasses are immobilized in a calcium alginate matrix (Thirunavukkarasu and Nithya, 2019; Aryal et al., 2019; Ahmad et al., 2019; Reddy et al., 2020). This allows the biomass to be retained in the column, significantly reducing its downstream recovery and increasing its thermo-mechanical stability.

With the further development of such immobilized systems, researchers have started developing much more complex systems by forming biocomposites with polymeric structures. These materials might provide the required protection for living microorganisms by resisting the operational factors. For instance, Paez-Velez et al. (2020) immobilized the bacterial strain *Lysinibacillus sphaericus* CBAM5 onto polycaprolactone micro-fibrous mats and alginate microcapsules to effectively capture gold from synthetic water samples.

The percentage composition of sodium alginate, porous volume of the polymeric beads, size of the immobilized beads, loading ratio, cell loading volume, pressure, and temperature are the influencing factors in determining the efficacy of the biosorption process. Ease of recovery, reusability, maintenance of live biomass, and potential to scale up the process are significant advantages of immobilized systems. On the other hand, additional diffusive resistances offered by the gel matrix might affect the performance of the sorptive capacity as the sorbate molecules can have difficulty reaching the accessible sites on the biosorbent surfaces.

Chemical and Genetic Modifications

Another strategy for improving the sorption capacity of biosorbents is the use of various chemicals to modify the functional properties of the biomass and alter their surface topography. Chemical treatments include esterification, graft polymerization, and treatment with chemical agents such as methanol, formaldehyde, triethyl phosphate, cationic surfactants, nitromethane, etc. (Ribeiro et al., 2019; Villen-Guzman et al., 2019; Kocaman et al., 2020; Zhang et al., 2020). These treatments are applicable to plant- and animal-derived biomasses and can even be employed with dead biomass.

For live biomasses, recombinant technology can be used. Native strains are screened and subjected to a molecular cloning process. The genetic elements are incorporated with the gene of interest through vector elements to improve resistance to pollutant toxicity and increase microbial accumulation/degradation. Further, the transformed microbial strains are screened for plasmid expression and maintained as recombinant strains. Several reports in the literature use these recombinant strains to remove heavy metals from wastewater (Tan et al., 2019; Dong et al., 2019; De Oliveira et al., 2020). The selection of strain, cloning vector, plasmid copy number, length of the gene of interest, expression vectors, and plasmid stability are the significant factors that affect the efficacy of the recombinant strain development process.

Despite the advantages of tailor-made strain development, desirable grafting of surface functional groups, and improved biosorptive capacity, the possibility of genetic mutations, tendency to become superbugs, resistance to microbicides, and toxicity of the added chemicals are potential challenges in applying such strategies in the biosorption process. Table 5.1 describes the reported literature on physicochemical pretreatment methods of various biosorbents.

Challenges in the Utilization of Biosorbents

Despite the use of biosorbents being a cost-effective and environmentally friendly technique, many challenges need to be tackled in their efficient use for wastewater treatment. Using versatile biosorbents in the continuous mode of large-scale operations needs to be comprehensively studied as the behavioral characteristics of these materials and their

Table 5.1 Biosorptive performances of biosorbents with physical and chemical treatments.

Biosorbent	Physical/Chemical treatment	Biosorbate	Biosorption capacity	Remarks	Reference
Lyophilized fungi (*Ophiostoma stenoceras* LLC), bacteria (*Pseudomonas veronii* ZW)	Amino-methylation	Ethyl acetate α-pinene n-hexane	620 mg/g 454 mg/g 374 mg/g	Modified fungal biosorbent was better than bacteria.	Cheng et al., 2019
Amberlite XAD-4 loaded with *Anoxybacillus kestanboliensis*	Heat treatment	Co(II) Hg(II)	24.3 mg/g 27.8 mg/g	Column can be used for over 35 cycles of biosorption/desorption without any loss in the biosorption.	Ozdemir et al., 2020
Yttrium/Sodium alginate (Y/SA) hydrogel	Chemical crosslinking of sodium alginate and yttrium ions	Congo red Weak acid brilliant blue RAW	1567 mg/g 1087 mg/g	Predominant adsorption due to electrostatic adsorption, hydrogen bonding and ion exchange reaction.	Li et al., 2020
Activated coastal waste of the zostera marina plant	Drying NaOH	Mn	120.6 mg/g	Optimum conditions: 10 mg, pH 6, Mn concentration 30 mg/L, and time 60 min. Pseudo second-order model Freundlich isotherm	Deniz et al., 2020
Living and dead biomass of *Rhodococcus* sp. HX-2	Sterilization and drying	Pb	88.74 mg/g (live) 125.5 mg/g (dead)	Dead biosorbents are better than live biosorbents.	Hu et al., 2020

(*Continued*)

Table 5.1 (Continued)

Biosorbent	Physical/Chemical treatment	Biosorbate	Biosorption capacity	Remarks	Reference
Coconut waste Black oak bark Wheat brans	Physical and chemical treatment	Pb, Cd, Hg, and Cr	263 mg/g 285 mg/g 400 mg/g 310 mg/g.	Mechanism: chemisorption, complexation, adsorption on surface, diffusion through pores, and ion exchange.	Alalwan et al., 2020
WSD@Fe3O4NPs-TA WSD@Fe3O4NPs-CA	Green microwave solvent-less technique and covalently binding of tartaric acid (TA) and citric acid (CA) to wood sawdust coated magnetic nanoparticles (WSD@ Fe3O4NPs)	Ca	18.4 mg/gWSD@ Fe3O4NPs-TA 27.2 mg/gWSD@ Fe3O4NPs-CA	Applied for fast reduction of calcium hardness in real water samples over 15–20 min and via two consecutive in situ batch operations.	Soliman et al., 2020
Pomegranate peel	Acid, alkali, formaldehyde and mixture treatment of formaldehyde and sulphuric acid	Cd Pb	97% 99%	Acid-treated pomegranate peel powder has better quality of adsorption of cadmium and lead from tannery wastewater.	Hameed et al., 2020
Biosorbent clay composite (CA-CMC/ Bt)	Carboxymethyl-chitosan/ sodium alginate grafted with amino-carbamate moiety/bentonite clay	Ni	159 mg/g	Hydrogel beads were observed to be very integrated and stable under a wide pH working range (from 2.0 to 12.0).	Shehzad et al., 2021
Soya bean adsorbent	No prior chemical treatment or modification	Pb and As	80% and 40%	Biosorption by soya bean adsorbent was exothermic	Gaur et al., 2018

Novel porous biosorbent (HF-D) based on Orange Peel (OP) OP-R 5 wt % HF-D 10 wt % HF-D 15 wt % HF-D	Fe treatment, Drying and Filtration	As	81.3 mg/g	HF-D showed well-ordered dendrimer-like structures and remarkable adsorption performance for the removal of arsenic. Biosorption of As was an adsorption-reduction process, including complex adsorption and inner reduction.	Meng et al., 2017
Haplophragma adenophyllum (HAB)	Drying and sieving (70 ASTM)	Gentian violet (GV) dye	13.21 mg/g	The optimum removal of GV by HAB was observed at pH 6.0.	Akram et al., 2021
Brassica nigra	No chemical treatment	Pb	53.476 mg/g	The best desorbing agent was HNO_3 (0.1N).	Romero et al., 2019
Date palm seeds (DS) Bleached date seed (BDS) Cellulose nano fibers from date seed (CNFs)	Physical and chemical treatment	Cu	22 mg/g 23 mg/g 34 mg/g	Cu adsorption was a spontaneous, entropy-increasing, endothermic process.	Maaloul et al., 2021
Amino-terminated hyper branched polymer (NH_2-HBP) and beta-cyclo dextrin (β-CD) onto cotton fibers	Chemical treatment	CR MB	350.8 mg/g 102.7 mg/g	Adsorption equilibrium was reached within 6 h for CR and within 4 h for MB. Adsorption mechanism: electrostatic attraction hydrogen-bonding and host-guest inclusion.	Yue et al., 2019
Dioscorea opposita Thunb. (DOT) modified with polyethylene imine (DOT@PEI)	Chemical treatment	Indigo carmine	344.83 mg/g	Optimum parameters: pH 2.0, 1.0 g/L dosage, temperature 20°C	Zhang et al., 2018

(Continued)

Table 5.1 (Continued)

Biosorbent	Physical/Chemical treatment	Biosorbate	Biosorption capacity	Remarks	Reference
Biosorbent activated de-oiled microalgae, *chlorella vulgaris* (AC)	H_2SO_4 treatment, drying and carbonization	Acid red 1	83.5%	The activation process transforms the biomass surface into a regular, small, homogeneous size that increases the surface area and enhances its adsorption capacity.	Mekkawi et al., 2021
Hybrid biosorbent (almond shell and *Pleurotus sajor caju*)	Crosslinking	Ni Pb	87% 85%	Hybrid biosorbent was extremely functional in the exclusion of Pb and Ni from wastewater.	Sattar, 2021
Cake-like biosorbents from tea waste, trimmed lawn grass, and *Nephrolepis cordifolia* leaves	One-step xanthation modification method combined with lyophilization	Pb Cu Cd	247.20 mg/g 85.80 mg/g 265.31 mg/g	Newly introduced $(C{=}S)-S-Na$ groups, which triggered ion exchange, complexation, and microprecipitation between heavy metal ions and functional groups.	Wang et al., 2021
Pine cone	Drying	Cd Cu Pb Cr	78.73 mg/g 33.55 mg/g 100.01 mg/g 57.36 mg/g	Pine cones are an efficient adsorbent to treat multimetal effluents for reuse.	Amar et al., 2021
Powdered activated carbon (AC) from *Stevia rebaudiana Bertoni* AC1-1 (0.046 mm) AC1-2 (0.039 mm).	Sodium hydroxide Solvent extraction carbonization	Malachite green	AC1-1: 74.87–284.45 mg/g AC1-2: 73.08–288.67 mg/g	No significant difference between AC1-1 and AC1-2.	Mullick et al., 2016
Zeolite 4A from Ethiopian kaolin	Physical and chemical treatment	Methylene blue	44.35 mg/g	Mechanism: electrostatic interaction, hydrogen bond, and n to π bond formation.	Belachew et al., 2021

Cosolvent-enhanced lignocellulosic fractionation (CELF)	Pretreatment and Mannich reaction combined to generate an aminated CELF lignin	Direct blue 1	502.7 mg/g	The incorporated amine functionalities as well as the increased specific surface area of the aminated CELF lignin resulted in adsorption efficiency.	Meng et al., 2020
Arachis hypogea shells, *Ziziphus mauritiana* sawdust, and *Cassia fistula* leaves	Physical treatment such as drying and sizing	Cr	*Arachis hypogea* shells, 4.32 mg/g *Ziziphus mauritiana* sawdust, 3.66 mg/g Cassia fistula leaves, 4.48 mg/g	Lagergren pseudo-first-order equation and pseudo-second-order equation. Chemical nature of the biosorption process.	Ahmad et al., 2017
Sugarcane bagasse cellulose-graft-poly stearyl methacrylate (SBC-g-PSMA)	Homogeneous graft copolymerization in the ionic liquid 1-butyl-3-methylimidazolium chloride ([BMIM]Cl)	Benzene Kerosene Diesel	10.24 g/g 9.76 g/g 9.74 g/g	Optimum conditions: the mass ratio of the monomer to sugarcane bagasse cellulose was 4:1 and the initiator to sugarcane bagasse cellulose ratio was 1:10, with a reaction time of 2 h.	Chen et al., 2015
Chitosan from shell dust of *Belamya javanica* snails (BSD)	Deproteinization using NaOH Demineralization with HCl treatment Deacetylation using NaOH and drying	Cu Cd Pb	147.48 mg/g 146.40 mg/g 138.57 mg/g	Adsorptive capacity: Pb < Cu, Cd^{2+}. The difference in absorption ability for each metal was due to the influence of the atomic radius and electronegativity.	Hardani et al., 2021

operational variables vary significantly in batch- and small-scale operation. As these biologically derived materials contain many phytoconstituents, the reactivity of such groups with sorbate molecules is uncertain. There are possibilities of cross-reactivity between the sorbate and sorbent, which might augment the toxicity of wastewater. Also, in the case of live microorganisms as biosorbents, the probability of mutated strains is significant, which might lead to the creation of superbugs. In addition, several operational constraints such as diffusion resistance due to continuous growth, substrate or oxygen depletion, plasmid stability and metabolic load with recombinant strains, etc., can contribute to the complexity of using live biomass. For instance, biofilm formation by algal strains on various parts of reacting vessels might change the rheological behavior of wastewater by increasing its viscosity.

The other significant challenge is the desorption of toxic contaminants or pollutants from the used biosorbent before it is subjected to disposal. Many studies have reported the desorption potential of batch-scale systems, and all the studies reported using chemicals such as HCl, NaOH, or acetic acid to desorb; very few studies utilized pressure-swing and temperature-swing desorption in continuous operation. The use of naive biosorbents is not preferable as releasing organic compounds into the suspending liquid can increase the BOD, COD, and total organic carbon. Hence, it is essential to pretreat the biosorbent materials before using them for wastewater treatment.

The selection of a pretreatment technique is vital as it determines the shelf-life, sorptive capability, and toxicity of the biosorbents. For instance, using dry heating can increase the storage period of the material and improve hydrophobic interactions on the surface by removing free and bound moisture. On the other hand, techniques like pyrolysis increase toxicity as they can emit volatile organic compounds at high temperatures; thus their applications are limited.

The use of chemicals such as acids or alkali and other chemical reagents is inevitable as recent research focuses on the synthesis of nano-sized biosorbents and biocomposite materials. The greatest threat in the use of nano-sized sorbents is their disposal. No investigations have been reported on good disposal practices for spent nano-sized biosorbents, and hence future studies must focus on this research area. Also, the lack of nationwide policies for the usage and disposal of nano-sized materials has led to the massive usage of such materials with no proper disposal (Thirunavukkarasu et al., 2020). These materials can also be derived from agricultural wastes and bio-based industrial discharge, which might improve the socio-economic viability of generating such huge amounts of waste into biosorbent preparation. But these wastes are also used as feedstock for the production of renewable energies such as biogas and bioethanol. Hence, the proper allocation of these wastes into environmental research and energy production is extremely challenging.

Conclusion

Adsorption using biologically derived materials is a simple, cost-effective, highly efficient, environmentally friendly, and sustainable green technology to remove pollutants from aqueous solutions. The versatile surface functional groups, ease of modification, plentiful availability, and renewable properties of the biosorbents makes them attractive to the scientific community involved in pollution research. Naive biosorbents tend to release organic compounds in the aqueous solution, which compete with metal ions for binding sites on the biosorbents, affecting the biosorption process's performance. Hence, naive biosorbents must be pretreated using various physical and chemical methods to prevent the release of organic compounds, increase the number of binding sites, and graft new surface functional compounds.

This chapter presented the biosorptive potential of such modified biosorbents to remove contaminants from wastewater. The chapter identified potential biosorption research avenues such as uncertainty in reactivity, the possibility of random mutations, the creation of superbugs, biofilm formation, poor understanding of scale-up analysis, disposal practices for spent biosorbents, feedstock shortages for renewable energy production, release of volatile organic compounds, use of toxic chemicals in pretreatment processes, etc., and these challenges should be scientifically addressed in future research.

References

Acheampong, M.A., Pakshirajan, K., Annachhatre, A.P. et al. (2013). Removal of Cu(II) by biosorption onto coconut shell in fixed-bed column systems. *Journal of Industrial and Engineering Chemistry* 19 (3): 841–848. doi:10.1016/j.jiec.2012.10.029.

Adewuyi, A. and Pereira, F.V. (2017). Underutilized Luffa cylindrica sponge: A local bio-adsorbent for the removal of Pb(II) pollutant from water system. *Beni-Suef University Journal of Basic and Applied Sciences* 6 (2): 118–126. doi:10.1016/j.bjbas.2017.02.001.

Ahluwalia, S.S. and Goyal, D. (2007). Microbial and plant derived biomass for removal of heavy metals from wastewater. *Bioresource Technology* 98 (12): 2243–2257. doi:10.1016/j.biortech.2005.12.006.

Ahmad, A., Bhat, A.H., and Buang, A. (2019). Enhanced biosorption of transition metals by living Chlorella vulgaris immobilized in Ca-alginate beads. *Environmental Technology* 40 (14): 1793–1809. doi:10.1080/09593330.2018.1430171.

Ahmad, A., Ghazi, Z.A., Saeed, M. et al. (2017). A comparative study of the removal of Cr(vi) from synthetic solution using natural biosorbents. *New Journal of Chemistry* 41 (19): 10799–10807. doi:10.1039/c7nj02026k.

Akram, M., Salman, M., Rehman, R. et al. (2021). Kinetic and isothermal investigations of cost-effective sorptive elimination of gentian violet dye from water using haplophragma adenophyllum biowaste. *Journal of Chemistry* 2021: 1–12. doi:10.1155/2021/5549536.

Aksu, Z. (2005). Application of biosorption for the removal of organic pollutants: a review. *Process Biochemistry* 40 (3-4): 997–1026. doi:10.1016/j.procbio.2004.04.008.

Alalwan, H.A., Kadhom, M.A., and Alminshid, A.H. (2020). Removal of heavy metals from wastewater using agricultural byproducts. *Journal of Water Supply: Research and Technology-Aqua* 69 (2): 99–112. doi:10.2166/aqua.2020.133.

Amar, M.B., Ben Amar, M., Walha, K. et al. (2021). Valorisation of pine cone as an efficient biosorbent for the removal of Pb(II), Cd(II), Cu(II), and Cr(VI). *Adsorption Science and Technology* 2021: 1–12. doi:10.1155/2021/6678530.

Amer, M., Nour, M., Ahmed, M. et al. (2019). The effect of microwave drying pretreatment on dry torrefaction of agricultural biomasses. *Bioresource Technology* 286: 121400. doi:10.1016/j.biortech.2019.121400.

Ao, W., Fu, J., Mao, X. et al. (2018). Microwave assisted preparation of activated carbon from biomass: A review. *Renewable and Sustainable Energy Reviews* 92: 958–979. doi:10.1016/j.rser.2018.04.051.

Arunachalam, T., Karpagasundaram, M., and Rajarathinam, N. (2017). Ultrasound assisted green synthesis of cerium oxide nanoparticles using Prosopis juliflora leaf extract and their structural, optical and antibacterial properties. *Materials Science-Poland* 35 (4): 791–798. doi:10.1515/msp-2017-0104.

Aryal, M. (2019). Calcium alginate entrapped Eupatorium adenophorum Sprengel stems powder for Chromium(VI) biosorption in aqueous medium. *Scientific Communication and Education* doi:10.1101/560680.

Belachew, N. and Hinsene, H. (2021). Preparation of zeolite 4a for adsorptive removal of methylene blue: optimization, kinetics, isotherm, and mechanism study. *Silicon.* doi:10.1007/s12633-020-00938-9.

Bolong, N., Ismail, A.F., Salim, M.R. et al. (2009). A review of the effects of emerging contaminants in wastewater and options for their removal. *Desalination* 239 (1-3): 229–246. doi:10.1016/j.desal.2008.03.020.

Chen, S., Zhang, X., Zhu, H. et al. (2015). Optimization synthesis conditions and characterization of oil biosorbent: sugarcane bagasse cellulose-graft-polystearylmethacrylate copolymer. *BioResources* 10 (1). doi:10.15376/biores.10.1.1357-1365.

Cheng, Y.-S., Zheng, Y., Yu, C.W. et al. (2010). Evaluation of high solids alkaline pretreatment of rice straw. *Applied Biochemistry and Biotechnology* 162 (6): 1768–1784. doi:10.1007/s12010-010-8958-4.

Cheng, Z., Feng, K., Su, Y. et al. (2020). Novel biosorbents synthesized from fungal and bacterial biomass and their applications in the adsorption of volatile organic compounds. *Bioresource Technology* 300: 122705. doi:10.1016/j.biortech.2019.122705.

Chojnacka, K. (2010). Biosorption and bioaccumulation – the prospects for practical applications. *Environment International* 36 (3): 299–307. doi:10.1016/j.envint.2009.12.001.

Crini, G. (2006). Non-conventional low-cost adsorbents for dye removal: A review. *Bioresource Technology* 97 (9): 1061–1085. doi:10.1016/j.biortech.2005.05.001.

Crini, G. and Badot, P.-M. (2008). Application of chitosan, a natural aminopolysaccharide, for dye removal from aqueous solutions by adsorption processes using batch studies: A review of recent literature. *Progress in Polymer Science* 33 (4): 399–447. doi:10.1016/j.progpolymsci.2007.11.001.

De Oliveira, V.H., De Oliveira, V.H., Ullah, I. et al. (2020). Bioremediation potential of Cd by transgenic yeast expressing a metallothionein gene from Populus trichocarpa. *Ecotoxicology and Environmental Safety* 202: 110917. doi:10.1016/j.ecoenv.2020.110917.

Demirbas, A. (2008). Heavy metal adsorption onto agro-based waste materials: A review. *Journal of Hazardous Materials* 157 (2-3): 220–229. doi:10.1016/j.jhazmat.2008.01.024.

Demirbas, A. (2009). Agricultural based activated carbons for the removal of dyes from aqueous solutions: A review. *Journal of Hazardous Materials* 167 (1-3): 1–9. doi:10.1016/j.jhazmat.2008.12.114.

Deniz, F. and Ersanli, E.T. (2020). An effectual biosorbent substance for removal of manganese ions from aquatic environment: a promising environmental remediation study with activated coastal waste of zostera marina plant. *BioMed Research International* 2020: 1–8. doi:10.1155/2020/7806154.

Digman, M.F., Shinners, K.J., Casler, M.D. et al. (2010). Optimizing on-farm pretreatment of perennial grasses for fuel ethanol production. *Bioresource Technology* 101 (14): 5305–5314. doi:10.1016/j.biortech.2010.02.014.

Dong, X.-B., Huang, W., Bian, Y.-B. et al. (2019). Remediation and mechanisms of cadmium biosorption by a cadmium-binding protein from Lentinula edodes. *Journal of Agricultural and Food Chemistry* 67 (41): 11373–11379. doi:10.1021/acs.jafc.9b04741.

Du, B., Sharma, L.N., Becker, C. et al. (2010). Effect of varying feedstock-pretreatment chemistry combinations on the formation and accumulation of potentially inhibitory degradation products in biomass hydrolysates. *Biotechnology and Bioengineering* 107 (3): 430–440. doi:10.1002/bit.22829.

El-Mekkawi, S.A., Abdelghaffar, R.A., Abdelghaffar, F. et al. (2021). Application of response surface methodology for color removing from dyeing effluent using de-oiled activated algal biomass. *Bulletin of the National Research Centre* 45 (1). doi:10.1186/s42269-021-00542-w.

Esteghlalian, A., Hashimoto, A.G., Fenske, J.J. et al. (1997). Modeling and optimization of the dilute-sulfuric-acid pretreatment of corn stover, poplar and switchgrass. *Bioresource Technology* 59 (2-3): 129–136. doi:10.1016/s0960-8524(97)81606-9.

Farooq, U., Kozinski, J.A., Khan, M.A. et al. (2010). Biosorption of heavy metal ions using wheat based biosorbents – A review of the recent literature. *Bioresource Technology* 101 (14): 5043–5053. doi:10.1016/j.biortech.2010.02.030.

Forgacs, E., Cserháti, T., and Oros, G. (2004). Removal of synthetic dyes from wastewaters: a review. *Environment International* 30 (7): 953–971.

Fosso-Kankeu, E. and Mulaba-Bafubiandi, A.F. (2014). Implication of plants and microbial metalloproteins in the bioremediation of polluted waters: A review. *Physics and Chemistry of the Earth, Parts A/B/C* 67–69: 242–252. doi:10.1016/j.pce.2013.09.018.

Fu, F. and Wang, Q. (2011). Removal of heavy metal ions from wastewaters: A review. *Journal of Environmental Management* 92 (3): 407–418. doi:10.1016/j.jenvman.2010.11.011.

Gadd, G.M. (2009). Biosorption: Critical review of scientific rationale, environmental importance and significance for pollution treatment. *Journal of Chemical Technology and Biotechnology* 84 (1): 13–28.

Gaur, N., Kukreja, A., Yadav, M. et al. (2018). Adsorptive removal of lead and arsenic from aqueous solution using soya bean as a novel biosorbent: equilibrium isotherm and thermal stability studies. *Applied Water Science* 8 (4): 98. doi:10.1007/s13201-018-0743-5.

Gogate, P.R. and Pandit, A.B. (2004). A review of imperative technologies for wastewater treatment I: oxidation technologies at ambient conditions. *Advances in Environmental Research* 8 (3–4): 501–551. doi:10.1016/s1093-0191(03)00032-7.

Gupta, V.K. and Suhas (2009). Application of low-cost adsorbents for dye removal – A review. *Journal of Environmental Management* 90 (8): 2313–2342. doi:10.1016/j.jenvman.2008.11.017.

Hai, F.I., Modin, O., Yamamoto, K. et al. (2012). Pesticide removal by a mixed culture of bacteria and white-rot fungi. *Journal of the Taiwan Institute of Chemical Engineers* 43 (3): 459–462. doi:10.1016/j.jtice.2011.11.002.

Hameed, M.A. (2020). Pomegranate peel treated with different chemicals as novel biosorbent of heavy metals such as PB, CD, from tanneries effluents. *Journal of Advanced Chemical Engineering* 10 (1). doi:10.35248/2090-4568.20.10.194.

Hu, X., Cao, J., Yang, H. et al. (2020). Pb2+ biosorption from aqueous solutions by live and dead biosorbents of the hydrocarbon-degrading strain Rhodococcus sp. HX-2 Mishra, Y.K. (ed.). PLOS ONE, 15 (1): e0226557. doi:10.1371/journal.pone.0226557.

Ibrahim, M.M., El-Zawawy, W.K., Abdel-Fattah, Y.R. et al. (2011). Comparison of alkaline pulping with steam explosion for glucose production from rice straw. *Carbohydrate Polymers* 83 (2): 720–726. doi:10.1016/j.carbpol.2010.08.046.

Jain, A.K., Gupta, V.K., Jain, S. et al. (2004). Removal of chlorophenols using industrial wastes. *Environmental Science and Technology* 38 (4): 1195–1200. doi:10.1021/es034412u.

Karthika, T., Thirunavukkarasu, A., Ramesh, S. (2010). Biosorption of copper from aqueous solutions using Tridax Procumbens. *Recent Research in Science and Technology* 2 (3): 86–91.

Kaushik, P. and Malik, A. (2009). Fungal dye decolourization: Recent advances and future potential. *Environment International* 35 (1): 127–141. doi:10.1016/j.envint.2008.05.010.

Khoramzadeh, E., Nasernejad, B., and Halladj, R. (2013). Mercury biosorption from aqueous solutions by Sugarcane Bagasse. *Journal of the Taiwan Institute of Chemical Engineers* 44 (2): 266–269. doi:10.1016/j.jtice.2012.09.004.

Kocaman, S. (2020). Synthesis and cationic dye biosorption properties of a novel low-cost adsorbent: coconut waste modified with acrylic and polyacrylic acids. *International Journal of Phytoremediation* 22 (5): 551–566. doi:10.1080/15226514.2020.1741509.

Kumar, R. and Wyman, C.E. (2009). Access of cellulase to cellulose and lignin for poplar solids produced by leading pretreatment technologies. *Biotechnology Progress* 25 (3): 807–819. doi:10.1002/btpr.153.

Kyzas, G., Fu, J., and Matis, K. (2013). The change from past to future for adsorbent materials in treatment of dyeing wastewaters. *Materials* 6 (11): 5131–5158. doi:10.3390/ma6115131.

Lakshmi, S.D., Avti, P.K., and Hegde, G. (2018). Activated carbon nanoparticles from biowaste as new generation antimicrobial agents: A review. *Nano-Structures and Nano-Objects* 16: 306–321. doi:10.1016/j.nanoso.2018.08.001.

Lesmana, S.O., Febriana, N., Soetaredjo, F.E. et al. (2009). Studies on potential applications of biomass for the separation of heavy metals from water and wastewater. *Biochemical Engineering Journal* 44 (1): 19–41. doi:10.1016/j.bej.2008.12.009.

Li, B. and Yin, H. (2020). Superior adsorption property of a novel green biosorbent yttrium/alginate gel beads for dyes from aqueous solution. *Journal of Polymers and the Environment* 28 (8): 2137–2148. doi:10.1007/s10924-020-01757-0.

Li, C., Knierim, B., Manisseri, C. et al. (2010). Comparison of dilute acid and ionic liquid pretreatment of switchgrass: Biomass recalcitrance, delignification and enzymatic saccharification. *Bioresource Technology* 101 (13): 4900–4906.

Ma, W. and Tobin, J.M. (2003). Development of multimetal binding model and application to binary metal biosorption onto peat biomass. *Water Research* 37 (16): 3967–3977. doi:10.1016/s0043-1354(03)00290-2.

MacDonald, D.G., Bakhshi, N.N., Mathews, J.F. et al. (1983). Alkali treatment of corn stover to improve sugar production by enzymatic hydrolysis. *Biotechnology and Bioengineering* 25 (8): 2067–2076. doi:10.1002/bit.260250815.

Maciel, G.M., de Souza, C.G.M., de Araújo, C.A.V. et al. (2013). Biosorption of herbicide picloram from aqueous solutions by live and heat-treated biomasses of Ganoderma lucidum (Curtis): Karst and Trametes sp. *Chemical Engineering Journal* 215-216: 331–338. doi:10.1016/j.cej.2012.09.127.

Manique, M.C., Faccini, C.S., Onorevoli, B. et al. (2012). Rice husk ash as an adsorbent for purifying biodiesel from waste frying oil. *Fuel* 92 (1): 56–61. doi:10.1016/j.fuel.2011.07.024.

Martín-Lara, M.A., Blázquez, G., Ronda, A. et al. (2013). Development and characterization of biosorbents to remove heavy metals from aqueous solutions by chemical treatment of olive stone. *Industrial and Engineering Chemistry Research* 52 (31): 10809–10819. doi:10.1021/ie401246c.

Mathialagan, T. and Viraraghavan, T. (2005). Biosorption of pentachlorophenol by fungal biomass from aqueous solutions: a factorial design analysis. *Environmental Technology* 26 (5): 571–580. doi:10.1080/09593332608618542.

Meng, F., Yang, B., Wang, B. et al. (2017). Novel dendrimerlike magnetic biosorbent based on modified orange peel waste: adsorption–reduction behavior of arsenic. *ACS Sustainable Chemistry and Engineering* 5 (11): 9692–9700. doi:10.1021/acssuschemeng.7b01273.

Meng, X., Scheidemantle, B., Li, M. et al. (2020). Synthesis, characterization, and utilization of a lignin-based adsorbent for effective removal of azo dye from aqueous solution. *ACS Omega* 5 (6): 2865–2877. doi:10.1021/acsomega.9b03717.

Maaloul, N., Oulego, P., Rendueles, M. et al. (2021). Novel biosorbents from tunisian date palm “bouhattam” seeds for copper(II). Ion adsorption. In: *Recent Advances in Environmental Science from the Euro-Mediterranean and Surrounding Regions*, 2e: 809–814. doi:10.1007/978-3-030-51210-1_127.

Morni, N.A.H., Radenahmad, N., Abu Bakar, M.S. et al. (2018). Potential of sewage sludge as energy recovery via gasification process. *7th Brunei International Conference on Engineering and Technology 2018 (BICET 2018)*. doi:10.1049/cp.2018.1528.

Mullick, A. and Neogi, S. (2016). Synthesis of potential biosorbent from used stevia leaves and its application for malachite green removal from aqueous solution: kinetics, isotherm and regeneration studies. *RSC Adv* 6 (70): 65960–65975. doi:10.1039/c6ra15225b.

Nemeş, L. and Bulgariu, L. (2016). Optimization of process parameters for heavy metals biosorption onto mustard waste biomass. *Open Chemistry* 14 (1): 175–187. doi:10.1515/chem-2016-0019.

Ngah, W.S.W. and Hanafiah M.A.K.M. (2008). Removal of heavy metal ions from wastewater by chemically modified plant wastes as adsorbents: a review. *Bioresource Technol* 99: 3935–3948.

Ngah, W.S.W., Wan Ngah, W.S., Teong, L.C. et al. (2011). Adsorption of dyes and heavy metal ions by chitosan composites: A review. *Carbohydrate Polymers* 83 (4): 1446–1456. doi:10.1016/j.carbpol.2010.11.004.

Nguyen, L.N., Hai, F.I., Yang, S. et al. (2014). Removal of pharmaceuticals, steroid hormones, phytoestrogens, UV-filters, industrial chemicals and pesticides by Trametes versicolor: Role of biosorption and biodegradation. *International Biodeterioration and Biodegradation* 88: 169–175. doi:10.1016/j.ibiod.2013.12.017.

Nguyen, Q.A., Tucker, M.P., Keller, F.A. et al. (2000). Two-stage dilute-acid pretreatment of softwoods. *Applied Biochemistry and Biotechnology* 84–86 (1-9): 561–576. doi:10.1385/abab:84-86: 1-9: 561.

Nithya, R., Sivasankari, C., Thirunavukkarasu, A. et al. (2018). Novel adsorbent prepared from bio-hydrometallurgical leachate from waste printed circuit board used for the removal of methylene blue from aqueous solution. *Microchemical Journal* 142: 321–328. doi:10.1016/j.microc.2018.07.009.

Ohring, M. and Kasprzak, L. (2015). An overview of electronic devices and their reliability. In: *Reliability and Failure of Electronic Materials and Devices* (eds. M. Ohring and L. Kasprzak), 1–38. Academic Press.

Ozdemir, S., Kılınç, E., and Fatih, S. (2020). A novel biosorbent for preconcentrations of Co(II) and Hg(II) in real samples. *Scientific Reports* 10 (1). doi:10.1038/s41598-019-57401-y.

Páez-Vélez, C., Castro-Mayorga, J.L., and Dussán, J. (2020). Effective gold biosorption by electrospun and electrosprayed bio-composites with immobilized Lysinibacillus sphaericus CBAM5. *Nanomaterials* 10 (3): 408. doi:10.3390/nano10030408.

Pearce, C. (2003). The removal of colour from textile wastewater using whole bacterial cells: a review. *Dyes and Pigments* 58 (3): 179–196. doi:10.1016/s0143-7208(03)00064-0.

Rafatullah, M., Sulaiman, O., Hashim, R. et al. (2010). Adsorption of methylene blue on low-cost adsorbents: A review. *Journal of Hazardous Materials* 177 (1-3): 70–80. doi:10.1016/j.jhazmat.2009.12.047.

Rajarathinam, N., Arunachalam, T., Raja, S. et al. (2020). Fenalan yellow G adsorption using surface-functionalized green nanoceria: An insight into mechanism and statistical modelling. *Environmental Research* 181: 108920. doi:10.1016/j.envres.2019.108920.

Ramrakhiani, L., Halder, A., Majumder, A. et al. (2017). Industrial waste derived biosorbent for toxic metal remediation: Mechanism studies and spent biosorbent management. *Chemical Engineering Journal* 308: 1048–1064. doi:10.1016/j.cej.2016.09.145.

Reátegui-Romero, W., Cadenas-Vásquez, W.J., King-Santos, M.E. et al. (2019). Evaluation of Pb (II) adsorption from aqueous solutions using brassica nigra as a biosorbent. *The Open Biotechnology Journal* 13 (1): 77–92. doi:10.2174/1874070701913010077.

Reddy, D.H.K., Harinath, Y. et al. (2010a). Biosorption of Pb(II) from aqueous solutions using chemically modified Moringa oleifera tree leaves. *Chemical Engineering Journal* 162 (2): 626–634. doi:10.1016/j.cej.2010.06.010.

Reddy, D.H.K., Ramana, D.K.V. et al. (2011). Biosorption of Ni(II) from aqueous phase by Moringa oleifera bark, a low cost biosorbent. *Desalination* 268 (1-3): 150–157. doi:10.1016/j.desal.2010.10.011.

Reddy, D.H.K., Seshaiah, K. et al. (2010b). Biosorption of Pb2 from aqueous solutions by Moringa oleifera bark: Equilibrium and kinetic studies. *Journal of Hazardous Materials* 174 (1-3): 831–838. doi:10.1016/j.jhazmat.2009.09.128.

Reddy, D.H.K., Seshaiah, K. et al. (2012). Optimization of Cd(II), Cu(II) and Ni(II) biosorption by chemically modified Moringa oleifera leaves powder. *Carbohydrate Polymers* 88 (3): 1077–1086. doi:10.1016/j.carbpol.2012.01.073.

Reddy, S. and Osborne, J.W. (2020). Biodegradation and biosorption of Reactive Red 120 dye by immobilized Pseudomonas guariconensis: Kinetic and toxicity study. *Water Environment Research* 92 (8): 1230–1241. doi:10.1002/wer.1319.

Ribeiro, V.R., Maciel, G.M., Fachi, M.M. et al. (2019). Improvement of phenolic compound bioaccessibility from yerba mate (Ilex paraguariensis) extracts after biosorption on Saccharomyces cerevisiae. *Food Research International* 126: 108623. doi:10.1016/j.foodres.2019.108623.

Saha, B.C., Iten, L.B., Cotta, M.A. et al. (2008). Dilute acid pretreatment, enzymatic saccharification, and fermentation of rice hulls to ethanol. *Biotechnology Progress* 21 (3): 816–822. doi:10.1021/bp049564n.

Salleh, M.A.M., Mahmoud, D.K., Karim, W.A.W. et al. (2011). Cationic and anionic dye adsorption by agricultural solid wastes: A comprehensive review. *Desalination* 280 (1-3): 1–13. doi:10.1016/j.desal.2011.07.019.

Sassner, P., Mårtensson, C.-G., Galbe, M. et al. (2008). Steam pretreatment of H2SO4-impregnated Salix for the production of bioethanol. *Bioresource Technology* 99 (1): 137–145. doi:10.1016/j.biortech.2006.11.039.

Sattar, A.A. (2021). Preparation of novel hybrid (almond shell and pleurotus sajor caju) biosorbent for the removal of heavy metals (nickel and lead) from wastewater. *Water Conservation and Management* 5 (1): 1–7.

Segovia-Sandoval, S.J., Ocampo-Pérez, R., Berber-Mendoza, M.S. et al. (2018). Walnut shell treated with citric acid and its application as biosorbent in the removal of Zn(II). *Journal of Water Process Engineering* 25: 45–53. doi:10.1016/j.jwpe.2018.06.007.

Shehzad, H., Ahmed, E., Din, M.I. et al. (2021). Facile synthesis of novel carboxymethyl-chitosan/sodium alginate grafted with amino-carbamate moiety/bentonite clay composite for effective biosorption of Ni (II) from aqueous solution. *Zeitschrift für Physikalische Chemie* 235 (5): 583–607. doi:10.1515/zpch-2019-1555.

Shuai, L., Yang, Q., Zhu, J.Y. et al. (2010). Comparative study of SPORL and dilute-acid pretreatments of spruce for cellulosic ethanol production. *Bioresource Technology* 101 (9): 3106–3114. doi:10.1016/j.biortech.2009.12.044.

Sills, D.L. and Gossett, J.M. (2011). Assessment of commercial hemicellulases for saccharification of alkaline pretreated perennial biomass. *Bioresource Technology* 102 (2): 1389–1398. doi:10.1016/j.biortech.2010.09.035.

Silverstein, R.A., Chen, Y., Sharma-Shivappa, R.R. et al. (2007). A comparison of chemical pretreatment methods for improving saccharification of cotton stalks. *Bioresource Technology* 98 (16): 3000–3011. doi:10.1016/j.biortech.2006.10.022.

Sivashankar, R., Thirunavukkarasu, A., Nithya, R. et al. (2020). Sequestration of methylene blue dye from aqueous solution by magnetic biocomposite: Three level Box–Behnken experimental design optimization and kinetic studies. *Separation Science and Technology* 55 (10): 1752–1765. doi:10.1080/01496395.2019.1607382.

Soliman, E.M., Ahmed, S.A., and Fadl, A.A. (2020). Microwave-enforced green synthesis of novel magnetic nano composite adsorbents based on functionalization of wood sawdust for fast removal of calcium hardness from water samples. *Water Environment Research* 92 (12): 2112–2128. doi:10.1002/wer.1383.

Soto, M.L., Domínguez, H., Núñez, M.J. et al. (1994). Enzymatic saccharification of alkali-treated sunflower hulls. *Bioresource Technology* 49 (1): 53–59. doi:10.1016/0960-8524(94)90173-2.

Srivastava, S., Agrawal, S.B., and Mondal, M.K. (2016). Animal wastes: an alternative adsorbent for removal of toxic heavy metals from industrial wastewater. *Journal of Scientific Research* 60: 65–72.

Sud, D., Mahajan, G., and Kaur, M. (2008). Agricultural waste material as potential adsorbent for sequestering heavy metal ions from aqueous solutions – A review. *Bioresource Technology* 99 (14): 6017–6027. doi:10.1016/j.biortech.2007.11.064.

Sun, Y. and Cheng, J. (2005). Dilute acid pretreatment of rye straw and bermuda grass for ethanol production. *Bioresource Technology* 96 (14): 1599–1606. doi:10.1016/j.biortech.2004.12.022.

Tadkaew, N., Hai, F.I., McDonald, J.A. et al. (2011). Removal of trace organics by MBR treatment: The role of molecular properties. *Water Research* 45 (8): 2439–2451. doi:10.1016/j.watres.2011.01.023.

Tan, L., Cui, H., Xiao, Y. et al. (2019). Enhancement of platinum biosorption by surface-displaying EC20 on Escherichia coli. *Ecotoxicology and Environmental Safety* 169: 103–111. doi:10.1016/j.ecoenv.2018.10.116.

Thirunavukkarasu, A., Muthukumaran, K., and Nithya, R. (2018). Adsorption of acid yellow 36 onto green nanoceria and amine functionalized green nanoceria: Comparative studies on kinetics, isotherm, thermodynamics, and diffusion analysis. *Journal of the Taiwan Institute of Chemical Engineers* 93: 211–225. doi:10.1016/j.jtice.2018.07.006.

Thirunavukkarasu, A. and Nithya, R. (2019). Response surface optimization of Cu (II) biosorption onto Candida tropicalis immobilized strontium alginate beads by Box-Behnken experimental design. *Journal of Environment and Biotechnology Research* 8 (2): 14–21.

Thirunavukkarasu, A. and Nithya, R. (2020). Adsorption of acid orange 7 using green synthesized CaO/CeO2 composite: An insight into kinetics, equilibrium, thermodynamics, mass transfer and statistical models. *Journal of the Taiwan Institute of Chemical Engineers* doi.org/10.1016/j.jtice.2020.04.007.

Thirunavukkarasu, A. and Nithya, R. (2021). The role of nanomaterials in wastewater treatment. *Sustainable Bioprocessing for a Clean and Green Environment.* doi:10.1201/9781003035398-7.

Thirunavukkarasu, A., Nithya, R., and Sivashankar, R. (2020). A review on the role of nanomaterials in the removal of organic pollutants from wastewater. *Reviews in Environmental Science and Bio/Technology* 19 (4): 751–778. doi:10.1007/s11157-020-09548-8.

Vijayaraghavan, J., Sardhar Basha, S.J., and Jegan, J. (2013). A review on efficacious methods to decolorize reactive azo dye. *Journal of Urban and Environmental Engineering*: 30–47. doi:10.4090/juee.2013.v7n1.030047.

Villen-Guzman, M., Gutierrez-Pinilla, D., Gomez-Lahoz, C. et al. (2019). Optimization of Ni (II) biosorption from aqueous solution on modified lemon peel. *Environmental Research* 179: 108849. doi:10.1016/j.envres.2019.108849.

Wang, J. and Chen, C. (2009). Biosorbents for heavy metals removal and their future. *Biotechnology Advances* 27 (2): 195–226. doi:10.1016/j.biotechadv.2008.11.002.

Wang, N., Qiu, Y., Hu, K. et al. (2021). One-step synthesis of cake-like biosorbents from plant biomass for the effective removal and recovery heavy metals: Effect of plant species and roles of xanthation. *Chemosphere* 266: 129129. doi:10.1016/j.chemosphere.2020.129129.

Witek-Krowiak, A. (2011). Analysis of influence of process conditions on kinetics of malachite green biosorption onto beech sawdust. *Chemical Engineering Journal* 171 (3): 976–985. doi:10.1016/j.cej.2011.04.048.

Witek-Krowiak, A. (2013). Application of beech sawdust for removal of heavy metals from water: biosorption and desorption studies. *European Journal of Wood and Wood Products* 71 (2): 227–236. doi:10.1007/s00107-013-0673-8.

Wyman, C.E., Dale, B.E., Elander, R.T. et al. (2009). Comparative sugar recovery and fermentation data following pretreatment of poplar wood by leading technologies. *Biotechnology Progress* 25 (2): 333–339. doi:10.1002/btpr.142.

Xu, J., Thomsen, M.H., and Thomsen, A.B. (2009). Pretreatment on corn stover with low concentration of formic acid. *Journal of Microbiology and Biotechnology* 19 (8): 845–850.

Yadanaparthi, S.K.R., Graybill, D., and von Wandruszka, R. (2009). Adsorbents for the removal of arsenic, cadmium, and lead from contaminated waters. *Journal of Hazardous Materials* 171 (1-3): 1–15. doi:10.1016/j.jhazmat.2009.05.103.

Yan, L., Yin, H., Zhang, S. et al. (2010). Biosorption of inorganic and organic arsenic from aqueous solution by Acidithiobacillus ferrooxidans BY-3. *Journal of Hazardous Materials* 178 (1-3): 209–217. doi:10.1016/j.jhazmat.2010.01.065.

Yue, X., Huang, J., Jiang, F. et al. (2019). Synthesis and characterization of cellulose-based adsorbent for removal of anionic and cationic dyes. *Journal of Engineered Fibers and Fabrics* 14: 155892501982819. doi:10.1177/1558925019828194.

Yusuff, A.S. (2019). Adsorption of hexavalent chromium from aqueous solution byLeucaena leucocephalaseed pod activated carbon: equilibrium, kinetic and thermodynamic studies. *Arab Journal of Basic and Applied Sciences* 26 (1): 89–102. doi:10.1080/25765299.2019.1567656.

Zhang, J., Chen, X., Zhou, J. et al. (2020). Uranium biosorption mechanism model of protonated Saccharomyces cerevisiae. *Journal of Hazardous Materials* 385: 121588. doi:10.1016/j.jhazmat.2019.121588.

Zhang, Y., Li, J., Zhao, J. et al. (2018). Utilization of modified Dioscorea opposita Thunb. as a novel biosorbent for the adsorption of indigo carmine in aqueous solutions. *RSC Advances* 8 (52): 30040–30048. doi:10.1039/c8ra05106b.

Zhao, Y., Wang, Y., Zhu, J.Y. et al. (2008). Enhanced enzymatic hydrolysis of spruce by alkaline pretreatment at low temperature. *Biotechnology and Bioengineering*.99 (6): 1320–1328. doi:10.1002/bit.21712.

Zhou, Y.-F. and Haynes, R.J. (2010). Sorption of heavy metals by inorganic and organic components of solid wastes: significance to use of wastes as low-cost adsorbents and immobilizing agents. *Critical Reviews in Environmental Science and Technology* 40 (11): 909–977. doi:10.1080/10643380802586857.

6

Characterization of the Biosorption Process

R. Sivashankar[1], A.B. Sathya[2], J. Kanimozhi[3], and B. Deepanraj[4]

[1] *Department of Chemical Engineering, Hindustan Institute of Technology and Science, Chennai, Tamilnadu, India*
[2] *Department of Biotechnology, Aarupadai Veedu Institute of Technology, Chennai, Tamilnadu, India*
[3] *Department of Biotechnology, Kalasalingam Academy of Research and Education, Krishnankoil, Tamilnadu, India*
[4] *Department of Mechanical Engineering, Jyothi Engineering College, Thrissur, Kerala, India*

Introduction

Environmental challenges have grown in tandem with the expansion of the industrial world. The earth is coping with water contamination as a prevalent type of pollution. Potential sources of water contamination include sewage and wastewater, oil emissions, agricultural waste, and radioactive elements.

Wastewater contain sa large number of heavy metals, particularly those discarded by industry. Heavy metals are one of the leading causes of water pollution. They cannot be degraded and accumulate in the environment. Heavy metals contaminate water that is inhabited by marine organisms and threaten not only aquatic life but also human health, as they bioaccumulate in the food chain. Arsenic, chromium, cadmium, copper, zinc, lead, and nickel are the primary water contaminants (Jaishankar et al., 2014).

Several methods for removing pollutants have been developed. Chemical precipitation, electrodialysis, ultrafiltration, reverse osmosis, and the industrial ion-exchange process are the traditional methods of removing heavy metals from industrial effluent (Sao, 2014). These treatment technologies have many drawbacks, including being costly, requiring considerable energy, and being ineffective (Simionato et al., 2014).

Biosorption technologies are an alternative, ecologically sound technique to eliminate toxic metals. Biosorption techniques are regarded as more sustainable and effective in the adsorption of heavy metals and dyes (Ali Redha, 2020). The term *biosorption* describes a series of physical and chemical adsorptive processes within a cell wall, including the exchange of ions, electrostatic interactions, complexation, chelation, and microprecipitation. These steps happen before anaerobic or aerobic biodegradation. Selectivity and efficiency are excellent (greater performance and less expensive). Biosorbents include microbial biomass (living and dead), plant- and animal-derived materials, industrial and agriculture byproducts, biopolymers, and so on (Derco and Vrana, 2018). Biosorbents are a

Biosorption for Wastewater Contaminants, First Edition.
Edited by Rangabhashiyam Selvasembian and Pardeep Singh.

less expensive and more efficient way to remove metallic elements from aqueous solution, especially heavy metals. The choice of the most effective biosorbent type is a significant challenge, given the wide range of promising and cheap biomaterials. Biosorbents should be developed/acquired to bind different metal ions with greater affinity(Macek and Mackova, 2011).

To achieve success with the sorption process, the principle of recovering and reusing sorbents is critical; but biosorption has lost ground in this respect. Biomass recuperation and reuse are vital; otherwise, sewage sludge management is complicated, and a permanent source of new sorbent biomass is required (Michalak et al., 2013). Parameters such as pH, temperature, metal ion concentration, biosorbent dose, and agitation rates influence biosorption. After removing heavy metals, biosorbents can be regenerated and reused, thus making the process economical (Kanamarlapudi et al., 2018).

The biosorption process has become more competitive for industrial applications due to benefits such as lower biosorbent costs, the ability to regenerate biosorbents, metal recovery, and minimal generation of toxic sludge. During the study and analysis of biosorbent functional groups that contribute to removing a specific metal and other pollutants, various analytical techniques such as titration, Raman spectroscopy, infrared spectroscopy, energy-dispersive X-ray spectroscopy, X-ray photoelectron spectroscopy, and X-ray absorption spectroscopy can be used (Javanbakht et al., 2014). This chapter reviews how biosorption and sorbents can be studied using various characterization methods.

Biosorption

Biosorption is a biological, physical, and chemical process employed by natural substances like plants and microbes to adsorb a product of interest such as metallic ions or dyestuffs (Gadd, 2009). There are two phases during a biosorption process: a solid phase known as the *biosorbent* (for example, plants, bacteria, or fungus) and a liquid phase known as the *sorbate*, typically containing water-based metallic ions (Abbas Ali et al., 2016). Biosorption capacity and equilibrium time are used to assess the biosorption potential for a metal ion by a biosorbent. The sorption capacity of biosorbent (mg/g) may be defined as the quantity of biosorbate (metal ions) adsorbed by biosorbent unit weight and stated by this equation:

$$q_e = \frac{(C_i - C_e)V}{m} \tag{6.1}$$

From the following equation, the percentage biosorption (R%) known as *metal biosorption efficacy* can be measured

$$R\% = \frac{C_i - C_e}{C_i} X100 \tag{6.2}$$

wherein q_e = number of metal ions adsorbed to the adsorbent (mg/g); C_i = metallic ion concentration present initially in the solution (mg/L); C_e = equilibrium concentration of the metal ions (mg/L); V = medium volume (L); and m = quantity of the utilized biomass in the sorption procedure (g) (Wei et al., 2016).

As the adsorption capacity increases, the biosorbent will adsorb more heavy metal ion(s). With the medium at optimum pH and temperature, maximal biosorption capacity may be reached. The adsorption capacity may be increased by optimizing parameters such as the equilibrium time between biosorbent and metal ion solution and dosage concentration of the biosorbent; however, the most significant variables are pH and temperature.

When selecting a biosorbent to remove a particular metallic ion, it is vital to think of biosorption capability. Biosorbents with a higher potential for biosorption are preferred because they can adsorb a greater amount of sorbate. Biosorbents with low biosorption capacities will eventually need to be replaced. Other factors to consider when selecting a biosorbent include the degree of sustainability with regard to any negative environmental implications of the biosorbent's use, speed of biosorption of the target species in terms of removal effectiveness, and the ability to regenerate and reuse the biosorbent (Ali Redha, 2020).

Characterization Methods

To evaluate the capacity of biomass to bind metal ions, sophisticated technologies such as scanning electron microscopy (SEM), X-ray photoelectron spectroscopy (XPS), Fourier-transform infrared spectroscopy (FTIR), energy dispersive X-ray analysis (EDX), nuclear magnetic resonance (NMR), X-ray fluorescence microscopy, transmission electron microscope (TEM), and others can be utilized. For macrostructural and microstructural characterization, simple approaches can be used. However, in many cases, advanced methods may be necessary to better understand structural property relationships. To develop a practical understanding of materials, characterization investigations are required. Characterization may also be used to connect structure and characteristics. The identification of materials includes a host of characterization methods; no single approach can give an in-depth analysis of the material to be characterized. Research conclusions must be drawn using two or three methods of characterization (Ananthapadmanaban, 2018).

Titration Technique

Titration techniques can readily estimate functional groups with acidic or fundamental qualities on a material surface and ion exchange. To classify surface groups on biosorbents, Boehm or potentiometric titration may also be utilized. Acidic sites are determined using the Boehm process, which involves combining small amounts (0.1 g) of adsorbent with 10 ml of a base such as 0.1M $NaHCO_3$, 0.05M Na_2CO_3, or 0.1M NaOH in a beaker. The beakers are sealed and agitated for 24 hours. After the solutions are filtered, 0.05M H_2SO_4 is used to titrate them.

In the same way, basic sites are found by mixing 10 ml of 0.1M HCl with 0.1 g of biosorbent, and the obtained mixture is titrated with 0.1M NaOH (Kalijadis et al., 2011). Potentiometric titration may be used to calculate the functional groups present on the adsorbent as well as their overall concentration in the biomass. The following is the protocol for the experiment: blank sample, i.e. 0.1–0.2 g of biomass dissolved in deionized water and 0.1M NaOH, is added until a pH of 11.5 is reached and then reversed until the solution pH reaches 2.5 with 0.1M HCl. Once stability is reached, the pH of the biomass dispersion must be recorded with each dose of the titrant. Water must be bubbled with argon for 3 hours before the experiment to eliminate any dissolved CO_2. The results of the biosorbent potentiometric titration process should next be matched to adsorption equations that consider the presence of one or more types of functional groups on the biomass surface (Michalak and Chojnacka, 2010).

Fourier Transform Infrared Spectroscopy

To investigate biosorption mechanisms, the sorbent functional groups involved in the biosorption process must be identified. FTIR is a technique used to acquire a solid, liquid, or gas infrared spectrum of absorption, emission, photoconductivity, and/or Raman scattering. The term *Fourier transform infrared spectroscopy* refers to the fact that a Fourier transform (mathematical process) is necessary to transform raw data into the actual spectrum. FTIR spectroscopy provides valuable information about the nature of bonds and identifies the various functional groups within the cell wall structure.

The magnitude of band shifts in raw and biomass loaded with metal shows the degree to which the functional group interacts with metal cations (Murphy et al., 2007). The three varieties of infrared regions are near-IR, mid-IR, and far-IR. Near IR ranging from 14,000 to 4000 cm^{-1} may be used to visualize the changes in vibrational and rotational speeds, as well as electron transitions. Mid-IR ranging from 4000 to 400 cm^{-1} is used to visualize the molecules that exhibit fundamental vibrations in fundamental vibrational levels. Far-IR ranging from 400 to 20 cm^{-1} is used to display variations in rotational energy levels. At the same time, an FTIR spectrometer gathers spectral data in a broad spectrum. This provides a significant advantage over a dispersive spectrometer that measures intensity over a limited number of wavelengths at a time. Some examples of stretching frequencies found in biomaterials' FTIR spectra are listed in Table 6.1. An FTIR image is shown in Figure 6.1.

Table 6.1 FTIR spectra of biomaterials.

Wave number (cm^{-1})	Bonds	Reference
3278	–NH stretching, –OH bonded,	(Sheng et al., 2004)
2930	(–CH) asymmetric stretch of aliphatic chains	(Pons et al., 2004)
1741	C=O stretch of COOH	(Fourest and Volesky, 1995)
1629	C=O asymmetric	(Fourest and Volesky, 1995)
1530	amide II	(Sheng et al., 2004)
1449	C=O symmetric	(Fourest and Volesky, 1995)
1369	$-SO_3$ asymmetric stretching	(Figueira et al., 1999)
1240	C–O stretch of COOH	(Fourest and Volesky, 1995)
1162	$-SO_3$ symmetric stretching	(Figueira et al., 1999)
1120	C–O (ether)	(Fourest and Volesky, 1995)
1035	C–O (alcohol)	(Sheng et al., 2004)
817	S=O stretch	(Figueira et al., 1999)

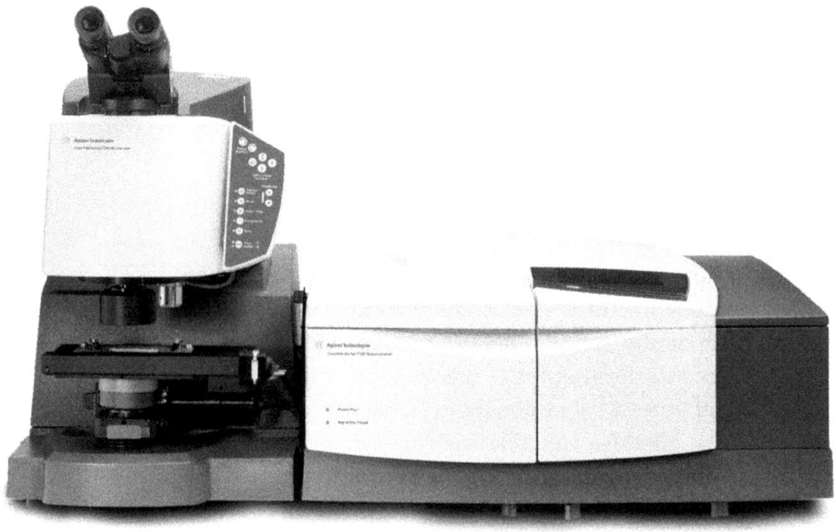

Figure 6.1 Fourier transforms infrared spectroscopy. *Source:* © 2021 Agilent Technologies, Inc.

The FTIR analysis is performed as follows. The particles are previously dried in a vacuum oven at 120 °C for 2 hours. At roughly 20 tons of pressure, potassium bromide powder is pressed into a clear disc that has high transmission over an infrared instrument range of 4000 to 400 cm^{-1}. To minimize scattering losses and distortions in the absorption band, the samples should be extremely finely ground. Prior to pressing, samples can be combined with KBr powder at concentrations ranging from 0.1 to 2% and their spectra collected in the KBr matrix. All spectra were adjusted for the typical gradual ATR rise in absorbance at lower wave numbers. The FTIR spectrums of *Gelidiium amansii* biomass samples were analyzed before and after Pb^{2+} biosorption to detect any alterations caused by interactions between the functional groups of the *Gelidiium amansii* biomass and the Pb^{2+} ions during biosorption.

The most prevalent acidic functional groups are carboxylic groups, and the presence of these binding sites is directly connected to the capacity of algae to adsorb. Carrageenan, the primary component of red algae, contains various binding sites for Pb^{2+} ions such as hydroxyl, carboxyl, amino, and sulfhydryl groups. The distinctive absorption peaks at 3435, 2923, 2525, 1802, 1624, 1471, 1417, 1082, 875, and 716 cm^{-1} were changed to 3438, 2923, 2525, 1802, 1624, 1419, 1036, 876, 719, and 546 cm^{-1} after lead biosorption by biomass. (El-Naggar et al., 2018). When FTIR spectra of *Ralstonia pickettii* HM8 were compared before and after heavy metal adsorption, it was discovered that the peak shape was nearly identical. The hydroxyl group's wavenumber altered from 3470 to 3420 cm^{-1} compared to *Ralstonia pickettii* HM8 before Mn(II) binding, and peak intensity at 3400 cm^{-1} dropped after Mn(II) binding. These findings suggested that Mn(II) could bind to hydroxyl groups. The absorption peaks of COO^- at 1400 cm^{-1} increased after adsorption of Mn(II), indicating that the carboxyl groups are engaged in the binding of Mn(II) onto its surface. Moreover, no significant variation in the peaks of the amide bonds was found following heavy metal adsorption, demonstrating that proteins do not affect adsorption. The data revealed that the most active groups for adsorption or complexation of Mn(II)were hydroxyl and carboxyl groups (Huang et al., 2018). It was found that primary and secondary amines (C–N), lignin aromatic (C=O), hydroxide (–OH), amide (N–H), C–H group, and aromatic compound groups appear on the surface of boiled wood chips (BWC), formaldehyde-treated wood chips (FWC) and wood chips treated with sulphuric acid(SWC) (Ofomaja et al., 2010). Further, the major functional group that contributes to copper adsorption is the hydroxyl group (–OH), according to Ofomaja et al. (2010) (Bansal et al., 2014).

Scanning Electron Microscopy with an Energy Dispersive X-ray Analytical System

The adsorbent surface morphology was studied using a scanning electron microscope (SEM). An SEM can produce images of solid samples and determine their elemental composition. The lower end of the magnification range for SEM is typically between 20X and 50X. The SEM is a powerful tool for studying biosorbent surface morphology before and after biosorption of metals ions (Reategui et al., 2010). It brilliantly observes surface phenomena such as swelling, aggregation, surface porosity, and molecular organization of the samples (Solangi et al., 2021). In this way, morphological variations of the biomass surface may be assessed: for example, alterations in the cell wall structure following metal ion binding. Further, an SEM provides useful knowledge about the distribution of constituents on the surface of the biomass in combination with energy dispersive spectrometer (EDS) techniques (Figueira et al., 1999).

The magnification power of an SEM is much higher than that of an optical microscope. However, an SEM gives only a qualitative assessment of the surface structure. The SEM and energy dispersive X-ray spectrometer (SEM/EDS) shown in Figure 6.2 can study and inspect samples at magnifications from 5X to 300,000X. The surface is evaluated for both the microstructure and fractures. An SEM analyzes small particles and/or areas inside a sample. Elemental analysis is conducted using an electron beam to measure the energy and intensity distribution of the X-ray signal emitted by the EDS. Samples are generally spattered with gold and held at an accelerating voltage of 10 kV in the SEM for about 15 seconds. Analysis of the primary composition of the adsorbent was performed using an EDS linked to the SEM.

Gelidium amansii biomass exhibited a homogeneous linked structure and a continuous surface. The process shows that *Gelidium amansii* biomass can adsorb and confiscate Pb^{2+} from aqueous media. As a result of the Pb^{2+} buildup caused by biosorption, the surface of the biomass became fragile and irregular, with light spots. The study evaluated the types of elements within the Pb^{2+} sample adhered to the *Gelidium amansii* biomass cell surface and validated the presence of Pb^{2+} using the EDS analysis. After adsorption of Pb^{2+} ions, a further optical absorption peak for Pb^{2+} is identified in the biomass, confirming *Gelidium amansii* biomass's participation in Pb^{2+} adsorption from an aqueous medium. (El-Naggar et al., 2018).

Ralstonia pickettii HM8 was cultivated in the absence of Mn(II) and subjected to 1000 mg/L of Mn(II) via SEM at 10000 magnification. According to the study, without Mn(II), *Ralstonia pickettii* HM8 was morphologically smooth, with a diameter of around

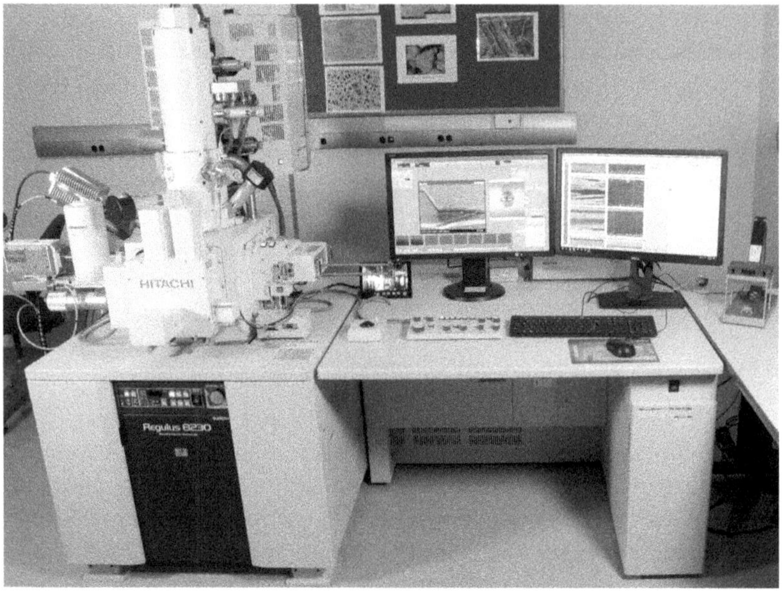

Figure 6.2 Scanning electron microscope attached to anenergy dispersive spectrometer. *Source:* © Hitachi High-Tech Corporation.

0.5 m. After metal biosorption, the cell-surface morphology changed considerably. Heavy metal ions appeared to weaken the surface of metal-loaded cells, which appeared blurry. After Mn(II) tension, the surface of HM8 cells became irregular and cracked, with wrinkles.

Since bacteria are small, they have a broad contact interface, making it easier for them to communicate with metals during the biosorption process. During metal biosorption, extracellular polymeric compounds are secreted, causing morphological changes (Huang et al., 2018). SEM images of boiled wood chips (BWC) surfaces before and after copper (II) ion biosorption specified the occurrence of copper (II) ions. It is assumed that irregular grooves and ridges are present in fibrous networks in the native BWC to provide heavy metal ions access to the biosorbent surface and, eventually, to the active sites.

EDS spectra of biosorbents equilibrated with metal ion solutions displayed distinctive peaks of copper (II) ions, indicating that these ions were truly adsorbed onto the biosorbents surface. Before examination, the biosorbents were coated with gold, which resulted in gold in the spectra. As a result, EDS analysis presented clear and supporting evidence for copper (II) ion-specific biosorption onto biosorbents (Ofomaja et al., 2010).

Magnetic pectin was synthesized from orange peel (Fe_3O_4@Pec-OPB) by Kadam et al. (2020). The Fe_3O_4@Pec-OPB biosorbent surface morphology was examined by SEM. The biosorbent formed exhibited a thin, nano-particulate, spherical morphology. Moreover, the reduction–precipitation approach produces a more distinct and uniform biosorbent compared to the synthesis of the same biological adsorbent utilizing co-precipitation (Kadam et al., 2020).

The surface morphology of synthesized adsorbent materials such as acid-modified *Salvinia molesta* (SM), iron oxide (IO), and magnetic biocomposite (MBC) were examined using an SEM at the desired magnification. This enables direct verification of changes on the material surface that occurred during the treatment process. The SEM micrographs of the IO and MBC were analyzed and found to be irregular in hierarchy because of aggregation and porous structured particles, with floccules, colonized shapes, and form. Microscopic investigations demonstrated a very rugged surface for all of the adsorbent materials (SM, IO, and MBC), offering a wide exposed surface for improved dye adsorption. The SEM image also reveals that the agglomerated structure of synthesized IO and MBC are extremely tiny spheroid crystals ranging in size from 0.916 to 1.0 μm for SM, 0.247 to 0.9 μm for IO, and 0.132 to 0.386 μm for MBC. All of these adsorbent materials exhibit a relatively homogenous structure, with a narrow distribution of agglomerate sizes except water hyacinth. The presence of well-developed pores on the surface resulted in a high surface area, which boosted the probability of dye entrapment and adsorption. The porous nature of the particle structure minimizes diffusion and makes mass transfer easier.

EDS images of acid-modified SM, IO, and MBC were analyzed, and it was observed that the elemental compositions were changed before and after the adsorption processes. The alteration in the elemental composition of the adsorbents may be caused by the interaction between the adsorbent function group and adsorbent reactive groups. Characteristic signals such as C and O weight percentage were increased after adsorption from 51.62 to 59.95 and 31.21 to 35.5. Similarly, the atomic percentage was also increased after adsorption from 51.26 to 55.1 and 41.28 to 43.09 (Sivashankar, 2017).

X-ray Photoelectron Spectroscopy Analysis

The quantitative X-ray spectroscopy (XPS) technique for chemical analysis is often referred to as *electron spectroscopy*, which allows the investigation of material surface chemistry. An XPS image is shown in Figure 6.3. The elementary composition, empirical formula, and electronic state of the elements in a material may be identified using this approach. XPS is frequently employed in sorption processes to acquire details of biosorption mechanisms and the oxidation state of adsorbed elements on biomaterial surfaces (Kebin et al., 2012).

XPS was also a valuable method for the analysis of adsorbent surface functional groups. The detection of binding energies and local chemical conditions allow for semi-quantitative study of all functional groups. Moreover, this method could be used to analyze elements such as oxygen, mercury, chlorine, nitrogen, and others.

The broad XPS scan spectrum is analyzed for natural reed, clean amine cross-connected reed (ACR), and nitrate-loaded ACR. New binding energies were allocated to the Cl2p and N1 core regions at 197 and 399 eV in the spectrum of clean ACR, indicating the grease of functional sites on the ACR surface. The N1 central area (from 394 to 408 eV) high-resolution spectrum was also analyzed. Two apparent highs with binding power of 398.7 and 401.8 eV, respectively, included the wider peak of N1 at 399 to 402 eV in clean ACR. N atoms linked to graphical C atoms (C–N) have been allocated the peak at 398.7 eV. N1s are, by comparison, allocated to $–NH^{3+}$ at 401.8 eV. The N1s spectrum at 399 to 402 eV showed that the contribution from neutral N was greater, and the contribution from positive N (–N/–N+ =.75) was comparatively small. Following the nitrate adsorption, the N1s in nitrate-laden ACR had a new high at a binding energy of 4049 eV. The laden nitrate was attributed to the new peak of the N1 core field. Neutral N and positive N radiation both decreased by 0.75 to 0.5; this meant the absorption of nitrate by the electrostatic attraction was regulated by tertiary amine groups in an ACR.

Figure 6.3 X-ray photoelectron spectroscopy. *Source:* © Western University.

X-Ray Diffraction Analysis

X-ray diffraction analysis (XRD), as shown in Figure 6.4, is a materials science technique used to identify a material's crystallographic structure. XRD works through the radiation of material with X-rays, after which the intensities and scattering angles of rays from the substance are measured. In XRD methods, X-rays strike the sample at various angles, resulting in diverse patterns of diffracted peaks that are connected to material crystallinity and unit cell structure (Veerakumar et al., 2017). The identification of materials based on their diffraction pattern is the main usage of XRD analysis.

XRD provides information about how the real structure is different from the ideal due to internal stresses and flaws, as well as phase identification. Crystals are regular atomic arrays, whereas X-rays may be seen as electromagnetic radiation waves. Crystal atoms scatter incident X-rays, principally through contact with the atoms' electrons. Crystal atoms scatter incoming X-rays, principally through electron interactions. This is called *elastic scattering*; the electron is referred to as the *scatterer*. A regular array of spherical waves isgenerated by the regular array of scatterers. Its large or small angle diffraction pattern distinguishes the XRD pattern. The unit cell and crystalline structure of the nanocomposite-based material are shown in wide-angle X-ray diffraction (WAXD) form, whereas the narrow X-ray diffraction (SAXD) pattern shows a material structure of around 10 Å or more. These waves cancel each other out in the majority of directions due to destructive interference, but they contribute constructively in a few specific directions, as established by Bragg's law

$$2d\sin\theta = n\lambda \tag{6.3}$$

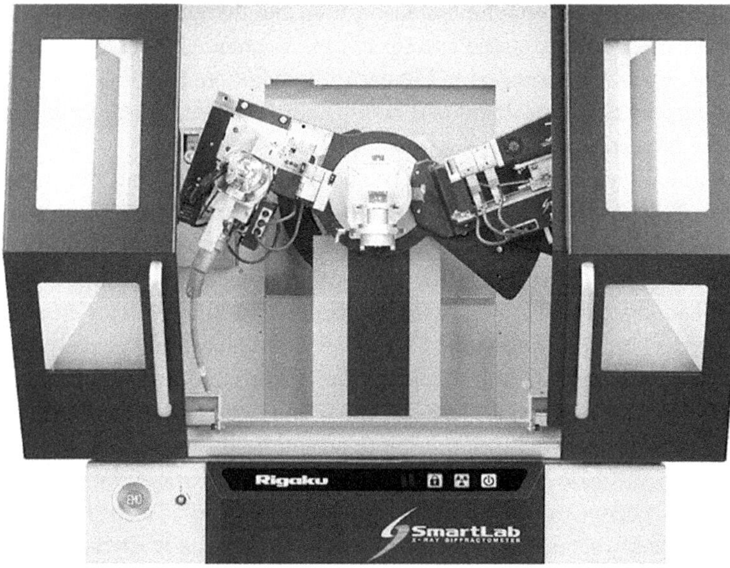

Figure 6.4 X-ray diffraction analysis.

where *d* is the distance between diffracting planes, the angle of incident is θ, *n* is the integer, and the beam wavelength is λ. The particular directions emerge as points in the so-called reflections diffraction pattern. As a result of the electromagnetic waves, X-ray diffraction patterns affect a regular array of scatterers. X-rays are used to produce the diffraction pattern because their wavelength is often the same order of magnitude as the spacing, *d*, between the crystal planes (1–100 Å).

Six principal lattice planes – (440), (511), (422), (400), (311), and (220) – were exhibited by the XRD pattern of the IO and MBC that are attributed to six diffraction peaks at 2θ=62.8°,57.2°,53.7°, 43.3°, 35.6°, and 30.2°, respectively. The four main diffraction peaks are maghemite or magnetite. Maghemite or magnetite may be attributed to the four primary diffraction peaks at 2θ=30.2°, 35.6°, 43.3°, and 57.2°. Additional peaks for MBC, which may be associated with hematite, are also found at 2θ=53.7° and 62.8°. In comparison, the peaks in XRD patterns of MBC with IO were shifted slightly to lower angles, indicating an interaction between biomass and metal oxide. The peak near the 28° reflection corresponds to the crystallographic (200) lattice plane. However, other peaks are also identified, which may be due to the abundance of such amorphous components in untreated, mineral acid-treated, and organic acid-treated biomass (SM). It may also be seen that the biosorption ability of the treated samples is greater than that of the untreated samples, suggesting that the treatment processes enhance the crystallinity of the biomass and may be caused by the elimination during the acid treatment of some amorphous components of SM (Sivashankar, 2017).

Brunauer-Emmett-Teller Analyzer

The surface of an adsorbent is typically a significant element in adsorption studies, i.e. the greater the surface, the more adsorption capacity. The large surface area of an absorbent material considerably improves organic contaminant adsorption, and therefore the Brunauer-Emmett-Teller (BET) surface area test is regarded as a significant technique for adsorbent surface defects. A BET surface analyzer is displayed in Figure 6.5. In addition, BET analysis may examine the volume and porosity of an adsorbent to determine an adsorbent's polar structure such as its mesosphere and microsphere.

BET determines the surface area of the total area (reactive surface) since all the pores absorb tiny gas molecules. Therefore the surface area of the BET is usually more than that of air permeability. The samples are dried in a vacuum using a nitrogen purge or at high temperatures. The volume of gas on the particle surface is determined at the boiling point of nitrogen (−196 °C). The amount of adsorbed gas is associated with the particles' overall surface area, including pores. The BET theory is the basis for the calculation. Typically, the adsorbed gas is utilized with nitrogen. Gas adsorption also permits the measurement of size and volume distribution of micropores (0.35–2.0 nm). The actual surface area of a particle is determined by the physical adsorption of a gas on the solid surface and the volume of adsorbate gas corresponding to a monomolecular layer on the surface. Weak interactions (Van der Waals forces) between the adsorbate gas molecules and adsorbed surface area of the test powder induce physical adsorption. The procedure is usually carried out at liquid nitrogen temperature. A volumetric or continuous flow technique may be used to determine the amount of gas adsorbed.

Jack fruit leaf is employed by Patel et al. (2016) to make fruit-based KOH activated carbon (JFAC). The BET analyzer showed that JFAC absorbents had a high 916.1 cm^2/g surface area, which means mesosphere architectures of pores are effective adsorbents to remove large molecular waste particles from wastewater (Patel et al., 2016). By using different biosorbents and deploying new technologies, the BET surface area improves. For example, the surface area for a crude banana peel was 0.65 m^2/g, but it improved to

Figure 6.5 Brunauer–Emmett–Teller Analyzer.

22.37 m^2/g following microwave treatment (Pathak and Mandavgane, 2015). It is impossible to completely understand the actions of adsorbents without BET studies.

Mercury intrusion porosimetry (MIP) is utilized to determine the surface area using a mercury porosimeter (Quantachrome, Poremaster 60 model). To measure different porosity parameters, data were taken from N_2 adsorption at 77k, CO_2 adsorption at 273k, and mercury-porosimetry studies. In the end, the total pore volume was estimated by the addition of the volume of each pore size and the average pore diameter of the surface area values, and the total pore volume.

The surface area was measured as 88.2028, 33.7704, and 1.7054 m^2/g for MBC, IO, and SM, respectively. The surface area of MBC is much larger than that of IO and SM. The surface area was found to decrease in the order MBC > IO > SM. The pore volume was measured as 0.1256, 0.1613, and 0.00557 cm^3/g for MBC, IO, and SM, respectively. The pore size was also measured as 25.689, 88.455, and 4.2853 Å. The volume and size of the pore of IO were found to be larger than other adsorbent particles (SM and MBC). Porous materials can be classified as having micropores (<50 nm in pore diameter) by the International Union of Pure and Applied Chemistry (IUPAC). As a result, according to IUPAC, the prepared adsorbents SM, IO, and MBC are micropores. Pores are typically microporous, with the exposed surface almost entirely contained within the microspores, which, once filled with adsorbate, leaves little or no exterior surface for additional adsorption. The Langmuir equation defines microporous material with a Type I isotherm that assumes monolayer adsorption (Sivashankar, 2017).

Thermal Stability Analyzer

Material thermal stability is generally analyzed through thermogravimetric analysis (TGA) and differential scanning calorimetry (DSC). TGA, as shown in Figure 6.6, is used to determine the thermogravimetric stability and decomposition of materials in specific temperature ranges. In TGA, a programmed heat rate is used to heat the sample, and the weight is monitored. The diagram is drawn from the weight loss to temperature. Until the temperature at which weight loss commences, the sample is thermally stable. A TGA analyzer was employed to heat samples at a rate of 10 °C/m between 30 and 900 °C. The biosorbent initially seemed stable from 30 to 150 °C and had a very low weight loss, which might be attributed to the loss of water molecules and decreased volatility of low-molecular-weight compounds. Significant degradation between 214 and 390 °C was observed, which is equivalent to a 55.2% loss in biosorbent weight. After 390 °C, organic matter deteriorated uniformly to 900 °C (Ahmad et al., 2020).

The weight loss was measured over the temperature range 100 to 940 °C for all of the prepared adsorbents. The loss of weight at 400 °C may be due to decomposition or degradation of the remaining cellulose and lignin. Polysaccharides such as cellulose and hemicellulose, as well as other organic polymers such as lignin, are significant components of biomasses, and their primary functional groups include CH, COOH, C–N, OH, and CH_2OH. Most lignocellulosic materials are composed of 30–70% cellulose, 25–40% hemicellulose, and 10–25% ligurine. The next event is hemicellulose deterioration at 400 °C, while lignin and cellulose begin to breakdown at 400 °C. It was also observed that 55.43% occurred at 600 °C. Above 600 °C, the biosorbent exhibits high ash content, and a further 13.75 % was observed at 910 °C. It was observed that no significant weight loss occurred in

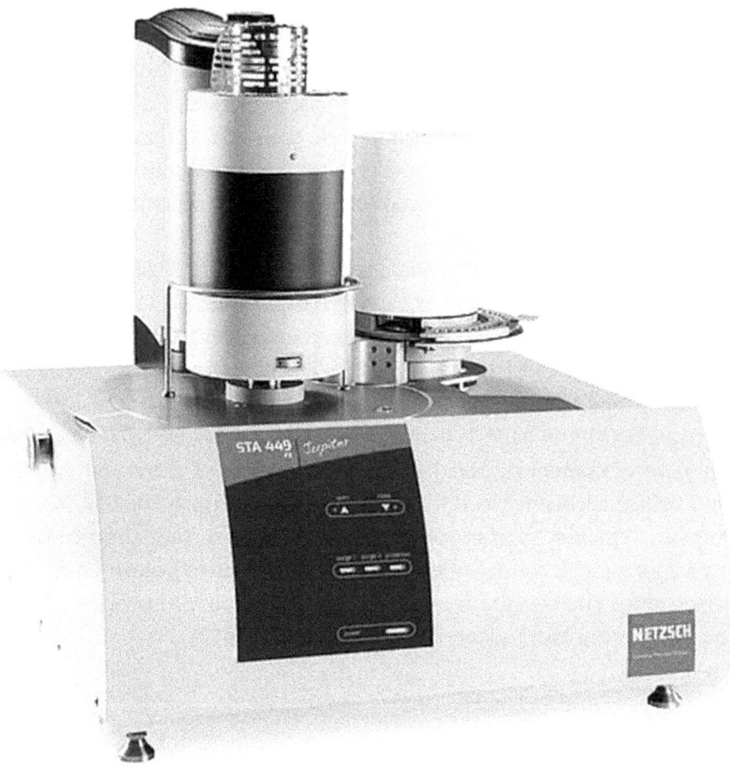

Figure 6.6 Thermo-gravimetric analyzer. *Source:* NETZSCH.

the TGA curve of the IO nanoparticles. The TGA curve showed that the weight loss for IO over the temperature range 100 to 940 °C was about 3.58%. This might be due to the loss of the remaining water and agents. A weight-loss step was seen at about 500 °C. This might be attributed partly to the decomposition of iron amorphous hydroxides and the creation of iron oxides. For an IO sample, the increase in weight over a broad range of temperatures is due to the oxidation of Fe_3O_4 in Fe_2O_3. The first loss of weight before 500 °C may result from removing leftover water and/or hydroxyls from the surface. The first weight loss for the MBC adsorbent before 500 °C was likely due to desorption of surface hydroxyls and/or adsorbed water and excess monomer. The excess monomers are due to the presence of carboxylic group of SM impregnated with metal oxides (Sivashankar, 2017).

Conclusion

Adsorption processes have evolved as a result of increased demand and needs. Adsorbent characterization is required in order to analyze adsorbent properties. This chapter presented an in-depth study of the characterization of adsorbent materials, including optical, thermal, physicochemical, and instrumental characterization. This comprehensive review of adsorbent characterization methods shows their usefulness and how they may be efficiently coupled and complementary. This helps researchers select the most suitable methodologies and allows them to evaluate their utilization more accurately.

References

Abbas Ali, A., Mohamed Sihabudeen, M., and Zahir Hussain, A. (2016). Biosorption of heavy metals by pseudomonas bacteria. *International Research Journal of Engineering and Technology* 3 (8): 1446–1450.

Ahmad, T., Belwal, T., Li, L. et al. (2020). Utilization of wastewater from edible oil industry, turning waste into valuable products: a review. *Trends Food Sci. Technol* 99: 21–33s.

Ali Redha, A. (2020). Removal of heavy metals from aqueous media by biosorption. *Arab Journal of Basic and Applied Sciences* 27 (1): 183–193. doi:10.1080/25765299.2020.1756177.

Ananthapadmanaban, D. (2018). Summary of some selected characterization methods of geopolymers. In: *Geopolymers and Other Geosynthetics*. Intech Open.

Bansal, M., Mudhoo, A., Garg, V.K., and Singh, D. (2014). Preparation and characterization of biosorbents and copper sequestration from simulated wastewater. *International Journal of Environmental Science and Technology* 11 (5): 1399–1412.

Derco, J. and Vrana, B. (2018). Introductory chapter: biosorption. In: *Biosorption*. InTech. doi:10.5772/intechopen.78961.

El-Naggar, N.E.A., Hamouda, R.A., Mousa, I.E. et al. (2018). Biosorption optimization, characterization, immobilization and application of Gelidium amansii biomass for complete Pb 2+ removal from aqueous solutions. *Scientific Reports* 8 (1): 1–19.

Figueira, M.M., Volesky, B., and Mathieu, H.J. (1999). Instrumental analysis study of iron species biosorption by Sargassum biomass. *Environmental Science and Technology* 33 (11): 1840–1846.

Fourest, E. and Volesky, B. (1995). Contribution of sulfonate groups and alginate to heavy metal biosorption by the dry biomass of Sargassum fluitans. *Environmental Science and Technology* 30 (1): 277–282.

Gadd, G.M. (2009). Biosorption: Critical review of scientific rationale, environmental importance and significance for pollution treatment. *Journal of Chemical Technology and Biotechnology* 84 (1): 13–28.

Huang, H., Zhao, Y., Xu, Z. et al. (2018). Biosorption characteristics of a highly Mn (II)-resistant *Ralstonia pickettii* strain isolated from Mn ore. *PloS One* 13 (8): e0203285.

Jaishankar, M., Tseten, T., Anbalagan, N. et al. (2014). Toxicity, mechanism and health effects of some heavy metals. *Interdisciplinary Toxicology* 7(2): 60–72. doi:10.2478/intox-2014-0009.

Javanbakht, V., Alavi, S.A., and Zilouei, H. (2014). Mechanisms of heavy metal removal using microorganisms as biosorbent. *Water Science and Technology* 69 (9): 1775–1787.

Kadam, A.A., Sharma, B., Saratale, G.D. et al. (2020). Super-magnetization of pectin from orange-peel biomass for sulfamethoxazole adsorption. *Cellulose* 27: 3301–3318.

Kalijadis, A., Vukcevic, M.M., Jovanović, Z.M. et al. (2011). Characterisation of surface oxygen groups on different carbon materials by the Boehm method and temperature-programmed desorption. *Journal of the Serbian Chemical Society* 76 (5): 757–768.

Kanamarlapudi, S.L.R.K., Chintalpudi, V.K., and Muddada, S. (2018). Application of biosorption for removal of heavy metals from wastewater. In: *Biosorption* (ed. J. Derco and B. Vrana). InTech. doi: 10.5772/intechopen.77315.

Kebin, L., Qinqin, W., Yan, D. et al. (2012). Characteristic and mechanism of Cr(VI) biosorption by buckwheat hull from aqueous solutions [J]. *Acta Chimica Sinica* 70 (07): 929–937.

Macek, T. and Mackova, M. (2011). Potential of biosorption technology. In: *Microbial Biosorption of Metals* (ed.: Kotrba, M. Mackova, and T. Macek). Springer. doi:10.1007/978-94-007-0443-5_2.

Michalak, I. and Chojnacka, K. (2010). Interactions of metal cations with anionic groups on the cell wall of the macroalga Vaucheria sp. *Engineering in Life Sciences* 10 (3): 209–217.

Michalak, I., Chojnacka, K., and Witek-Krowiak, A. (2013). State of the art for the biosorption process – a review. *Applied Biochemistry and Biotechnology* 170 (6): 1389–1416.

Murphy, V., Hughes, H., and McLoughlin, P. (2007). Cu (II) binding by dried biomass of red, green and brown macroalgae. *Water Research* 41 (4): 731–740.

Ofomaja, A.E., Unuabonah, E.I., and Oladoja, N.A. (2010). Competitive modeling for the biosorptive removal of copper and lead ions from aqueous solution by Mansonia wood sawdust. *Bioresour Technol* 101: 3844–3852.

Patel, S.P., Patel, M.T.A.V., and Patel, J.V. (2016). Study of KOH impregnated jack fruit leaf based carbon as adsorbent for treatment of wastewater contaminated with nickel. *Int J. Recent Trends Eng. Resour* 2: 219–229.

Pathak, P.D. and Mandavgane, S.A. (2015). Preparation and characterization of raw and carbon from banana peel by microwave activation: application in citric acid adsorption. *J. Environ. Chem. Eng* 3: 2435–2447.

Pons, M.N., Le Bonté, S., and Potier, O. (2004). Spectral analysis and fingerprinting for biomedia characterisation. *Journal of Biotechnology* 113 (1–3): 211–230.

Reategui, M., Maldonado, H., Ly, M., and Guibal, E. (2010). Mercury (II) biosorption using Lessonia sp. kelp. *Applied Biochemistry and Biotechnology* 162 (3): 805–822.

Sao, K. (2014). A review on heavy metals uptake by plants through biosorption. *International Proceedings of Economics Development and Research* 75 (17): 78–83. doi:10.7763/IPEDR.

Sheng, P.X., Ting, Y.P., Chen, J.P., and Hong, L. (2004). Sorption of lead, copper, cadmium, zinc, and nickel by marine algal biomass: characterization of biosorptive capacity and investigation of mechanisms. *Journal of Colloid and Interface Science* 275 (1): 131–141.

Simionato, J.I., Guerra Villalobos, L.D., Keller Bulla, M. et al. (2014). Application of chitin and chitosan extracted from silkworm chrysalides in the treatment of textile effluents contaminated with remazol dyes. *Acta Scientiarum. Technology Maringá* 36 (4): 693–698. doi:10.4025/actascitechnol.v36i4.24428.

Sivashankar, R. (2017). Synthesis and characterization of magnetically modified biosorbent for sequestration of dyes from aqueous solution. PhD dissertation, National Institute of Technology Calicut, India.

Solangi, N.H., Kumar, J., Mazari, S.A. et al. (2021). *Development of fruit waste derived bio-adsorbents for wastewater treatment: A review. Journal of Hazardous Materials* 125848.

Veerakumar, P., Thanasekaran, P., Lin, K.-C., and Liu, S.-B. (2017). Biomass derived sheet-like carbon/palladium nanocomposite: an excellent opportunity for reduction of toxic hexavalent chromium. *ACS Sustain. Chem. Eng* 5: 5302–5312.

Wei, W., Wang, Q., Li, A. et al. (2016). Biosorption of Pb (II) from aqueous solution by extracellular polymeric substances extracted from Klebsiella sp. J1: Adsorption behavior and mechanism assessment. *Scientific Reports* 6 (1): 10.

7

Isotherm and Kinetic Modeling Analysis of Water Decontamination through Biosorption

Subramanyam Busetty[1], *Ramprasad Chandrasekaran*[1], *and Srihari Vedartham*[2]

[1] *School of Civil Engineering, SASTRA Deemed University, Thanjavur, Tamilnadu, India*
[2] *NICMAR, Hyderabad Campus at Shamirpet, Hyderabad, Telangana, India*

Adsorption Equilibrium Analysis

Basics of Adsorption Equilibrium

Adsorption is the deposition of concentration at a surface – an affinity of adsorbate molecules (a gaseous or liquid component) to an adsorbent surface. The relation between adsorbate and adsorbent mainly consists of atomic molecular forces developing a permanent dipole, induced dipole, and quadruple electrostatic effects, popularly known as Van der Wall's forces of attraction. When adsorption occurs between adsorbate and adsorbent, the forces become balanced.

Adsorption is an exothermic process. The heat evolved during the adsorption of a solute onto a solid from a liquid is strongly dependent on the source and history of the solid adsorbent (Do, 1998).

Models of Adsorption Equilibrium

Different adsorption isotherm models have been proposed, some hypothetical and some of a more experimental nature. Most of these derived equations are effective over relatively small pressure ranges but sometimes do not fit the investigational data when tested with the maximum range of comparative pressures. Only those that are generally used for the surface of biosorbents are outlined.

Many researchers in the field of biosorptive removal of water pollutants have used the Langmuir and Freundlich isotherm equations to represent correlation coefficient results when using activated carbon and organic contaminants. In an attempt to determine (Boyd et al., 1988; Brigatti et al., 1995) the adsorption potential of soil and agro soil-based adsorbents for removing organic phenol, Subramanyam and Das (2012) analyzed using 14 isotherm models, ranging from single-parametric to multi-parametric models of the system (up

Biosorption for Wastewater Contaminants, First Edition.
Edited by Rangabhashiyam Selvasembian and Pardeep Singh.

to five parameters). The results show that, in general, the accuracy of models to fit experimental data improves as the degrees of freedom increase. The correlation coefficients, average percentage error, and student t-test were used to determine the best-fitting isotherm model and understand the mechanism involved with different types of sorbate-sorbent systems. Among the two-parameter models, three-parameter models, and four-parameter isotherm modelling, the Temkin isotherm model, Langmuir-Freundlich isotherm model, Fritz- Schlunder model, and Baudu model were found to be the best-fitting models. This study emphasizes the importance of simultaneous solution of multi-parametric equations rather than solving their linearized forms, as commonly used by contemporary researchers (Darrowski et al., 2005).

The commonly used isotherm studies have attempted to categorize the fitness of selected raw and treated biomass. The study of biosorption of pollutants at a steady concentration and constant temperature is referred to as an *adsorption isotherm*. It usually describes a quantifiable association between the strength of the biosorbate in solution (C_e, mg/l) and its biosorbate concentration over the biosorbent (q_e, mg/g), as shown in Table 7.1 (Ozkaya, 2006). The type of isotherm curve depends on factors such as the nature, concentration, and number of pollutants; type of biosorbent; characteristics of the aqueous medium; and other environmental factors.

El Alouani et al. (2018) used a geopolymer based on fly ash for the adsorption of methylene blue by studying adsorbent dosages and initial adsorbate concentration. They identified the extreme adsorption capacity q_e (mg/g) as 37.04. Namasivayam and Kavitha (2002) used coir pith as an adsorbent to adsorb Congo red and determined the maximum adsorption capacity q_e as 2.6 (mg/g) under optimized pH and dosage conditions. A Langmuir isotherm was identified as the best-fit isotherm or kinetic model. Bharathi and Ramesh (2012) worked on *Citrullus lanatus* rind as an adsorbent to adsorb crystal violet and found 11.9 (mg/g) as the maximum adsorption capacity. Langmuir and pseudo-second-order were considered the best-fit isotherm or kinetic model. Khraisheh et al. (2018) used a nano-composite material to adsorb cadmium under optimized pH, contact time, adsorbent dosage, and initial concentration and identified the maximum adsorption capacity q_e (mg/g) as 148.32. The best-fit isotherm or kinetic model was Langmuir. Zhang et al. (2017) also worked on a cadmium removal study using phosphate-modified activated bamboo as an adsorbent with various optimized parameters such as pH, contact time, adsorbent dosage, and initial concentration; the maximum adsorption capacity q_e (mg/g) was 202.55. Langmuir and pseudo-second-order were the best fit isotherm or kinetic model. Bulgariu and Bulgariu (2018) focused on lead adsorption by using soy waste as an adsorbent; the maximum adsorption capacity q_e (mg/g) was 0.4089 mmol/g using optimized parameters such as pH, contact time, adsorbent dosage, and initial concentration. Freundlich and pseudo-second-order were the best-fit isotherm or kinetic models. Zhang et al. (2019) carried out an extensive study on selenium (IV) removal using zeolite synthesis from low-calcium fly ash as an adsorbent under specified conditions including pH, contact time, adsorbent dosage, and initial concentration. The maximum adsorption capacity q_e (mg/g) was 4.16, and the best-fit isotherm or kinetic models were Freundlich and pseudo-second-order.

Ba et al. (2018) used activated carbon from olive waste as an adsorbent to separate chromium using optimized parameters such as pH, contact time, adsorbent dosage, and initial

Table 7.1 Reported maximum biosorption capacities for different adsorbents and adsorbates.

Adsorbate	Adsorbent	Parameter optimized	Best-fit isotherm and kinetic model	Maximum biosorption capacity q_e (mg/g)	Reference
Methylene blue	Fly ash-based geopolymer	Adsorbent dosages and initial adsorbate concentration	Langmuir and pseudo-second-order	37.04	El Alouani et al., 2018
Congo red	Coir pith	pH and dosage	Langmuir	2.6	Namasivayam and Kavitha, 2002
Crystal violet	Citrulluslanatus rind	NA	Langmuir and pseudo-second-order	11.9	Bharathi and Ramesh, 2012
Cadmium	Nano-composite	pH, contact time, adsorbent dosage, initial concentration	Langmuir	148.32	Khraisheh et al., 2018
Cadmium	Phosphate-modified activated bamboo	pH, contact time, adsorbent dosage, initial concentration	Langmuir and pseudo-second-order	202.55	Zhang et al., 2017
Lead	Soy waste	pH, contact time, adsorbent dosage, initial concentration	Freundlich and pseudo-second-order	0.4089 mmol/g	Bulgariu and Bulgariu, 2018
Selenium (IV)	Zeolite synthesis from low-calcium fly ash	pH, contact time, adsorbent dosage, initial concentration	Freundlich and pseudo-second-order	4.16	Zhang et al., 2019
Chromium	Activated carbon from olive waste	pH, contact time, adsorbent dosage, initial concentration	Langmuir, and pseudo-second-order	14.3–74.9	Ba et al., 2018
Nickel	Sewage sludge	pH, contact time, adsorbent dosage temperature, initial ion concentration	Freundlich, and pseudo-second-order	11.53	Khelifi et al., 2018
Surfactants	Four different soils	pH, contact time, adsorbent dosage	Langmuir model	4–28.6	Ramprasad and Philip, 2017

(*Continued*)

Table 7.1 (Continued)

Adsorbate	Adsorbent	Parameter optimized	Best-fit isotherm and kinetic model	Maximum biosorption capacity q_e (mg/g)	Reference
Heavy metals	Chitosan	NA	Langmuir model	1–2	Renge et al., 2012
Malachite green	Neem sawdust	Agitation time, dye concentration, adsorption dose, pH, temperature	Langmuir model	4.354	Khattri and Singh, 2009
Arsenic V	Activated-Bauxsol-coated sand	NA	Langmuir, and pseudo-second-order	2.14	Genç-Fuhrman et al., 2005
Copper	Red mud	NA	NA	19.72	Apak et al., 1998
Phenol	Red mud	pH, contact time, adsorbent dosage	NA	4.12	Gupta and Ali, 2006

concentration and found a maximum adsorption capacity q_e (mg/g) between 14.3 and 74.9. Langmuir and pseudo-second-order were the best-fit isotherm or kinetic models. Khelifi et al. (2018) studied nickel removal by using sewage sludge as an adsorbent with optimized parameters such as pH, contact time, adsorbent dosage temperature, and initial ion concentration. The value 11.53 was the maximum adsorption capacity q_e (mg/g). Freundlich and pseudo-second-order were the best-fit isotherm or kinetic models. Ramprasad and Philip (2017) carried out research to remove surfactants by using four different soils with different pH, contact time, and adsorbent dosage parameters and found that the maximum adsorption capacity q_e (mg/g) varied between 4 and 28.6. The Langmuir model was the best-fit isotherm or kinetic model. Khattri and Singh (2009) worked on chitosan as an adsorbent to separate heavy metals. The maximum adsorption capacity q_e (mg/g) was 1–2, and the Langmuir model was the best-fit isotherm or kinetic model. Khattri and Singh (2009) carried out an extensive study on malachite green by using neem sawdust as the adsorbent and considering agitation time, dye concentration, adsorption dose, pH, and temperature as basic optimized parameters; 4.354 was the maximum adsorption capacity q_e (mg/g). The best-fit isotherm or kinetic model was Langmuir. Genç-Fuhrman et al. (2005) used activated-Bauxsol-coated sand as an adsorbent to remove arsenic V and determined the maximum adsorption capacity q_e (mg/g) to be 2.14. Apak et al. (1998) worked on adsorbing copper by using red mud as an adsorbent. The maximum adsorption capacity q_e (mg/g) was 19.72. Gupta and Ali (2006) studied phenol removal using optimized conditions such as pH, contact time, adsorbent dosage and found 4.12 to be the maximum adsorption capacity q_e (mg/g).

In addition, several other generally recognized mathematical relationships have been developed to describe the equilibrium distribution of a solute between the dissolved (liquid) and adsorbed (solid) phases. To enhance the design of an adsorption system for the separation of adsorbates by adsorption, it is important to develop the most suitable correlation for the balancing equilibrium curves. Various isotherm models mentioned in the preceding paragraphs are used to explain the balanced nature of adsorption.

Two-Parameter Model

Langmuir Isotherm Model (Langmuir, 1918)

The equation for the popular biparameter Langmuir isotherm (Langmuir, 1917) is as follows (Weber and Chakravarthy, 1974)

$$q_e = \frac{x}{m} = \frac{q_m b C_e}{1 + bC_e} \tag{7.1}$$

where x = mass of solute biosorbed (mg); m = mass of biosorbent (g); C_e = final aqueous phase concentration (mg/l), q_e = amount of solute adsorbed per unit weight of adsorbent at equilibrium (mg/g), q_m = extreme adsorption capacity (mg/g), and b = constant describing the adsorption free energy (l/mg). The Langmuir model generally is better to explain the capacity of adsorption of gases onto solids, whereas the Freundlich model is superior for the adsorption of liquid solutions (Cooney, 1999). The simplified form of the Langmuir equation is given as

$$\frac{C_e}{q_e} = \frac{C_e}{q_m} + \frac{1}{bC_e} \tag{7.2}$$

Table 7.2 The Langmuir separation factor.

R_L Value	Type of process
$R_L > 1$	Unfavorable
$R_L = 1$	Linear
$0 < R_L < 1$	Favorable
$R_L = 0$	Irreversible

The Langmuir isotherm can also be expressed by a separation factor (R_L), given by the following equation (Weber and Chakravarthy, 1974).

$$R_L = \frac{1}{1 + b\ C_i} \tag{7.3}$$

where C_i = initial concentration of phenol in mg/l and b = Langmuir constant in l/g. The separation factor R_L indicates the nature of the adsorption process, as given in Table 7.2.

Freundlich Isotherm Model

The Freundlich model is by far the most commonly used isotherm model for wastewater treatment. It has been reported that data for adsorption involving adsorbates in a liquid phase are best fitted using the Freundlich model (Cooney, 1999). The two-parameter Freundlich model relates the sorbed phase concentration to an equilibrium concentration of the adsorbate according to the following equation

$$\frac{x}{m} = q_e = k_f C_e^{1/n} \tag{7.4}$$

where x = mass of solute adsorbed (mg); m = mass of adsorbent (g); C_e = final aqueous phase concentration (mg/L); k_f, n = empirical constants for the adsorption capacity and intensity of adsorption measuring parameters; q_e is the amount of pollutant biosorbed per unit mass of biosorbent; and C_e is the equilibrium concentration (mg/l). The readings for the constant k_f represent a water phase equilibrium concentration of 1 mg/l (i.e. $C_e = 1$) when the equation is changed to the form $x/m = k_f$, and thus, k is the adsorptive capacity and has the units w/w or mg/g for the example. The Freundlich model assumes that the distribution of energy for adsorption sites is exponential (Cooney, 1999).

The adsorption and desorption rates vary with the adsorption energy associated with the sites, and there is a chance of the formation of more than one monomolecular layer of adsorptive coverage. The Freundlich model also does not require that the surface coverage approach a constant value corresponding to one complete monolayer as C_e increases. At high concentrations, the equation fails to fit the experiment (Cooney, 1999). In such cases, the logarithmic form of the Freundlich equation can be expressed as follows:

$$\log\frac{x}{m} = \log q_e = \log k_f + \frac{1}{n}\log C_e \tag{7.5}$$

The linear form has a slope with the value $1/n$, and the Y-axis intercept is log k. The smaller the slope, the better the adsorption over the entire range of concentrations, whereas a sharper slope indicates the best adsorption at greater concentrations. A larger value of k_f indicates a greater adsorptioncapacity. The Freundlich isotherm equation assumes that the adsorbent has a heterogeneous surface composed of adsorption sites with different adsorption potentials.

Modified Langmuir – 1

$$q_e = \frac{bT^{-n}C_e}{1+bC_e} \tag{7.6}$$

where q_e and C_e have their usual meanings (as given above), T is the Kelvin temperature, and the constant b indicates the free energy of adsorption (l/mg).

Modified Langmuir – 2

$$q_e = \frac{bC_e}{1+bC_e}\left(\left(\frac{1+\sigma^2(1-bC_e)}{2(1+bC_e)^2}\right)\right) \tag{7.7}$$

where q_e and C_e have their usual meanings (as described above) and σ is the constant of the isotherm connected to the degree of adsorption.

Dubinin–Radushkevich Isotherm

The Dubinin-Radushkevich (D-R) isotherm is often (Dubinin, 1960) applied to determine porosity and the apparent free energy of adsorption. Hence, to evaluate these parameters, the Dubinin-Radushkevich isotherm is used in the following form

$$q_e = K_{DR}\exp\left[b\left(RTln\left(1+\frac{1}{C_e}\right)\right)^2\right] \tag{7.8}$$

where q_e = amount adsorbed at equilibrium (mg/g); C_e = adsorbed concentration at equilibrium (mg/l); K_{DR} = sorption free energy; and b = isotherm constant for Dubinin-Radushkevich related to the degree of sorption.

Temkin Model

The Temkin isotherm model shows that adsorbate interactions cause the heat of adsorption within molecules to decrease linearly. The Temkin isotherm is usually expressed in the following form (Vijayaraghavan et al, 2006)

$$Q_e = \frac{RT}{b}(lnAC_e) \tag{7.9}$$

where q_e = amount adsorbed at equilibrium (mg/g); C_e = adsorbed concentration at equilibrium (mg/l); and parameters A and b = isotherm constants.

Three-Parameter Models

Langmuir-Freundlich Isotherm (Kobe and Corrigan, 1952)

$$q_e = \frac{q_{mLF}\left(K_{LF}C_e\right)^{mLF}}{1+\left(K_{LF}C_e\right)^{mLF}} \tag{7.10}$$

where q_e = amount adsorbed at equilibrium (mg/g); q_{mLF} = Langmuir-Freundlich higher adsorption capacity (mg/g); C_e = adsorbed concentration at equilibrium (mg/L); K_{LF} = constant of equilibrium for a heterogeneous substance; and m_{LF} = Langmuir-Freundlich heterogeneity parameter, usually between 0 and 1. At low adsorbate concentrations, the Langmuir-Freundlich isotherm model can be successfully reduced to the Freundlich isotherm and thus disobeys Henry's law. At maximum adsorbate concentrations, it performs like sorption of monolayers and exhibits Langmuir isotherm characteristics.

Fritz-Schlunder Equation

The specific model using the Fritz-Schlunder equation is as follows (Yaneva et al., 2013)

$$q_e = \frac{q_{mFS}K_{FS}C_e}{1+q_{mFS}C_e^{\propto}} \tag{7.11}$$

where q_e = amount adsorbed at equilibrium (mg /g); C_e = concentration of the adsorbate at equilibrium (mg/L); q_{mFS} = Fritz-Schlunder higher adsorption capacity (mg/g); K_{FS} = Fritz–Schlunder equilibrium constant (L/mg); and α = Fritz–Schlunder model exponent.

Radke-Prausnitz Model

The Radke-Prausnitz model equation is as follows (Vijayaraghavan et al., 2006)

$$q_e = \frac{q_{mRP}K_{RP}C_e}{\left(1+K_{RP}C_e\right)^{mRP}} \tag{7.12}$$

where q_e = amount adsorbed at equilibrium (mg/g); q_{mRP} = Radke-Prausnitz higher adsorption capacities (mg/g); C_e = equilibrium concentration of the adsorbate (mg /l); K_{RP} = Radke-Prausnitz equilibrium constant; and m_{RP} = Radke–Prausnitz model exponents.

Toth Model

Toth has carried out extensive studies on the Langmuir equation to minimize the error between experimental analysis data and projected values of adsorption equilibrium data. His equation is well-matched to multilayer adsorption and is similar to the BET isotherm, which is a distinct type of Langmuir isotherm that has very instructive validity. The Toth model equation is given as (Toth, 2000)

$$q_e = \frac{C_e q_{mT}}{\left(\frac{1}{K_T}+C_e^{mT}\right)^{\frac{1}{mT}}} \tag{7.13}$$

where q_e = amount adsorbed at equilibrium (mg/g); C_e = concentration of the adsorbate at equilibrium (mg/l); q_{mT} = Toth higher adsorption capacity (mg/g); K_T = Toth equilibrium constant; and mT = Toth model exponent.

Jossens Model

The specific Jossens model is explained by the energy distribution (interaction between adsorbate and adsorbent on adsorption sites) (Hamdaoui and Naffrechoux, 2007) as follows

$$C_e = \frac{q_e}{H}\exp\left(Fq_e^p\right) \tag{7.14}$$

where q_e = amount adsorbed at equilibrium (mg/g); C_e = adsorbate concentration at equilibrium (mg/L); and parameters F, H, and p are Jossens model constants. H and F are mainly depending on temperature. At lower capacities, this equation can be reduced to Henry's law.

Redlich-Peterson Model

The Redlich-Peterson (1959) model is a three-parameter model:

$$q_e = \frac{x}{m} = \frac{AC_e}{1+BC_e^{\beta}} \tag{7.15}$$

The linearized form is

$$\ln\left[\frac{AC_e}{q_e}-1\right] = \ln B + \beta\ \ln\ C_e \tag{7.16}$$

where x = moles of adsorbate adsorbed (mg); m = mass of adsorbent (g); C_e = final sorbate concentration (mg/l); and parameter β lies between 0 and 1. Whereas A has the units mg/g, B has the units l/g (Redlich and Peterson, 1959).

Plotting *ln[(AC_e/q_e) - 1] against ln C_e* produces a straight line with a slope of β and corresponding intercept ln B. Due to three unknown parameters, equation plotting is not applicable. Therefore, a minimization procedure is used (for the maximum R^2 correlation coefficient) between the theoretical data for q_e predicted by the above equation and the experimental data:

$$q_e = \frac{RT}{b_T}\ \ln\left(K_T\right) + \frac{RT}{b_T}\ \ln\left(C_e\right) \tag{7.17}$$

where $1/b_T$ = potential of adsorption and K_T = Temkin isotherm constant. The explicit factor of the Temkin isotherm considers the interfaces between adsorbing species and the adsorbate. This isotherm assumes that

- The heat of adsorption of all molecules decreases linearly.
- Binding energies with a uniform distribution will characterize adsorption up to some binding energy values (Kim et al., 1997).

Sips Model

The Sips isotherm model equation is expressed as follows

$$q_e = \frac{\left(K_S C_e\right)^{\beta}}{\left[1+\left(\alpha_S C_e\right)^{\beta}\right]} \tag{7.18}$$

where K_S (l/g) and α (l/mg)$^{\beta}$ = Sips constants; and β = exponent constant, which lies between 1 and 0.

Four-Parameter Models

Fritz-Schlunder Model

$$q_e = \frac{AC_e^{\propto}}{1 + BC_e^{\beta}} \tag{7.19}$$

where constants A and B = Fritz-Schlunder factors; and α and β = Fritz-Schlunder equation exponents.

Baudu Model

$$q_e = \frac{q_m b_0 C_e^{(1+x+y)}}{1 + b_0 C_e^{(1+x)}} \tag{7.20}$$

where q_m = Baudu higher adsorption capacity (mg/g); b_0 = constant of equilibrium; and x and y = Baudu factors.

Five-Parameter Model

Fritz-Schlunder Model

$$q_e = \frac{q_{mFSS} K_1 C_e^{m_1}}{1 + K_2 C_e^{m_2}} \tag{7.21}$$

where q_{mFSS} = Fritz-Schlunder maximum adsorption capacity (mg/g); and K_1, K_2, m_1, and m_2 = Fritz-Schlunder parameters.

Adsorption Kinetics

A product's cost, performance, and mode of application are always concerns, to control process efficiency. Therefore, the adsorption capacity and sufficient contact time are two of the most important factors to be considered in biosorption. Biosorption equilibrium is developed when the sorbate concentration in the bulk solution is in energetic balance with that of the interface. The capacity of a sorbent can be analyzed under an equilibrium, as shown in Table 7.3. However, thermodynamic data can envisage only the final state of a system from an initial non-equilibrium mode. Hence it is important to estimate how sorption rates are dependent on the sorbate concentration in solution and how rates are affected by sorption capacity or by the character of the sorbent. Kinetic analysis shows that the uptake rate of solute, which governs the residence time necessary to complete the sorption reaction, may be established. Researchers have used various kinetic models with pseudo-first-order (Vasanth et al., 2007) and pseudo-second-order models (Ho and McKay, 1999).

Shafique et al. (2012) carried out an extensive study on pine leaves to adsorb the primary heavy metal, arsenic. The pH and temperature were maintained at 5.0 and 25 °C, respectively. The biomass dosage adsorbent 20.0 g/L and pollutant concentration 0–10 mg/L as adsorbate were

Table 7.3 Comparison of maximum biosorption capacities and kinetic rate constants from various precursors.

Adsorbate	Bioadsorbent or Biosorbent	pH and temperature maintained	Sorbent and sorbate concentrations	Adsorption kinetics explored	Rate constant	Maximum adsorption capacity (mg/g)	Reference
Arsenic	Pine leaves	pH 5.0 and temperature 25 °C	Biomass dosage 20.0 g/L; pollutant concentrations 0–10 mg/L	Pseudo-first-order and pseudo-second-order kinetic models	0.166/min, pseudo-first-order; 21.45 g/mg/minute, pseudo-second-order	0.0232	Shafique et al., 2012
Carbofuran	Rice husk biochar	pH 5.0 and temperature 30 °C	Biomass dosage 1.0 g/L; pollutant concentrations 0–50 mg/L	Pseudo-first-order kinetic models	0.11/min	24.6	Vithanage et al., 2016
Chromium	*Sargassummuticum*	pH 2.0 and temperature 20 °C	Biomass dosage 2.0 g/L; pollutant concentrations 10–150 mg/L	Pseudo-second-order	2 g/mg/minute	196.1	Bermúdez et al. 2012
Chromium	*Sericin-alginate* particles	pH 3.5 and temperature 40 °C	Biomass dosage 10.0 g/L; pollutant concentrations 1–442 mg/L	Pseudo-first-order and pseudo-second-order kinetic models; film diffusion, surface diffusion, pore diffusion	0.0082/min, pseudo-first-order; 0.122 g/mmol/min, pseudo-second-order; $1.77e^{-7}$ m/s, film diffusion; 0.0011/min, surface diffusion; $7.2e^{-4}$ mmol/g/min$^{0.5}$, pore diffusion	5.3	Andrade et al., 2016
Chromium	*Salvinianatans*	pH 4.0 and temperature 25 °C	Biomass dosage 2.0 g/L; pollutant concentrations 1–208 mg/L	Bohart-Adams model	0.0301 L/mmol/min	17.2	Lima et al., 2016

(*Continued*)

Table 7.3 (Continued)

Adsorbate	Bioadsorbent or Biosorbent	pH and temperature maintained	Sorbent and sorbate concentrations	Adsorption kinetics explored	Rate constant	Maximum adsorption capacity (mg/g)	Reference
Chromium	Hydrolyzed olive cake	pH 6.0 and temperature 25 °C	Biomass dosage –1.0 g/L; pollutant concentrations 50–200 mg/L	Thomas, Yoon-Nelson, and dose-response models	0.191 mL/min, Thomas; 0.04/min, Yoon-Nelson; 1.633, dose-response	2.251	Fernández-González et al., 2018
Cadmium	Rice husks	pH 7 and temperature 15 °C	Biomass dosage 0.1 g/L; pollutant concentrations 10–250 mg/L	Pseudo-second-order kinetic model	0.5 g/mg/min	137.16	Qu et al., 2018
Cadmium	Dealginated residue	pH 3.5 and temperature 30 °C	Biomass dosage 2.00 g/L; pollutant concentrations 56–2025 mg/L	Pseudo-first, pseudo-second-order, MTEF, intra-particle diffusion, Boyd model	0.090/min, pseudo-first-order; 0.185 g/mmol/min, pseudo-second-order; 0.066/min, MTEF; 0.144 mmol/g/min$^{0.5}$, intra-particle diffusion; 5.792e^{-5} sq.cm/min, Boyd	48.3	Nishikawa et al., 2018
Cadmium	Straw biochar	pH 5.5 and temperature 25 °C	Biomass dosage 1.25 g/L; pollutant concentrations 2–200 mg/L	Pseudo-first-order, pseudo-second-order kinetic models	0.012/min, pseudo-first-order; 0.042 g/mg/min, pseudo-second-order	81.10	Li et al. 2017
Cadmium	*A. Auricular* waste	pH 5.0 and temperature 20 °C	Biomass dosage 2.0 g/L; pollutant concentrations 50–150 mg/L	Pseudo-first-order, pseudo-second-order kinetic models	–0.02/min, pseudo first-order; 0.0097 g/mg/min, pseudo-second-order	35.51	Song et al. 2017

Cadmium	*Ulva lactuca sp.* waste	pH 5.0 and temperature 20 °C	Biomass dosage 8.0 g/L; pollutant concentrations 20–540 mg/L	Pseudo-first-order, pseudo-second-order, intra-particle diffusion model	0.0077/min, pseudo-first-order; 0.0749 g/mg/min, pseudo-second-order; 0.2388 mg/g/min$^{0.5}$, intra-particle diffusion	41.9	Bulgariuand Bulgariu 2016
Cephalexin	Walnut shell activated	pH NA and temperature 30 °C	Biomass dosage 6.0 g/L; pollutant concentrations 0–100 mg/L	Adams-Bohart, Thomas, Yoon-Nelson kinetic models	4.13e5 L/mg/min, Adams-Bohart; 7.46e^{-5} mL/min/mg, Thomas; 1.12e^{2}/min, Yoon-Nelson	139.98	Nazari et al., 2016
Copper	Biopolymer	pH 5.5 and temperature 30 °C	Biomass dosage 0.02 g/L; pollutant concentrations 0–10 mg/L	Pseudo-second-order kinetic model	3.175 g/mg/min	389.92	Zhou et al., 2016
Copper	*Sargassumsp*	pH 5.0 and temperature 30 °C	Biomass dosage 1.33 g/L; pollutant concentrations 0–445 mg/L	Thomas model	1.46e^{-3} L/mmol/min	94.24	Barquilha et al. 2017
Copper	Dealginated residue	pH 5.0 and temperature 40 °C	Biomass dosage 2.0 g/L; pollutant concentrations 30–1271 mg/L	Yoon-Nelson model	0.027 L/mmol/min	227.5	de Freitas et al., 2018
Diclofenac	Cocoa shell activated	pH 2.0 and temperature 25 °C	Biomass dosage 2.5 g/L; pollutant concentrations 0–150 mg/L	Pseudo-first-order, pseudo-second-order kinetic models	0.497/min, pseudo-first-order; 0.0035 g/mg/min, pseudo-second-order	51.63	Saucier et al., 2015

(*Continued*)

Table 7.3 (Continued)

Adsorbate	Bioadsorbent or Biosorbent	pH and temperature maintained	Sorbent and sorbate concentrations	Adsorption kinetics explored	Rate constant	Maximum adsorption capacity (mg/g)	Reference
Diphenhydramine	Montmorillonite clay	pH 5.0 and temperature 22 °C	Biomass dosage 5.0 g/L; pollutant concentrations 0–1000 mg/L	Pseudo-second-order kinetic model	77 g/mmol/h	0.66 mmol/g	Li et al., 2011
Ibuprofen	Activated bamboo waste	pH 3.0–4.0 and temperature 25 °C	Biomass dosage 2.0 g/L; pollutant concentrations 0–120 mg/L	Pseudo first-order, pseudo-second-order, Elovich equation, liquid film diffusion	0.036 g/mg/min, pseudo first-order; 0.012 g/mg/min, pseudo-second-order; 1.02e^{19} mg/g/min$^{0.5}$, Elovich equation; 2.49/min, liquid film diffusion	61.6	Reza et al., 2014
Ibuprofen	Activated rice husks	pH 6.3 and temperature 30 °C	Biomass dosage 2.4 g/L; pollutant concentrations 0–100 mg/L	Pseudo-first-order, pseudo-second-order, intra-particle diffusion kinetic models	0.117 g/mg/min, pseudo-second-order best fit	42.7	Alvarez-Torrellas et al., 2016
Lead	*L. puntacta*	pH 4.6 and temperature 25 °C	Biomass dosage 4.0 g/L; pollutant concentrations 5–2000 mg/L	Pseudo first-order, pseudo-second-order kinetic models	0.0056/min, pseudo-first-order; 0.1732 g/mg/min, pseudo-second-order	12.14	Tang et al. 2017
Lead	Hydrolyzed olive cake	pH 6.0 and temperature 25 °C	Biomass dosage –1.0 g/L; pollutant concentrations 50–200 mg/L	Thomas, Yoon-Nelson, dose-response models	0.07 mL/min, Thomas; 0.015/min, Yoon-Nelson; 3.919, dose-response	12.683	Fernández-González et al., 2018

Mercury	Sugarcane bagasse	pH 4.0 and temperature 30 °C	Biomass dosage 5.0 g/L; pollutant concentrations 0–76 mg/L	Pseudo first-order, pseudo-second-order kinetic models	0.0575/min, pseudo-first-order; 4.46 g/mg/min, pseudo-second-order	14.7	Khoramzadeh et al., 2013
Mercury	Mercapto-grafted rice straw	pH 5.0 and temperature 50 °C	Biomass dosage 1.0 g/L; pollutant concentrations 0–400 mg/L	Pseudo-first-order, pseudo-second-order, Elovich models	0.0025/min, pseudo-first-order; 0.0042 g/mg/min, pseudo-second-order; 7.79 mg/g/min, Elovich	45.3	Song et al., 2013
Nickel	*Sargassumsp*	pH 5.0 and temperature 30 °C	Biomass dosage 1.33 g/L; pollutant concentrations 0–411 mg/L	Pseudo-first-order, pseudo-second-order, diffusion models	0.032/min, pseudo first-order; 0.063 g/mmol/min, pseudo-second-order; 4.768e^{-5} sq.cm/min, diffusion model	53.59	Barquilha et al. 2017
Nickel	Modified pine bark	pH 5.0 and temperature 25 °C	Biomass dosage 5.0 g/L; pollutant concentrations 0–154 mg/L	Pseudo-first-order, pseudo-second-order kinetic models	3.134 l/min, pseudo first-order; 0.756 g/mg/min, pseudo-second-order	7.22	Arim et al., 2018
Nickel	Dealginated residue	pH 3.5 and temperature 50 °C	Biomass dosage 2.0 g/L; pollutant concentrations 29–1057 mg/L	Pseudo-first-order, pseudo-second-order, extra mass transfer resistance, Boyd models	0.699/min, pseudo first-order; 4.242 g/mmol/min, pseudo-second-order; 0.211 m/sec, extra mass transfer resistance; 2.41e^{-6} sq.cm/min, Boyd	61.1	Moino et al. 2019

(*Continued*)

Table 7.3 (Continued)

Adsorbate	Bioadsorbent or Biosorbent	pH and temperature maintained	Sorbent and sorbate concentrations	Adsorption kinetics explored	Rate constant	Maximum adsorption capacity (mg/g)	Reference
Phenol	Commercial activated carbon	pH 6.5 and temperature 25 °C	Biomass dosage 2.0 g/L; pollutant concentrations 0–100 mg/L	Pseudo-first-order, pseudo-second-order, intra-particle diffusion models	0.054/min, pseudo-first-order; 0.19 g/mg/min, pseudo-second-order; 0.052 mg/g/min$^{0.5}$, intra-particle diffusion	45	Nourmoradi et al., 2016
Zinc	Hydrolyzed olive cake	pH 6.0 and temperature 25 °C	Biomass dosage –1.0 g/L; pollutant concentrations 0–21 mg/L	Thomas, Yoon-Nelson, dose-response models	0.112 mL/min, Thomas; 0.023/min, Yoon-Nelson; 2.247, dose-response	5.036	Fernández-González et al., 2018
Zinc	*Arthrospira platensis*	pH 6.0 and temperature 22 °C	Biomass dosage 0.5 g/L; pollutant concentrations 0–10 mg/L	Pseudo-second-order kinetic model	0.99 g/mg/min	0.92	Zinicovscaia et al., 2018

maintained. The pseudo-first-order and pseudo-second-order kinetic models were well-suited, and the rate constants (k) 0.166/min (pseudo-first-order) and 21.45 g/mg/min (pseudo-second-order) were recorded. The maximum adsorption capacity q_e (mg/g) was 0.0232. Vithanage et al. (2016) worked on rice husk biochar as an adsorbent to adsorb carbofuran with an adsorbent dosage of 1.0 g/L and adsorbate concentrations 0–50 mg/L with pH 5.0 and temperature 30 °C. The maximum adsorption capacity q_e (mg/g) was 24.6. The pseudo-first-order kinetic models were explored, and the rate constant (k) was 0.11/min. Bermúdez et al. (2012) used *Sargassummuticum* as an adsorbent with a dosage of 2.0 g/L to adsorb chromium in the range of 10 to 150 mg/L. The optimum conditions were pH 2.0 and temperature 20 °C. The maximum adsorption capacity q_e (mg/g) was 196.1. The adsorption pseudo-second-order kinetics revealed the rate constant (k) to be 2 g/mg/min.

Andrade et al. (2016) mainly focused on *Sericin-alginate* particles as adsorbents with a biomass dosage of 10.0 g/L to remove chromium as a pollutant at concentrations of 1–442 mg/L. During their study, the optimum conditions wr found to be pH 3.5 and temperature 40 °C. The maximum adsorption capacity *qe* (mg/g) was 5.3. The pseudo-first-order and pseudo-second-order kinetic models, film diffusion, surface diffusion, and pore diffusion revealed rate constant values of 0.0082/min, 0.122 g/mmol/min, $1.77e^{-7}$ m/s, 0.0011/min, and $7.2e^{-4}$ mmol/g/min$^{0.5}$, respectively. Lima et al. (2016) carried out their studies on *Salvinianatans* with biomass dosage 2.0 g/L to adsorb chromium as a pollutant with concentrations of 1–208 mg/L at the best pH of 4.0 and temperature 25 °C and found the maximum adsorption capacity q_e (mg/g) to be 17.2. The Bohart-Adams model provided the best adsorption kinetics with the rate constant (k) 0.0301 L/mmol/min. Fernández-González et al. (2018) also worked on chromium as a pollutant with concentrations 50–200 mg/L, using Hydrolyzed olive cake as an adsorbent with a dosage of 1.0 g/L under the optimum conditions of pH 6.0 and temperature 25 °C; the maximum adsorption capacity q_e (mg/g) was 2.251. The adsorption kinetics revealed that the Thomas, Yoon-Nelson, and dose-response models were the best, with a the rate constant (k) of 0.191 mL/min, 0.04, and 1.633, respectively. Qu et al. (2018) used a rice husk (as biomass) dosage of 0.1 g/L for removing cadmium concentrations 10–250 mg/L under the optimum conditions of pH 7 and temperature 15 °C, with a best maximum adsorption capacity q_e (mg/g) of 137.16. The adsorption kinetics study revealed that the pseudo-second-order kinetic model was best, with a rate constant (k) of 0.5 g/mg/min.

Nishikawa et al. (2018) used dealginated residue as biomass with dosage 2.00 g/L to adsorb cadmium as a pollutant at concentrations of 56–2025 mg/L under optimum conditions of pH 3.5 and temperature 30 °C. They found a maximum adsorption capacity (mg/g) of 48.3. The adsorption kinetics study revealed that the pseudo-first-order, pseudo-second-order, MTEF, intra-particle diffusion, and Boyd models were best, with rate constant (k) 0.090/min, 0.185 g/mmol/min, 0.066/min, 0.144 mmol/g/min$^{0.5}$, and $5.792e^{-5}$ sq.cm/min, respectively. Li et al. (2017) used straw biochar as biomass with a dosage of 1.25 g/L to separate cadmium as a pollutant at concentrations of 2–200 mg/L under experimental conditions of pH 5.5 and temperature 25 °C and found a maximum adsorption capacity q_e (mg/g) of 81.10. The pseudo-first-order and pseudo-second-order kinetic models showed a rate constant (k) of 0.012/min and 0.042 g/mg/min, respectively. Song et al. (2017) also remove cadmium concentrations of 50–150 mg/L on *A. Auricular* waste with dosage 2.0 g/L as an adsorbent at conditions of pH 5.0 and temperature 20 °C. The maximum adsorption capacity q_e (mg/g) was 35.51. The pseudo-first-order and pseudo-second-order kinetic models were studied with a rate constant (k) of –0.02/min and 0.0097 g/mg/min, respectively.

Bulgariu and Bulgariu (2016) also worked on removing cadmium concentrations of 20–540 mg/L by using *Ulva lactuca sp.* Waste as an adsorbent with a dosage of 8.0 g/L at pH 5.0 and temperature 20 °C gave a maximum adsorption capacity q_e (mg/g) of 41.9. The pseudo-first-order, pseudo-second-order, and intra-particle diffusion models gave a rate constant (k) of 0.0077/min, 0.0749 g/mg/min, and 0.2388 mg/g/min, respectively.

Nazari et al. (2016) used activated walnut shells as an adsorbent with a dosage of 6.0 g/L to remove cephalexin as a pollutant at concentrations of 0–100 mg/L at 30 °C and found a maximum adsorption capacity q_e (mg/g) of 139.98. The kinetic study showed the Adams-Bohart, Thomas, and Yoon-Nelson kinetic models suited with a rate constant (k) of $4.13e^{-5}$/L/mg/min, $7.46e^{-5}$ mL/mg/min, and $1.12e^{2}$/min, respectively. Zhou et al. (2016) used biopolymer as biomass with a dosage of 0.02 g/L to separate copper at concentrations of 0–10 mg/L under optimum conditions of pH 5.5 and temperature 30 °C and found a maximum adsorption capacity q_e (mg/g) of 389.92. The rate constant (k) was 3.175 g/mg/min for the pseudo-second-order model as the best adsorption kinetic. Barquilha et al. (2017) used the *Sargassumsp*as adsorbent with a biomass dosage of 1.33 g/L to remove copper at concentrations of 0–445 mg/L and found 94.24 to be the maximum adsorption capacity q_e (mg/g). The adsorption kinetics study revealed that the Thomas model rate constant (k) was $1.46e^{-3}$ L/mmol/min. De Freitas et al. (2018) also worked on copper concentrations of 30–1271 mg/L by using dealginated residue as an adsorbent with a dosage of 2.0 g/L under optimum conditions of pH 5.0 and temperature 40 °C and found the maximum adsorption capacity q_e (mg/g) to be 227.5. During the adsorption kinetics study, the Yoon-Nelson model rate constant (k) was 0.027 L/mmol/min. Saucier et al. (2015) used activate cocoa shells as an adsorbent with a dosage of 2.5 g/L to separate diclofenac concentrations of 0–150 mg/L at pH 2.0 and temperature 25 °C and found 51.63 to be the maximum adsorption capacity q_e (mg/g). The adsorption kinetics showed a rate constant (k) of 0.497/min for the pseudo-first-order model and 0.0035 g/mg/min for the pseudo-second-order model. Li et al. (2011) identified Montmorillonite clay as an adsorbent with a dosage of 5.0 g/L to separate diphenhydramine concentrations of 0–1000 mg/L under optimum conditions of pH 5.0 and temperature 22 °C and found the maximum adsorption capacity q_e (mg/g) to be 0.66 mmol/g. the pseudo-second-order kinetics model showed the rate constant (k) to be 77 g/mmol/h.

Reza et al. (2014) used activated bamboo waste as an adsorbent with a dosage of 2.0 g/L to remove Ibuprofen concentrations of 0–120 mg/L using pH 3.0–4.0 and temperature 25 °C and obtained a maximum adsorption capacity q_e (mg/g) of 61.6. The adsorption kinetics study showed that the rate constant (k) was 0.036 g/mg/min (pseudo first-order), 0.012 g/min (pseudo-second-order), $1.02e^{19}$ mg/g/min$^{0.5}$ (Elovich equation), and 2.49/min (liquid film diffusion). Alvarez-Torrellas et al. (2016) also worked on Ibuprofen concentrations of 0–100 mg/L using activated rice husks as an adsorbent with a dosage of 2.4 g/L under optimum conditions of pH 6.3 and temperature 30 °C and obtained a maximum adsorption capacity q_e (mg/g) of 42.7. The adsorption kinetics showed a 0.117 g/mg/min pseudo-second-order best fit. Tang et al. (2017) worked on lead concentrations of 5–2000 mg/L using *L. puntacta as* biomass with a dosage 4.0 g/L under conditions of pH 4.6 and temperature 25 °C and obtained a maximum adsorption capacity q_e (mg/g) of 12.14. The kinetic study showed that the rate constant (k) was 0.0056/min (pseudo-first-order) and 0.1732 g/mg/min (pseudo-second-order). Fernández-González et al. (2018) also worked on lead at concentrations of 50–200 mg/L using hydrolyzed olive cake as an adsorbent with a dosage of 1.0 g/L under optimum conditions of pH 6.0 and temperature 25 °C and found the maximum adsorption capacity q_e (mg/g) to be 12.683. The adsorption kinetics showed that the Thomas, Yoon-Nelson, and dose-response models were best, with a rate constant (k) of 0.07 mL/min, 0.015, and 3.919, respectively. Khoramzadeh et al. (2013) used a sugarcane bagasse dosage of 5.0 g/L for removing mercury concentrations of 0–76 mg/L at pH 4.0 and temperature 30°C and obtained a maximum adsorption capacity q_e (mg/g) of 14.7. The adsorption kinetic rate constant (k) was 0.0575/min (pseudo-first-order) and 4.46 g/mg/min (pseudo-second-order). Song et al. (2013) also worked to remove mercury concentrations of 0–400 mg/L using Mercapto-grafted rice straw as an adsorbent with a dosage of 1.0 g/L at optimum pH 5.0 and temperature 50 °C. The maximum adsorption capacity q_e (mg/g) was 45.3. The adsorption kinetic rate constant (k) was 0.0025/min (pseudo-first-order), 0.0042 g/mg/min (pseudo-second-order), and 7.79 mg/g/min (Elovich). Barquilha et al. (2017) utilized sargassumspas as an adsorbent with a dosage of 1.33 g/L to remove nickel concentrations of 0–411 mg/L and found a maximum adsorption capacity q_e (mg/g) of 53.59 under optimum conditions of pH 5.0 and temperature 30 °C. The adsorption kinetic rate constant (k) was 0.032/min (pseudo-first-order), 0.063 g/mmol/min (pseudo-second-order), and 4.768e^{-5} sq.cm/min (diffusion model). Arim et al. (2018) used modified pine bark as an adsorbent with a dosage of 5.0 g/L to separate nickel concentrations of 0–154 mg/L under optimum conditions of pH 5.0 and temperature 25 °C and obtained a maximum adsorption capacity q_e (mg/g) of 7.22. The adsorption kinetic rate constant (k) was 3.134 l/min (pseudo-first-order) and 0.756g/mg/min (pseudo-second-order).

Moino et al. 2019 carried their analysis on dealginated residue with a dosage of 2.0 g/L to remove nickel pollutant concentrations ranging from 29 to 1057 mg/L and obtained a maximum adsorption capacity q_e (mg/g) of 61.1 The adsorption kinetic rate constant (k) was 0.699/min (pseudo-first-order), 4.242 g/mmol/min (pseudo-second-order), 0.211 m/sec (extra mass transfer resistance), and 2.41e^{-6} sq.cm/min (Boyd model). Nourmoradi et al. (2016) worked on phenol concentrations of 0–100 mg/L with an adsorbent commercial activated carbon dosage of 2.0 g/L under optimum conditions of pH 6.5 and temperature 25 °C and obtained a maximum adsorption capacity q_e (mg/g) of 45. The adsorption kinetic rate constant (k) was 0.054/min (pseudo-first-order), 0.19 g/mg/min (pseudo-second-order), and 0.052 mg/g/min$^{0.5}$ (intra-particle diffusion). Fernández-González et al. (2018) also worked on zinc pollutant concentrations of 0–21 mg/L using hydrolyzed olive cake as an adsorbent with a dosage of 1.0 g/L under optimum conditions of pH 6.0 and temperature 25 °C and found a maximum adsorption capacity q_e (mg/g) of 5.036. The adsorption kinetics showed that the Thomas, Yoon-Nelson, and dose-response models were best, with a rate constant (k) of 0.112 mL/min, 0.023/min, and 2.247, respectively. Zinicovscaia et al. (2018) studied *Arthrospira platensis* as an adsorbent with a dosage of 0.5g/L to adsorb zinc concentrations ranging from 0 to 10 mg/L under optimum conditions of pH 6.0 and temperature 22 °C and obtained a maximum adsorption capacity q_e (mg/g) of 0.92. The adsorption kinetics rate constant (k) was 0.99 g/mg/min for the pseudo second-order kinetic model.

Pseudo-First-Order Kinetics

The adsorption of pollutants from the liquid phase to a solid phase can be explained by a reversible kinetic equilibrium developed between the solid phase and the solution. Assuming non-dissociating molecular adsorption of phenols on soil and soil-agro-based adsorbent particles, the adsorption phenomenon can be explained as a diffusion-controlled process (Fogler, 1998). The Lagergren rate equation of is the most widely used for the adsorption of a solute from a solvent (Tseng et al., 2003; Ho, 2004; Weber and Chakravorti, 1974):

$$\frac{dq_t}{dt} = k_t\left(q_e - q_t\right) \tag{7.22}$$

where k_t (1/min) = pseudo-first-order rate constant; and q_e and q_t = sorption capacity at equilibrium and time t (mg/g). After integration by applying the conditions $q_t = 0$ at $t = 0$ and $t = t$, $q_t = q_t$, Eq. (7.22) becomes

$$\log\left(q_e - q_t\right) = \log q_e - \frac{k_t}{2.303}t \tag{7.23}$$

The value of k_t can be calculated from the plots of log $(q_e - q_t)$ vs. t for different concentrations of phenol.

Pseudo-Second-Order Kinetics

The chemisorption kinetic rate equation for the pseudo-second-order model (Ho and McKay, 1999) is expressed as follows

$$\frac{dq_t}{dt} = k_2\left(q_e - q_t\right)^2 \tag{7.24}$$

where k_2 is the second-order rate constant (g/mg/min). After integration by applying the conditions $q_t = 0$ at $t = 0$ and $t = t$, $q_t = q_t$, Eq. (7.24) becomes

$$\frac{t}{q_t} = \frac{1}{k_2 q_e^2} + \frac{t}{q_e} \tag{7.25}$$

This equation can be written as

$$\frac{t}{q_t} = \frac{1}{h} + \frac{t}{q_e} \tag{7.26}$$

where $h = k_2 q_e^2$; plotting t/q_t vs. t shows a linear relationship. Values of k_2 and q_e can be calculated from the intercept and slope of the plots of t/q_t vs. t.

The Elovich Equation

The Elovich equation is a useful models and can be expressed as follows (Aharoni and Tompkins, 1970)

$$\frac{dq_t}{dt} = \alpha \exp\left(-\beta q_t\right) \tag{7.27}$$

where α = initial adsorption rate (mg/g/min); and β is the desorption constant (g/mg) during any one experiment.

To simply the Elovich equation, assume that $\alpha\beta t >> t$, and apply boundary conditions $q_t = 0$ at $t = 0$ and $q_t = q_t$ at $t = t$. Equation (7.27) becomes

$$q_t = \frac{1}{\beta}\ln(\alpha\beta) + \frac{1}{\beta}\ln(t) \tag{7.28}$$

The graph of q_t vs. $\ln(t)$ produces a linear association with a slop of $(1/\beta)$ and an intercept of $(1/\beta)\ln(\alpha\beta)$. The parameter $1/\beta$ specifies the number of sites available for adsorption.

Avrami Kinetic Equation

Most researchers use the Lagergren kinetic equations to study adsorption kinetics. Sometimes it is not possible to estimate fractional kinetic parameters using existing kinetic models. The Avrami kinetic equation is a good kinetic model given possible changes of the sorption rates in the function of initial concentration (mg/l) and adsorption time (t), as well as determining fractional kinetic orders (Lopes et al., 2003):

$$q_t = q_e\left\{1 - \exp\left[-(K_{AV}t)\right]^{n_{AV}}\right\} \tag{7.29}$$

$$\frac{q_t}{q_e} = \left\{1 - \exp\left[-(K_{AV}t)\right]^{n_{AV}}\right\} \tag{7.30}$$

where $\alpha = \frac{q_t}{q_e}$ = adsorption fraction at time t; K_{AV} = Avrami kinetic constant; and n_{AV} = constant that can be relational to the adsorption mechanism. The linearized form of this equation is

$$\ln(-ln(1-\alpha)) = nlnK_{AV} + nlnt \tag{7.31}$$

Sorption Diffusion Models

In adsorption theory, the concept of kinetics between individual atoms, ions, or molecules of adsorbate and the surface or interface depends on the forces and their relations. Basic equilibrium studies can provide information for modeling the adsorption of an adsorbate onto an adsorbent. For the adsorption of phenol, which has large molecules and a long contact time to equilibrium, the adsorption mechanism is likely to be diffusion control. Diffusion adsorption is usually controlled by external film resistance and/or internal diffusion mass transport or intra-particle diffusion. The diffusion process involves four migration steps to transport the adsorbate from the bulk fluid phase to the specific sites within the interior surface of the pores. There are three steps for adsorption on the exterior surface of the particle, which is usually very small and ignored with respect to the adsorption on the surface of the pores.

The following mass-transfer mechanisms are involved during adsorption:

Bulk transport refers to as the transport of the adsorbate in the solution phase to the vicinity of the particle. In a well-agitated batch system, bulk mass transfer is so rapid that it is not considered the rate-limiting step.

Film transport refers to the diffusion of the adsorbate through a film, which causes resistance to mass transfer at the surface of the adsorbent particle.

Intra-particle transport refers to the mass diffusion of the adsorbate within the adsorbent particle. It involves a two-step mechanism for diffusing the adsorbate: pore diffusion and surface diffusion.

The overall rate of adsorption is controlled by the slowest step, which is film transport (external diffusion) or intra-particle transport (internal diffusion). These two rate-limiting steps were developed by several workers.

Several researchers have used the film diffusion model to determine the external mass-transfer coefficients for the sorption of phenol. Keinath and Weber (1968) reported that the intra-particle mass transfer coefficient for dinitro-o-sec-butyl phenol varies with the initial concentration and diameter of the particle. Distribution profiles in branched pore models for gas adsorption were reported by Do (1998), Hu (1999), and Hu et al. (1999). Mathews and Weber (1976) developed a solid-phase diffusion model based on finite difference approximations for any arbitrary isotherm and extended the model to multicomponent systems. Several researchers, including Lee et al. (1980), Weber and Liu (1980), Thacker et al. (1981), and McKay and Allen (1984), successfully applied this approach to solutes and mixtures in batch reactors and fixed beds. In their studies, Weber and Morris (1962), Crank (1977), McKay et al. (1981), and Mathews and Weber (1998) developed another type of single-resistance model based on intra-particle diffusion. The model correlates a linear relationship between the sorption uptake (q_t) and the square root of time ($t^{0.5}$).

Calculating the External Mass Transfer Coefficient

The adsorption rate is represented as follows

$$\frac{dC_t}{dt} = -K_f S_A \left(C_t - C_s\right) \tag{7.32}$$

with these initial conditions:

$$\text{At } t = 0,\ C_t = C_0 \tag{7.33}$$

The particle is assumed to be spherical, so the external surface area per unit volume of solution S_A can be calculated as follows:

$$S_A = \frac{A_s}{V} = \frac{6M_S}{d_p \rho_b} \tag{7.34}$$

Integrating Eq. 7.34 and applying boundary conditions $(0,C_0)$ to (t,C_t),

$$\ln\left({C_t}/{C_0} \right) = -K_f S_A t \tag{7.35}$$

A plot of $\ln\left({C_t}/{C_0} \right)$ vs. t gives a straight line with slope $K_f S_A$. Knowing S_A, the value of the external diffusion coefficient (K_f) can be estimated.

Intra-Particle Diffusion Control

The theory of diffusion, especially in isotropic substances, assumes that the rate of transfer of a diffusing substance through a unit area of a section (F) is directly proportional to the concentration gradient measured normal to the section. This is given by Fick's first law of diffusion:

$$F = -D\frac{\partial C}{\partial x} \tag{7.36}$$

The fundamental equation of diffusion, where there is a concentration gradient along the x-axis, is given by Fick's second law of diffusion:

$$\frac{\partial C}{\partial t} = D\frac{\partial^2 C}{\partial x^2} \tag{7.37}$$

Based on these two diffusion laws, a mathematical derivation was derived by Boyd et al. (1947) (cited by Crank, 1977; Srivastava et al., 2006), which was valid under the experimental conditions used. With the diffusion rate controlling adsorption on particles with a spherical shape, the solution to the simultaneous set of differential and algebraic equations leads to

$$\frac{q_t}{q_e} = 6\left(\frac{Dt}{R^2}\right)^{0.5}\left\{\pi^{-0.5} + 2\sum_{n=1}^{\infty} ierfc\frac{nR}{Dt^{0.5}}\frac{nR}{Dt^{0.5}}\right\} - 3\frac{Dt}{R^2} \tag{7.38}$$

where D can be replaced by the diffusion coefficient D_f:

$$\frac{q_t}{q_e} = 6\left(\frac{D_f t}{\pi R^2}\right)^{0.5} t^{0.5} \tag{7.39}$$

Equation 7.38 can be written as follows:

$$F(t) = \frac{q_t}{q_e} = 1 - \frac{6}{\pi^2}\sum_{n=1}^{\infty}\frac{1}{n^2}\exp\left(-\frac{\mathrm{D_i}\pi^2 \mathrm{n}^2 \mathrm{t}}{\mathrm{r}^2}\right) \tag{7.40}$$

Equation 7.40 converges asymptotically for a time since the higher terms of the summation become small and are negligible, so it can be rewritten as

$$1 - \frac{q_t}{q_e} = \frac{6}{\pi^2}\exp\left(-\frac{D_i \pi^2 z^2 t}{r^2}\right)t \tag{7.41}$$

and

$$1 - \frac{q_t}{q_e} = \frac{6}{\pi^2}\exp(-k')t \tag{7.42}$$

k' is the overall rate constant as

$$k' = -\frac{D_i \pi^2 z^2 t}{r^2} \tag{7.43}$$

Taking the logarithm on both sides of Eq. (7.42),

$$\ln\left(1-\frac{q_t}{q_e}\right)=\ln\frac{6}{\pi^2}-k't \tag{7.44}$$

A plot of $ln(1-q_t/q_e)\ln\left({}^{C_t}\!/_{C_0}\right)$ vs. t gives a straight line with slope k′, and using Eq. (7.43), the internal diffusion coefficient (D_i) can be calculated.

Natarajan and Khalaf Equation

Natarajan and Khalaf (cited by Raji et al., 1986) developed a relation between the initial concentration and the concentration at any time t. The equation is given as

$$\log\left(\frac{C_o}{C_t}\right)=\left(\frac{k_N}{2.303}\right)t \tag{7.45}$$

where C_0 and C_t are the concentrations of phenol (in mg/l) at time zero (initial concentration) and at a time t, respectively. A linear plot of log (C_0/C_t) vs. time was drawn, and the slope gives the k_N value.

Bhattacharya and Venkobachar Equation

The Bhattacharya and Venkobachar kinetic equation (1984) develop a relation between phenol uptake onto the adsorbent and time t. The equation is given as

$$\log\left[1-U(T)\right]=\left(\frac{k_B}{2.303}\right)t \tag{7.46}$$

where

$$U(T)=\frac{C_o-C_t}{C_o-C_e} \tag{7.47}$$

where C_e = equilibrium adsorbate concentration (ppm); and C_i and C_t = concentrations of adsorbate (mg/l) at time zero (initial concentration) and time t, respectively.

Power Function Equation

The kinetic power function equation (Goswami and Ghosh, 2005) developed a relationship between phenol accumulation and the sorbent at time t. The equation is given as

$$\log\ q_t=\log\ a+b\ \log\ t \tag{7.48}$$

A linearized plot of log q_t vs. log t describes the constants of power functions a nd b. The parameter a signifies the initial rate and indicates the y-intercept of the plot of log q_t vs. log t. The parameter b is the plot slope and the reaction of the rate constant.

Bangham's Equation

Kinetic information can be further utilized to check whether pore diffusion is a rate-controlling step in the adsorption system using Bangham's equation (Aharoni et al., 1979)

$$\log.\log\left(\frac{C_o}{C_o - q_t m}\right) = \log\left(\frac{k_o m}{2.303V}\right) + \alpha.\log(t) \quad (7.49)$$

where m = mass of adsorbent per liter of solution (g/l); q_t = amount of adsorbate adsorbed by the adsorbent at time t (mg/g); V = volume of the solution (ml); C_0 = initial concentration of the adsorbate in solution (mg/l); and α (< 1) and k_0 (ml/(g/l)) are constants.

Boyd Model

Boyd developed a kinetic model for film diffusion. It is based on a single-resistance model to study the rate-limiting step of the kinetics. The equation is based on the boundary layer. The boundary surrounding the adsorbent has a greater effect on the diffusion of the adsorbate (Boyd et al., 1947). Equation (7.41) is applied to the Boyd model as follows:

$$F = 1 - \frac{6}{\pi^2}\sum_{1}^{\infty}\left(\frac{1}{n^2}\right)\exp\left(-n^2 B_t\right) \quad (7.50)$$

where F = fraction of adsorbate adsorbed at time t (q_t) to adsorbate adsorbed at an infinite time ($q\infty$); ($t\infty$ > 24 hours for better results). B_t is a mathematical function of F:

$$F = \frac{q_t}{q_\infty} \quad (7.51)$$

It is hard to estimate appropriate values of B_t using Eq. 7.51; B_t can be calculated using the integrated Fourier transform:

$$0 \le F \le 0.85: B_t = 2\pi - \frac{\pi^2 F}{3} - 2\pi\left(1 - \frac{\pi F}{3}\right)^{1/2} \quad (7.52)$$

$$0.86 \le f \le 1: B_t = -0.4977 - \ln(1 - F) \quad (7.53)$$

The graph of B_t vs. t helps to predict the rate-controlling step. If the graph approximates y = mx passing through the origin, the rate-controlling step is intra-particle diffusion; otherwise, the film diffusion model dominates the process. Most studies conclude that film diffusion is the limiting step during the initial stages of the adsorption process, followed by intra-particle diffusion when particles reach the surface of the adsorbent.

References

Aharoni, C. and Tompkins F.C. (1970). Kinetics of adsorption and desorption and the Elovich equation. In: *Advance in Catalysis and Related Subjects* (ed. D.D. Eley, H. Pines, and P.B. Weisz), 1–49. Academic Press.

Alouani, M.E.L., Alehyen, S., Achouri, M.E.L., and Taibi, M. (2018). Removal of cationic dye—methylene blue—from aqueous solution by adsorption on fly ash-based geopolymer. *Journal of Materials and Environmental Sciences* 9 (1): 32–46.

Álvarez-Torrellas, S., Muñoz, M., Zazo, J.A. et al. (2016). Synthesis of high surface area carbon adsorbents prepared from pine sawdust-Onopordum acanthium L. for nonsteroidal anti-inflammatory drugs adsorption. *Journal of Environmental Management* 183: 294–305.

Apak, R., Güçlü, K., and Turgut, M.H. (1998). Modeling of copper (II), cadmium (II), and lead (II) adsorption on red mud. *Journal of Colloid and Interface Science* 203 (1): 122–130.

Arim, A.L., Guzzo, G., Quina, M.J., and Gando-Ferreira, L.M. (2018). Single and binary sorption of Cr (III) and Ni (II) onto modified pine bark. *Environmental Science and Pollution Research* 25 (28): 28039–28049.

Ba, S., Alagui, A., and Hajjaji, M. (2018). Retention and release of hexavalent and trivalent chromium by chitosan, olive stone activated carbon, and their blend. *Environmental Science and Pollution Research* 25 (20), 19585-19604.

Barquilha, C.E.R., Cossich, E.S., Tavares, C.R.G., and Silva, E.A. (2017). Biosorption of nickel (II) and copper (II) ions in batch and fixed-bed columns by free and immobilized marine algae Sargassum sp. *Journal of Cleaner Production* 150: 58–64.

Bermúdez, Y.G., Rico, I.L.R., Guibal, E. et al. (2012). Biosorption of hexavalent chromium from aqueous solution by Sargassum muticum brown alga. *Application of statistical design for process optimization. Chemical Engineering Journal* 183: 68–76.

Bharathi, K.S. and Ramesh, S.T. (2012). Biosorption of crystal violet from aqueous solution by Citrullus lanatus (watermelon) rind. *Journal of Environmental Research and Development* 7 (1A), 321-329.

Bhattacharya, A.K and Venkobachar, C (1984). Removal of cadmium (II) by low cost adsorbents *Journal of Environmental Engineering* 110: 110–122.

Boyd, G.E., Schubert, J., and Adamson, A.W. (1947). The exchange adsorption of ions from aqueous solutions by organic zeolites; ion-exchange equilibria. *J Am Chem Soc* 69 (11): 2818–29.

Boyd, S.A, Morland, M.M., and Chiou, C.T (1988). Sorption characteristics of organic compounds on hexadecyl trimethyl ammonium-semectite. *Soil Sci. Soc. Am. J.* 52: 652–65.

Brigatti, M., Corrodini, F., Franchini, G. et al. (1995). Interaction between montmorillonite and pollutants from industrial wastewaters: exchange of Zn2+ and Pb2+ from aqueous solutions. *Applied Clay Science:* 383–395.

Bulgariu, D. and Bulgariu, L. (2016). Potential use of alkaline treated algae waste biomass as sustainable biosorbent for clean recovery of cadmium (II) from aqueous media: batch and column studies. *Journal of Cleaner Production* 112: 4525–4533.

Bulgariu, L. and Bulgariu, D. (2018). Functionalized soy waste biomass-A novel environmental-friendly biosorbent for the removal of heavy metals from aqueous solution. *Journal of Cleaner Production* 197: 875–885.

Cooney, D.O. (1999). *Adsorption Design for Wastewater Treatment.* Lewis Publishers, CRC Press LLC.

Crank, J. (1977). *The Mathematics of Diffusion*, vol. 84. Oxford Clarendon Press.

da Silva, T.L., da Silva, A.C. Jr., Vieira, M.G.A. et al. (2016). Biosorption study of copper and zinc by particles produced from silk sericin–alginate blend: evaluation of blend proportion and thermal cross-linking process in particles production. *Journal of Cleaner Production* 137: 1470–1478.

Dabrowski, A., Podkościelny, P., Hubicki, Z., and Barczak, M. (2005). Adsorption of phenolic compounds by activated carbon – a critical review. *Chemosphere* 58 (8):1049-70.

De Andrade, J.R., Da Silva, M.G.C., Gimenes, M.L., and Vieira, M.G. (2016). Equilibrium and thermodynamic studies on adsorption of trivalent chromium by sericin-alginate particles prepared from Bombyx Mori cocoons. *Chemical Engineering Transactions* 52: 169–174.

de Freitas, G.R., da Silva, M.G.C., and Vieira, M.G.A. (2019). Biosorption technology for removal of toxic metals: a review of commercial biosorbents and patents. *Environmental Science and Pollution Research* 26 (19): 19097–19118.

Do, D.D. (1998). *Adsorption Analysis: Equilibria and Kinetics*. Imperial College Press.

Dubinin, M.M. (1956). The potential theory of adsorption of gases and vapors for adsorbents with energetically nonuniform surfaces. *Chem. Rev.* 60 (2): 235–241.

Fernández-González, R., Martín-Lara, M.A., Iáñez-Rodríguez, I., and Calero, M. (2018). Removal of heavy metals from acid mining effluents by hydrolyzed olive cake. *Bioresource Technology* 268: 169–175.

Fogler, H.S. (1998). *Elements of Chemical Reaction Engineering*, 3e. Prentice-Hall.

Genç-Fuhrman, H., Bregnhøj, H., and McConchie, D. (2005). Arsenate removal from water using sand–red mud columns. *Water Research* 39 (13): 2944–2954.

Goswami, S. and Ghosh, U.C. (2005). Studies on adsorption behaviour of Cr(VI) onto synthetic hydrous stannic oxide. *Water SA* 31 (44): 57–602.

Gupta, V.K. and Ali, I. (2006). Removal of 2, 4-dinitrophenol from wastewater by adsorption technology: a batch and column study. *International Journal of Environment and Pollution* 27 (1–3): 104–120.

Hamdaouia, O. and Naffrechoux. E. (2007). Modeling of adsorption isotherms of phenol and chlorophenols onto granular activated carbon Part II. *Models with more than two parameters, Journal of Hazardous Materials* 147: 401–411.

Ho, Y.S. (2004). Citation review of Lagergren kinetic rate equation on adsorption reactions. *Scientometrics* 59 (1): 171–177. doi:10.1023/b:scie.0000013305.99473.cf.

Ho, Y.S. and McKay, G. (1999). Pseudo-second order model for sorption processes. *Process Biochemistry* 34 (5): 451–465. doi:10.1016/S0032-9592(98)00112-5.

Hu, X.J., Qiao, S., and Do, D.D. (1999). Multi-component adsorption kinetics of gases in activated carbon: Effect of pore size distribution. *Langmuir* 15: 6428–6437.

Hu, X.J. (1999). Multi-component adsorption equilibrium of gases in Zeolites: Effect of pore size distribution. *Chemical Engineering Communication* 174: 201–214.

Khattri, S.D. and Singh, M.K. (2009). Removal of malachite green from dye wastewater using neem sawdust by adsorption. *Journal of Hazardous Materials* 167 (1–3), 1089-1094.

Khelifi, O., Nacef, M., and Affoune, A.M. (2018). Nickel (II) adsorption from aqueous solutions by physico-chemically modified sewage sludge. *Iranian Journal of Chemistry and Chemical Engineering (IJCCE)* 37 (1): 73–87.

Khoramzadeh, E., Nasernejad, B., and Halladj, R. (2013). Mercury biosorption from aqueous solutions by Sugarcane Bagasse. *Journal of the Taiwan Institute of Chemical Engineers* 44 (2): 266–269. doi:10.1016/j.jtice.2012.09.004.

Khraisheh, M., Al-Ghouti, M.A., and AlMomani, F. (2020).: putida as biosorbent for the remediation of cobalt and phenol from industrial waste wastewaters. *Environmental Technology and Innovation* 20: 101148.

Kim, J.M.Y., Jae, K.P., and Tuncer, B.E. (1997). Sorption of organic compounds in the aqueous phase onto tire rubber. *Journal of Environmental Engineering* 23 (9): 827–835.

Koble, R. and Corrigan, T.E. (1952). Adsorption isotherms for pure hydrocarbons, *Industrial and Engineering Chemistry* 44 (2): 383–387.

Langmuir, I. (1917). The constitution and fundamental properties of solids and liquids. *II. Liquids. Journal of the American Chemical Society* 39 (9): 1848–1906.

Langmuir, I. (1918). The adsorption of gases on plane surfaces of glass, mica and platinum. *Journal of the American Chemical Society* 40 (9): 1361–1403.

Lee, M.C., Thacker, W.E., Snoeyink, V.L., and Crittenden, J.C. (1980). Mathematical modeling of humic substances removal with activated carbon beds. In: *ASCE Environmental Engineering Conference Proceedings*, 387–393.

Li, B., Yang, L., Wang, C.Q. et al. (2017). Adsorption of Cd (II) from aqueous solutions by rape straw biochar derived from different modification processes. *Chemosphere* 175: 332–340.

Li, Z., Chang, P.H., Jiang, W.T. et al. (2011). Removal of diphenhydramine from water by swelling clay minerals. *Journal of Colloid and Interface Science* 360 (1): 227–232.

Lima, L.K.S., Silva, M.G.C., and Vieira, M.G.A. (2016). Study of binary and single biosorption by the floating aquatic macrophyte Salvinia natans. *Brazilian Journal of Chemical Engineering* 33 (3): 649–660.

Lopes, E.C.N, Anjos, F.S.C., Vieira, E.F.S, and Cestari A.R. (2003). An alternative Avrami equation to evaluate kinetic parameters of the interaction of Hg (II) with thin chitosan membranes. *J. Colloid Interface Sci.* 263: 542–547.

Mathews, A.P. and Weber, W.J. (1998). Effect of external mass transfer and intra-particle diffusion on adsorption rates in slurry reactors. *American Institute of Chemical Engineers Symposium Series* 73: 91–017.

McKay, G. and Allen, S.J. (1984). Pore diffusion model for dye adsorption onto peat in batch adsorbents. *The Canadian Journal of Chemical Engineering* 62: 340–345.

McKay, G., Otterburn, M.S., and Sweeney, A.G. (1981). Kinetics of colour removal from effluent using activated carbon. *Journal of the Society of Dyers and Colourists* 96 (11): 576–579.

Moino, B.P., Costa, C.S.D., Carlos da Silva, M.G., and Vieira, M.G.A. (2020). Reuse of the alginate extraction waste from Sargassum filipendula for Ni (II) biosorption. *Chemical Engineering Communications* 207 (1): 17–30.

Moradi, M., Fazlzadehdavil, M., Pirsaheb, M. et al. (2016). Response surface methodology (RSM) and its application for optimization of ammonium ions removal from aqueous solutions by pumice as a natural and low cost adsorbent. *Archives of Environmental Protection* 42: 2: 33–43.

Namasivayam, C. and Kavitha, D. (2002). Removal of congo red from water by adsorption onto activated carbon prepared from coir pith, an agricultural solid waste. *Dyes and Pigments* 54: 47–58.

Nazari, G., Abolghasemi, H., and Esmaieli, M. (2016). Batch adsorption of cephalexin antibiotic from aqueous solution by walnut shell-based activated carbon. *Journal of the Taiwan Institute of Chemical Engineers* 58: 357–365.

Nishikawa, E., da Silva, M.G.C., and Vieira, M.G.A. (2018). Cadmium biosorption by alginate extraction waste and process overview in life cycle assessment context. *Journal of Cleaner Production* 178: 166–175.

Nourmoradi, H., Avazpour, M., Ghasemian, N. et al. (2016). Surfactant modified montmorillonite as a low-cost adsorbent for 4-chlorophenol: Equilibrium, kinetic and thermodynamic study. *Journal of the Taiwan Institute of Chemical Engineers* 59: 244–251.

Ozkaya, B. (2006). Adsorption and desorption of phenol on activated carbon and a comparison of isotherm models. *Journal of Hazardous Materials* 129 (1–3): 158–163.

Qu, J., Meng, X., Jiang, X. et al. (2018). Enhanced removal of Cd (II) from water using sulfur-functionalized rice husk: characterization, adsorptive performance and mechanism exploration. *Journal of Cleaner Production* 183: 880–886.

Raji, C., Manju, G.N., and Amirudhan, T.S. (1986). Removal of heavy metal ions from water using sawdust-based activated carbon. *Indian J. Eng. Mater. Sci.* 4 (6): 254–260.

Ramprasad, C. and Philip, L. (2017). Sorption of surfactants and personal care products in Indian soils. *International Journal of Environmental Science and Technology* 14 (4): 853–866.

Redlich, O. and Peterson, D.L. (1959). A useful adsorption isotherm. *Journal of Physical Chemistry* 63: 1024–30.

Renge, V.C., Khedkar, S.V., and Pande, S.V. (2012). Removal of heavy metals from wastewater using low cost adsorbents: a review. *Sci. Revs. Chem. Commun* 2 (4): 580–584.

Reza, R.A., Ahmaruzzaman, M., Sil, A.K., and Gupta, V.K. (2014). Comparative adsorption behavior of ibuprofen and clofibric acid onto microwave assisted activated bamboo waste. *Industrial and Engineering Chemistry Research* 53 (22): 9331–9339.

Saucier, C., Karthickeyan, P., Ranjithkumar, V. et al. (2017). Efficient removal of amoxicillin and paracetamol from aqueous solutions using magnetic activated carbon. *Environmental Science and Pollution Research* 24 (6): 5918–5932.

Shafique, U., Ijaz, A., Salman, M. et al. (2012). Removal of arsenic from water using pine leaves. *Journal of the Taiwan Institute of Chemical Engineers* 43 (2): 256–263.

Song, S.T., Saman, N., Johari, K., and Mat, H. (2013). Removal of Hg (II) from aqueous solution by adsorption using raw and chemically modified rice straw as novel adsorbents. *Industrial and Engineering Chemistry Research* 52 (36): 13092–13101.

Song, T., Liang, J., Bai, X. et al. (2017). Biosorption of cadmium ions from aqueous solution by modified Auricularia Auricular matrix waste. *Journal of Molecular Liquids* 241: 1023–1031.

Srivastava, V.C., Swamy, M.M., Mall, I.D. et al. (2006). Adsorptive removal of phenol by bagasse fly ash and activated carbon: Equilibrium, kinetics and thermodynamics. *Colloids and Surfaces A: Physicochemical and Engineering Aspects* 272 (1–2): 89–104.

Subramanyam, B. and Ashutosh, D. (2012). Adsorption isotherm modeling of phenol onto natural soils applicability of various isotherm models *Int. J. Environ. Res.* 6 (1): 265–276.

Tang, J., Chen, C., Chen, L.et al. (2017). Effects of pH, initial Pb 2+ concentration, and polyculture on lead remediation by three duckweed species. *Environmental Science and Pollution Research* 24 (30), 23864-23871.

Thacker, W.E., Snoeyink, V.L., and Crittenden, J.C. (1981). Modeling of activated carbon and coal gasification char adsorbents in single solute and bi-solute systems. *University of Illinois, Urbana-Champaign, Water Resources Center, research report.* 161: 1–166.

Toth, J. (2000). Calculation of the BET-compatible surface area from any type I isotherms measured above the critical temperature. *J. Colloid Interf. Sci.* 225: 378–383.

Tseng, R.L., Wu, F.C., and Juang, R.S. (2003). Liquid-phase adsorption of dyes and phenols using pinewood based activated carbons. *Carbon* 41: 487–495.

Vijayaraghavan, K., Padmesh, T.V.N., Palanivelu, K., and Velan, M. (2006). Biosorption of nickel(II) ions onto Sargassum wightii: Application of two-parameter and three-parameter isotherm models. *Journal of Hazardous Materials* B133: 304–308.

Vithanage, M., Mayakaduwa, S.S., Herath, I. et al. (2016). Kinetics, thermodynamics and mechanistic studies of carbofuran removal using biochars from tea waste and rice husks. *Chemosphere* 150: 781–789.

Weber, W.J. Jr. and Liu, K.T. (1980). Determination of mass transport parameters for fixed bed adsorbers. *Chem. Eng.Comm* 3: 49–60.

Yaneva, Z.L., Koumanova, B.K., and Georgieva, N.V. (2013). Linear and nonlinear regression methods for equilibrium modelling of p-nitrophenol biosorption byrhizopus oryzae: comparison of error analysis criteria. *Journal of Chemistry*: 1–10.

Zhang, X., Li, X., Zhang, F. et al. (2019). Adsorption of Se (IV) in aqueous solution by zeolites synthesized from fly ashes with different compositions. *Journal of Water Reuse and Desalination* 9 (4): 506–519.

Zhang, Z., Zheng, Z., Cai, J. et al. (2017). Effect of cadmium on oxidative stress and immune function of common carp (Cyprinus carpio L.) by transcriptome analysis. *Aquatic Toxicology* 192: 171–177.

Zhou, Y., Xia, S., Zhang, J. et al. (2017). Insight into the influences of pH value on Pb (II) removal by the biopolymer extracted from activated sludge. *Chemical Engineering Journal* 308: 1098–1104.

Zhou, Y., Xia, S., Zhang, Z. et al. (2016). Associated adsorption characteristics of Pb (II) and Zn (II) by a novel biosorbent extracted from waste-activated sludge. *Journal of Environmental Engineering* 142 (7): 04016032.

Zhou, Y., Zhang, Z., Zhang, J., and Xia, S. (2016). Understanding key constituents and feature of the biopolymer in activated sludge responsible for binding heavy metals. *Chemical Engineering Journal* 304: 527–532.

Zinicovscaia, I., Cepoi, L., Povar, I. et al. (2018). Metal uptake from complex industrial effluent by cyanobacteria Arthrospira platensis. *Water, Air, and Soil Pollution* 229 (7): 1–10.

8

Dynamic Biosorption for Removal of Wastewater Contaminants

Arunachalam Thirunavukkarasu and Rajarathinam Nithya

Department of Industrial Biotechnology, Government College of Technology, Coimbatore, Tamilnadu, India

Introduction

Biosorption or biomass-mediated sorption is a metabolically independent process in which solute molecules (sorbate) diffuse passively across the solid-liquid interface and concentrate on a biomass's surface (biosorbent). Through this process, solute molecules, usually pollutants or contaminants, can be translocated from the bulk liquid solution (wastewater) to the surfaces of the solid biosorbents. As this process is not energy-dependent, the phase transfer of sorbate molecules is entirely mediated by molecular diffusion, and with some modifications to the batch operational conditions, it can also be facilitated with advection currents. Hence, the difference in the concentration gradients among solid and liquid phases is the primary driving force for the biosorption process (Karthika et al., 2010; Thirunavukkarasu and Nithya 2021).

When the solute-enriched liquid phase contacts the surface-active biomasses, a change of molecular dynamics occurs at the solid-liquid interface. With the relative affinity toward the surface functional moieties of the biomasses, the sorbate molecules tend to move across the phase and form weak and reversible physical bonds or strong and irreversible chemical interactions on the surface of the biosorbent. The reversible physical interactions could be weak Van der Waals attractive forces, dipole–dipole interactions, dispersion interactions, or hydrogen bonding among the sorbate-biosorbent molecules, with a low-enthalpy change of about 20 kJ/mol. Irreversible chemical interactions between sorbate and biosorbent molecules are associated with high-enthalpy changes of about 200 kJ/mol (Hardiljeet et al., 2011). Despite the term *sorption*, which refers to both two-dimensional surface adsorption and three-dimensional interphase absorption, it is ideal to consider biosorption to mean *bioadsorption* from the extensive literature reports.

On the other hand, biosorption can be viewed as an enrichment technique in which pollutants or contaminants (sorbate) are concentrated on a localized or confined region (biosorbent). This process reduces huge volumes of pollutant-containing wastewater and

Biosorption for Wastewater Contaminants, First Edition.
Edited by Rangabhashiyam Selvasembian and Pardeep Singh.

thus facilitates further waste-management techniques. Hence, biosorption can be effectively employed in the secondary phase of the wastewater treatment process.

Fundamentals of Biosorption

Biosorbates

Biosorbates are a wide range of solute molecules present in solution in varied forms such as ionic species, particulates, colloids, precipitates, etc. These solute molecules show different affinities toward a diverse range of biosorbents. Scientific researchers select a biosorbent for which the biosorbate possesses a high partition coefficient. Also, a proper understanding of the establishment of equilibrium conditions at the solid-liquid interface provides greater insights into the mechanistic aspects of the biosorption process.

In the past two decades, biosorption studies of a wide range of metals, colloids, organometals, inorganic compounds, and organic compounds such as dyes, fluorides, pharmaceuticals, etc., using various classes of biosorbents have been reported in the literature. Most of these research works were extremely focused on metals and organic compounds, as these materials are often considered highly toxic and recalcitrant molecules in wastewater. Hence, this chapter presents scholarly works on the effective removal of metals and organic compounds using natural, synthetic, and semi-synthetic classes of biosorbents.

Metals

Three out of four elements in the periodic table are grouped as metals, and most of these metals have been considered as sorbates in the past two decades of scientific research. Toxicity, recalcitrance, secondary market value, recovery potential, radioactivity, etc., are the primary reasons for their consideration in biosorption research (Nithya et al., 2021a). From the literature reports, lead, copper, mercury, cadmium, chromium, arsenic, radionuclides of cobalt, strontium, uranium, thorium, etc., are the most widely studied metals (Gadd 2009, Karthika et al., 2010; Gadd and Fomina, 2011). Based on their ionic and chemical properties, these metals are either cationic or anionic, or possess a wide range of oxidation states or existing complexes. In some cases, these metals exist in modified forms such as hydroxylated forms or are complexed with other compounds. All of these metallic compounds are unavoidable in the mining, metallurgy, automobile, paint, and other industries. In addition to cationic removal, several studies were conducted to remove toxic anions such as arsenic, selenium, chromium, molybdenum, and vanadium (Michalak et al., 2013). These toxic metal(oids) are removed primarily using conventional approaches including ion exchange, solvent extraction, and precipitation technologies. Based on literature reports, biosorption has proven to be an effective alternate approach (Michalak et al., 2013).

Organic Compounds

Organic compounds are a class of chemical compounds containing at least one carbon-hydrogen bond. A wide range of these compounds are found in everyday life, and eventually, they are released into the ecosystem. Most organic compounds are degraded by nature's microbiota, and the rate and extent of such degradation depend on the complexity of the compounds' chemical structure. In some cases, degraded products pose a significant threat to the environment due to their toxicity. Certain groups of human-made compounds called *xenobiotics* are highly resistant to microbial degradation and highly recalcitrant, as they are chemically complex and inaccessible to microbial actions. Compounds such as dyes, pesticides, phenolic compounds, etc., are very difficult to degrade: they are persistent, resistive to aerobic degradation, and stable to most oxidizing agents even in low concentrations (Aksu 2005; Thirunavukkarasu et al., 2020b; Nithya et al., 2021b).

Conventional strategies such as precipitation, electrolytic methods, extraction, ion exchange, etc. suffer from many limitations including high cost, the need for skilled labor, scaling issues, solid waste generation, etc. However, biosorption has proven its potential to remove various acidic and basic dyes, phenolic compounds, and other organic compounds from wastewater.

Biosorbents

Biosorbents are materials of biological origin that can capture inorganic and organic compounds from wastewater with diverse functional groups on their surfaces (Gadd, 2009; Dhankhar and Hooda, 2011). Biosorbents may be derived from microbes (archea, bacteria, cyanobacteria, fungi, filamentous fungi, algae, yeasts, etc.), seaweed (macroalgae), plant and agricultural wastes (leaves, bark, stems, roots, fruit, vegetables, peels, seeds, rice straw, baggases, wheat bran, sugar beet pulp, soybean hulls, etc.), industrial discharge (fermented waste, sludge, etc.), and marine materials (shells, chitosan, cellulose, etc.) (Park et al., 2010; Dhankhar and Hooda, 2011). The vast quantity of biomass generated due to deforestation also paves the way for a search for sustainable uses of such wastes in wastewater treatment technologies.

Abundance, diverse functionalities, low cost, recycled green products, and versatility are the primary factors for adopting such biomass-based materials in the biosorption process. The sorptive capacity of these biosorbents varies considerably even within the same biomass as it depends on a wide range of factors including the physicochemical properties of the solution microenvironment, operational conditions, and the type of surface pretreatment. A proper selection of biosorbents from the large pool of biomass-based materials is crucial for researchers, and desirable properties of ideal biosorbents are listed here:

- Non-edible biomaterials
- Easily available and abundant
- Non-toxic given a wide range of chemicals, temperatures, and pressures
- Microbes that are dead or attenuated
- Easy maintenance of microbial growth in the case of living biomass
- Inexpensive and nontoxic bio-based industrial discharges
- Plant-derived biomass with low or zero economic importance

Factors Affecting Biosorption

The solution pH is the most important factor in the biosorption process as it affects the solution chemistry of the contaminants or pollutants and the surface functional moieties of the biosorbents, and it determines the relative affinity of coexisting ions in the solution (Vijayaraghavan and Yun, 2008; Thirunavukkarasu et al., 2018). An alkaline pH environment can remove cationic metals or basic dyes, and an acidic pH favors the removal of anionic metals or acidic dyes. In an alkaline microenvironment, hydrogen ions and cationic species compete for binding sites, and metals ions will form hydroxides and thus reduce their binding potential. Similarly, in an acidic environment, hydroxyl ions and anionic species competitively bind on the surface, resulting in poor binding of negatively charged molecules on the biosorbent.

The ionic strength of the sorbate solution is another important factor determining the performance of the biosorption process. A higher ionic concentration in the solution reduces the sorptive removal as ions compete with the sorbate ions for sorbents' surface binding pockets (Thjrunavukkarasu et al., 2018; Barquilha et al., 2019). This tendency limits the applications of effluents that contain a complex mixture of cationic species.

The sorptive performance of biosorbents is also influenced by the initial concentration of sorbate ions. At low concentrations, more surface functional groups are available in the biosorbent, resulting in a rapid rate of solute transfer at the interface. On the other hand, at higher concentrations, removal efficiency is affected by the limited availability of active sites.

The presence of other molecules or ions can also influence the removal capacity as they compete for binding surfaces on the biosorbents. For instance, cation-loaded biosorbents can increase the uptake of cationic pollutants due to pH buffering effects (Fourest et al., 1994). Likewise, cationic-modified biosorbents positively attract the counter ionic anion species by enhancing electrostatic interactions (Gadd, 2009). The physicochemical properties, presence of functionalized groups on the surface, type of pretreatment, amount or dosage, and particle size distribution of the biosorbents are other significant factors affecting the overall performance of the biosorption process (Gadd, 2009; Karthika et al., 2010; Park et al., 2010; Kotrba, 2011; Li and Tao, 2013; Arunachalam et al. 2017).

Temperature is defined as the average kinetic energy of the molecules, and when it is increased, the net intrinsic energy of the sorbate molecules increases, resulting in improved collision frequency. Likewise, an increase in the temperature positively influences the potential of the biosorbent by increasing its surface properties and its action on the biosorbates. Another explanation is that the increased temperature of live biosorbents results in increased metabolic action inside the cell, and hence the uptake of contaminants or pollutants is also quite high. However, if the temperature goes beyond its optimal value, the cell wall is disrupted, resulting in the loss of sorption capacity (Dai et al., 2019; Hu et al., 2020). In some instances, a higher temperature also affects the surface integrity of the biosorbent molecules and reduces their overall sorptive potential (Park et al., 2010; Thirunavukkarasu et al. 2018).

Operational Modes of Biosorption

Biosorbates can contact solid biosorbents via different modes such as batch and continuous flow. In the continuous mode of operation, the biosorbents are solid matrices or *beds*, and these beds can be moving, fixed, or operated under fluidized conditions. Table 8.1 compares the batch and continuous modes of biosorption operation, and it is evident that fixed-bed biosorption is the optimal choice for removing a wide range of contaminants or pollutants from wastewater.

Table 8.1 Comparison of batch and continuous flow operations.

Batch system	Continuous flow system
Constant volume.	Variable volume.
Easy to characterize.	Difficult to comprehensively understand.
Concentration of biosorbate at any point inside the batch system is assumed as constant.	Concentration of biosorbate varies along the length of the column.
No or less mass-transfer resistance across the phase boundary.	Greater mass-transfer resistance due to the non-ideal behavior of the liquid.
No pressure drop is observed.	Pressure drop plays a significant role in the column integrity and its performance.
Agitation prevents the formation of a stagnant or dark zone.	No agitation is feasible, so the formation of a stagnant or dark zone is inevitable.
Temperature can be easily controlled as the volume is low.	Temperature control is challenging.
To understand the equilibrium and kinetics of the biosorbent, batch mode is preferable.	To treat a large volume of samples, continuous bed flow is preferable.
Applicable only for small-scale wastewater treatment.	Can be applied to industrial-scale treatment.
Low contaminant load in the sample is preferable.	High contaminant load can be treated.
Loading or charging of biosorbent is not required during the sorption process.	Loading or charging the biosorbent is essential during the sorption process.
Biosorbents can be easily removed with simple filtration or magnetic techniques.	Continuous regeneration of the biosorbent bed is essential.
Initial solution pH, initial biosorbate concentration, biosorbent dose, temperature, agitation speed, and contact time are the significant parameters affecting the performance of the process.	Initial biosorbate concentration, bed height, flow rate, temperature, and contact time are the vital parameters affecting column efficiency.
The performance of the batch system can be assessed with the adsorption capacity from the equilibrium models.	The performance of the continuous fixed bed system can be assessed with breakthrough curves.
Cheap and economical as it requires no accessory devices	Relatively expensive as it has to be accompanied by a flow-regulating device and thermostat, etc.

Batch Biosorption

Batch-mode biosorption is characterized by a constant-volume system with neither input nor output, as the reacting system is charged initially with the sorbate solution and sorbent species. Sampling is done at fixed intervals, followed by a quantitative determination of the sorbate molecules using appropriate techniques. As a batch system is simple to operate and has a constant volume, most research is carried out using this mode to investigate the sorptive potential of various sorbents (Nithya et al., 2018; Rajarathinam et al., 2020; Thirunavukkarasu and Nithya 2020). The batch mode of operation is also easy to control or regulate and is completely characterized; thus it is the mode of choice for the preliminary analysis of biosorption.

Sorption experiments were carried out in batch reactors to understand the equilibrium and kinetics of biosorbent toward targeted biosorbate species. It is also easier to manage multiple batch systems simultaneously to look for possible errors in manual operation and check the redundancy of the performance data. Despite these advantages, batch systems can never be used for large volumes of samples, which is an essential requirement for wastewater treatment.

Dynamic Biosorption

In fixed-bed column studies, the biosorbents are preprocessed and packed as a homogenous matrix in a column of the desired length. The sample containing the contaminants or pollutants is made to pass through the biosorbent column at a predefined volumetric flow rate. The relative affinity of the sorbate toward the functional moieties on the surface of the biosorbents at the solid-liquid interface determines the efficiency of the biosorption process; hence, sufficient time must be provided for the interaction of the biosorbate and biosorbent (Thirunavukkarasu et al., 2018; Thirunavukkarasu et al., 2020). In addition, operational parameters such as the length of the bed and flow rate of the sample may influence the efficacy of the process.

The performance of this mode of column studies can be viewed using sorptive characteristic curves called *breakthrough curves*. Usually, these curves are plotted with the time of operation on the horizontal axis and contaminant concentration on the vertical axis. The underlying mechanism in such continuous-flow sorption is axial dispersion, in which biosorbates move from the bulk liquid phase and are bound onto the surface moieties of the biosorbent by establishing physicochemical interactions at their interface. Resistive forces such as film-diffusion resistance and intra-particle diffusion resistance control the rate of dispersive motion of the biosorbates, and hence the fixed bed should be designed to facilitate the molecular motion of the biosorbates.

The mass transfer zone (MTZ) is a key variable related to breakthrough curves and fixed-bed adsorption capacity. At the start of the continuous fixed-bed biosorption process, a predefined volume of feed (wastewater) is allowed to flow at a specified flow rate into the biosorptive column. The feed water wets the dry sorptive column, and due to the relative affinity of the biosorbates toward the surface functional groups, biosorption occurs rapidly and effectively on the top layers of the biosorbents. The rapidity of the solute transport is due to the availability of a greater number of active pockets on the surface of the biosorbents and the fact that few biosorbate molecules were initially allowed. In other words, the MTZ is attained on the top layers of the biosorbent column (Figure 8.1). As a result, no biosorbate molecules are available in the runoff stream at the first stage, and the concentration of biosorbate (C) in the effluent is zero. As the process progresses, the top layer becomes saturated and attains equilibrium with the biosorbate molecules, so no functional groups are available to grab the biosorbates. These layers of biosorbents are saturated regions or unavailable beds. Hence, the MTZ needs to move further down the length of the column to continue the solute transport in increasing concentrations.

Moving the MTZ and forming saturated regions and diffusional resistance along the length of the column significantly reduces the sorptive capacity of the bed. The length of the unused bed increases down the column, increasing the ratio of the effluent

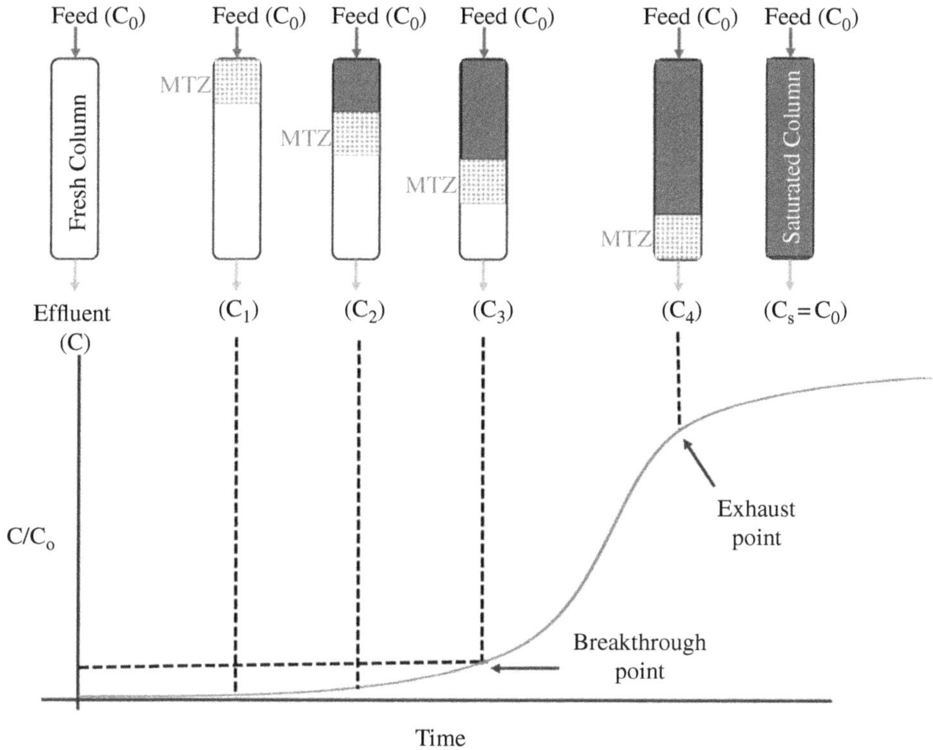

Figure 8.1 Illustration of MTZ movement along the length of the column, showing the breakpoint time and exhaust time.

concentration to the inflow concentration. The magnitude of the ratio increases further, and when it attains the EPA threshold value of the contaminants or pollutants that is the breakthrough point of the sorption process, the corresponding time is defined as the *breakthrough time.* The sorption process continues, and when the bed is completely saturated or the MTZ is unavailable, the ratio of effluent concentration to inflow concentration becomes one: this point is defined as the *saturation zone.* Hence, the breakthrough curves for various contaminants or pollutants are S-shaped with varying degrees of steepness (Chowdhury et al., 2013; Shafeeyan et al., 2014; Hasanzadeh et al., 2016).

If the length of the MTZ in a fixed-bed sorption column is assumed to be uniform, the time taken by the bed to establish the MTZ (*tx*), the time required for the exchange of sorbate ions in the MTZ region, the rate of movement of the MTZ along the bed length, and the fraction of usable bed available in the total length of the column are the key parameters in designing an optimal column for the biosorption process (Gupta and Ali, 2012; Crittenden and Thomas, 1998). In the biosorptive mode of wastewater treatment, the effects of various factors such as initial biosorbate concentration, initial pH, flow rate of the biosorbate into the column, bed height, particle size of the biosorbent in the solid matrix, and temperature of the column were determined and analyzed with the help of breakthrough curves. Every factor significantly affects the performance of the biosorption process. For instance, an increase in the biosorbate concentration results in the early onset of breakthrough and exhaustion. The higher the initial concentration, the greater the concentration difference between the liquid and solid phase, and hence the greater the magnitude of the mass-transfer coefficient. A further increase in the sorbate concentration results in decreased sorptive potential as the column functional groups become unavailable at the interface (Moyo et al., 2017). Similarly, a continuous fixed-bed column with greater biosorbate flow rates will attain an early breakthrough point as the residence time of the biosorbate-biosorbent interaction for solute transport at the interface is not adequate.

Continuous column operation with a lower flow rate results in higher removal efficiency as sufficient interaction time is provided for the sorbate molecules to transit across the phase at the interface (Lopez-Cervantes et al., 2017; Ahmad and Hameed 2009; Sheng et al., 2018). On the other hand, an increase in the bed length or height may result in a longer breakthrough and exhaustion time. This can be explained by the fact that the increased bed height provides greater surface area, and as a result, more functional active moieties can attract biosorbate molecules from the bulk liquid phase (Fathi et al., 2014; Teutscherova et al., 2018). A similar observation can be made for the biosorbent particle size: as the particle size becomes smaller, the surface area increases, which increases the biosorptive capability of the biosorbent.

The higher the temperature, the slower the onset of breakthrough and exhaustion times. A high temperature facilitates biosorbate transit by reducing the diffusional resistance and opens the pores on the surface of the biosorbent in the column, thus increasing the performance of the column operation. Table 8.2 lists the various scholarly reports from the literature that have studied these factors in removing a wide range of contaminants or pollutants from wastewater.

Models of Dynamic Biosorption

Models are the mathematical representations of any physicochemical process and give the desired output as a function of input variable(s) (Thirunavukkarasu and Nithya, 2011; Thirunavukkarasu et al., 2019; Sivashankar et al., 2020). As biosorption can be considered a physicochemical process, modeling approaches are the optimal choice for predicting key variables such as biosorbent capacity, usable bed capacity, bed regeneration time, etc. These models can also provide greater insights into the mechanism of the sorptive removal process. From the literature, diverse mathematical models have been employed by assuming that biosorbate removal processes are mediated by axial dispersion, with counteractive resistance offered by the intra-particle and stagnant liquid film surrounding the biosorbent at the interface.

The Thomas model, Adams-Bohart model, Clark model, Yoon-Nelson model, Wolborska model, and modified dose-response model are the most widely used models reported in the literature to analyze column performance in the removal of contaminants from wastewater. The most extensively studied Thomas model provides key variables such as the sorptive rate coefficient and biosorbate concentration on the biosorbent surface. The Langmuir sorptive equilibrium and second-order reversible rate kinetics are the underlying assumptions of the Thomas model. Using gas-charcoal adsorptive system, Adams and Bohart proposed a relationship between the ratio of sorbate concentration to the initial concentration and time in a continuous flow mode. This model implies that the sorption rate is purely a

Table 8.2 Biosorptive removal of contaminants or pollutants from wastewater using fixed-bed columns.

S.No	Biosorbent	Biosorbate	Operational parameters	Models	Adsorption capacity	Reference
1	Activated charcoal prepared from neem leaf powder (Ac-NLP)	Pb, Cu, Cr, Zn, Ni, Cd	Flow rate, initial concentration, bed height	Thomas model	Pb: 205.6 mg/g; Cu: 185.8 mg/g; Cd: 154.5 mg/g; Zn: 133.3 mg/g; Ni: 120.6 mg/g; Cr: 110.9 mg/g	Patel 2020
2	Modified wheat bran	Cr, Cu, Cd	Bed height, initial concentration, flow rate	Mass-transfer model	Cr: 10.41 mg/g; Cu: 8.32 mg/g; Cd: 1.78 mg/g	Renu et al., 2020
3	Acorn shell from *Quercus crassipes Humb. Bonpl. (QCS)*	Total chromium, Cr(VI)	Influent solution pH, influent flow rate, QCS bed height, influent Cr(VI) concentration	Dose-response model	181.56 mg/g; 110.35 mg/g	Aranda-García et al., 2020
4	Two commercial granular activated carbons (E-GAC and H-GAC)	Sevoflurane	Bed depth, gas flow rate, inlet sevoflurane concentration	Adams-Bohart, Thomas, Yoon-Nelson models	60% removal	Ang et al., 2019
5	Hydrolyzed olive cake	Cr, Cu, Pb, Zn	Inlet metal concentration	Thomas, Yoon-Nelson, dose-response models	Cr: 3.08 mg/g; Cu: 5.17 mg/g; Pb: 13.21 mg/g; Zn: 5.51 mg/g	Fernández-González et al., 2019
6	Sugarcane bagasse	Pb and Cd	Single-component and two-component systems	Yoon-Nelson, Thomas models	Single-component – Pb: 91% removal; Cd: 90% removal. Two-component – Pb: 90% removal; Cd: 92% removal	Vera et al., 2019

(*Continued*)

Table 8.2 (Continued)

S.No	Biosorbent	Biosorbate	Operational parameters	Models	Adsorption capacity	Reference
7	Compressible Prussian blue/polyurethane sponges	Cs	Flow rate, initial cesium ions concentration	Thomas, Yoon-Nelson models	68.2 mg/g	Chang et al., 2018
8	*Enteromorpha prolifera* (EP)	Methylene blue (MB)	Bed depth, influent flow rate, pH, MB initial concentration, column shape	Thomas, Yoon-Nelson, Bohart–Adams, bed depth service time (BDST) models	348.06 mg/g	Zhong et al., 2018
9	Organic modified spent grains (OSGs)	Arsenate, As(V); arsenite, As(III)	Influent flow rate and arsenic ions initial concentration	Thomas, Adams-Bohart, Yoon-Nelson kinetic models	As(V): 6.0 mg/l; As(III): 3.0 mg/l	Yunnen et al 2017
10	Bone Char	Mn	Initial concentration	Column Adsorption Model	—	Maria et al., 2017
11	*Chrysanthemum indicum* flower waste raw (CIF-R); *Chrysanthemum indicum* flower waste bio char (CIF-BC)	Ni	Bed height, flow rate, initial metal ion concentration	Adams-Bohart, Thomas, Yoon-Nelson models	14.02 mg/g; 29.44 mg/g	Vilvanathan et al., 2017
12	Raw and treated date palm leaves, orange peel waste biomass	Cu, Pb	Bed height, initial metal concentration, particle size, flow rate	Thomas, Yoon-Nelson models	60%	Amin et al., 2017
13	Chitosan–glutaraldehyde	Direct blue 71	Bed height, inlet direct blue 71 concentration, and feed flow rate	Adams-Bohart, Thomas, BDST models	343.59 mg/g	López-Cervantes et al., 2017

14	Multi-metal binding biosorbent (MMBB)	Cd, Cu, Pb, Zn	Influent flow rate, metal concentration, bed depth	Thomas, dose-response, Yoon-Nelson models	Cd: 38.25 mg/g; Cu: 63.37 mg/g; Pb: 108.12 mg/g; Zn : 35.23 mg/g	Abdolali et al., 2017
15	Eucalyptus bark	Methylene blue (MB)	Inlet MB dye flow rate, initial MB dye concentration, adsorbent bed height	Thomas, Yoon–Nelson, BDST models	45.24 mg/g; 48.29 mg/g; 49.55 mg/g	Afroze et al., 2016
16	Sesame waste biomass	Cd	Flow rate, initial Cd concentration, bed height	Thomas, Yoon-Nelson models	22.88 mg/g	Cheraghi et al., 2016
17	Sugarcane bagasse	Pb, Cu	Bed height, flow rate	Thomas, Adams-Bohart, Yoon-Nelson models	Pb: 4.54 mg/g; Cu: 3.98 mg/g	Duga et al., 2016
18	Immobilized *Bacillus thuringiensis*	Cu, Mn	Flow rate, bed height	BDST, Thomas, Yoon-Nelson models	14.23 mg/g; 25.27 mg/g	Khosravi et al., 2015
19	Chemically carbonized rubber wood sawdust	Pb	Flow rate, bed depth, concentration, pH	Thomas, Yoon-Nelson, Adams-Bohart models	38.56 mg/g	Biswas et al., 2015
20	Algae, *Cystoseira indica*	Thorium(IV)	pH, initial concentration	Clark, mass-transfer models	151.3 mg/g; 195.7 mg/g; 120.6 mg/g	Riazi et al., 2014
21	*Artocarpus heterophyllus* jackfruit leaf powder *(JLP)*	Crystal violet (CV)	Bed height, flow rate	Thomas, BDST models	43.39 mg/g	Saha et al., 2012
22	Acid-treated *Penicilium sp*	Reactive red	Initial dye concentration, adsorbent dose, flow rate	Thomas model	150.95 mg/L; 64.84% removal	Zhang et al., 2011

(*Continued*)

Table 8.2 (Continued)

S.No	Biosorbent	Biosorbate	Operational parameters	Models	Adsorption capacity	Reference
23	*Cystoseira indica*	Uranium	Initial concentration	Mass-transfer model	287.21 mg/g; 371.39 mg/g; 359.70 mg/g; 78% removal	Ghasemi et al., 2011
24	Scales of *Catla catla* fish	Cu	Fixed volume of solution	Mass-transfer model	—	Das et al., 2008
25	Dried *Rhizopus arrhizus*	Gemacion (Procion) red H-E7B (GR), Gemazol turquise blue G (GTB), Gemactive (reactive) black HFGR (GB)	Flow rate, inlet dye concentration	Thomas, Yoon-Nelson models	GR: 1007.8 mg/g; GTB: 823.8 mg/g; GB: 635.7 mg/g	Aksu et al., 2007
26	Palm fruit bunch and maize cob	Fe, Mn	Bed height, solution flow rate	BDST model	—	Nassar et al., 2003
27	Ca-loaded *Laminaria japonica*	Cu, Pb	Initial metal concentration	Mass-transfer model	Cu: 1.22 mmol/g; Pb: 1.08 mmol/g	Lee et al., 2000
28	*Rhizopus arrhizus,* and *Mucor miehei*	Cu	Flow rate	Two-parameter fixed-bed adsorption model	300 mmol/g	Brady et al., 1999

function of the sorbate concentration and the residual capacity of the sorption process (Dorado et al., 2014). On the other hand, the Yoon-Nelson model proposes that the declining rate of sorption is linearly proportional to the biosorbate sorption and breakthrough on the biosorbents, and it is highly independent of the type, nature, and physical features of the biosorption bed. Clark proposed breakthrough models based on the assumption that the sorption kinetics follow the Freundlich isotherm and the flow behavior is of piston type in continuous fixed-bed column studies. By ignoring dispersive effects, mass-transfer equations were developed to define the performance of the sorption process. The Wolborska model also describes a relationship between concentration distribution and sorption capacity for a continuous flow system, and it has limited applicability for low biosorbate concentration ranges. The modified dose-response model was intended to prevent errors resulting from using the Thomas model, specifically at higher or lower time zones on the breakthrough curve (Lee et al., 2015; Biswas and Mishra, 2015). Table 8.3 depicts the column biosorption models along with their mathematical expressions.

Challenges in Dynamic Biosorption

Despite biosorption being an effective strategy for removing pollutants from wastewater, it generates a considerable amount of used solid biosorbents; and a lack of effective regeneration potential has resulted in limited applications on an industrial scale. In addition, this technique is applicable only for liquid samples with little or no suspended solids or colloids; hence, for typical industrial wastewater, pretreatment techniques need to be carried out to facilitate the sorption process (Thirunavukkarasu et al., 2020). This solute-removal process can be considered a secondary or tertiary treatment option; the efficacy of the other treatment processes can significantly influence the overall recovery process. Further, pretreating solutions in continuous columns faces some significant challenges, including the particle size distribution of the biosorbent, carrier selection, diffusional resistance, pressure-drop effects, channeling behavior, formation of stagnant or black zones, poor temperature control, etc.

The biosorbent particle size is one of the key variables in the column design, as the size distribution determines the kind of packing: regular or random. The primary difference between regular and random packing is the magnitude of the column's porosity and void volume, which can be directly correlated with effective mass transfer at the solid-liquid interface. If the particles are too small to be packed into the column, an inert carrier can be used to encapsulate or embed the biosorbent molecules to prevent their elution with the sorbate runoff.

Carrier selection is another design factor for continuous flow studies as it determines the efficacy of the column operation and the solute removal process. Encapsulated or immobilized biosorbents should be accessible to the liquid phase containing biosorbates; the carrier should facilitate lateral movement. Using porous media such as water-insoluble gels (e.g. calcium alginate) is the optimal choice, and such carriers must be non-reactive with the sorbent and sorbate molecules.

In addition to diffusive resistance from the sorbent molecules and their surrounding liquid film, the presence of inert carriers creates additional resistance to the flow of sorbate

Table 8.3 Column biosorption models.

Model	Mathematical function (linear model)	Graphical plot	Estimated parameter
Thomas model (TM)	$\ln\left[\left(\frac{C_0}{C_t}\right)-1\right]=\frac{k_{TH}q_0m}{Q}-\frac{k_{TH}C_0V_{eff}}{Q}$	$\ln\left[\left(\frac{C_0}{C_t}\right)-1\right]$ vs. t	k_{TH} and q_0
Adam-Bohart model (ABM)	$\ln\left[\left(\frac{C_0}{C_t}\right)-1\right]=K_{AB}N_0\frac{Z}{u}-K_{AB}C_{tt}$	$\ln\left[\left(\frac{C_0}{C_t}\right)-1\right]$ vs. t	K_{AB} and N_0
Bed depth service time model (BDST)	$t=\frac{N_0}{C_0F}Z-\frac{1}{k_{BDST}C_0}\ln\left[\left(\frac{C_0}{C_t}\right)-1\right]$	Bed height vs. time	k_{BDST} and N_0
Yoon-Nelson model (YNM)	$\ln\left(\frac{C_t}{C_0-C_t}\right)=k_{YN}t-\tau k_{YN}$	$\ln\left(\frac{C_t}{C_0-C_t}\right)$ vs. t	k_{YN} and τ
Clark model (CM)	$\frac{C_t}{C_0}=1/\left(1+Ae^{-rt}\right)^{\left[1/n-1\right]}$	$\frac{C_t}{C_0}$ vs. t	A and r
Wolborska model (WM)	$\ln\left(\frac{C_t}{C_0}\right)=\frac{\beta C_0t}{N_0}-\frac{\beta Z}{U}$	$\ln\left(\frac{C_t}{C_0}\right)$ vs. t	β and N_0
Modified dose response model (MDRM)	$\ln\left(\frac{C_t}{C_0-C_t}\right)=a\ln\left(C_0Qt\right)-a\ln\left(q_{mdr}m\right)$	$\ln\left(\frac{C_t}{C_0-C_t}\right)$ vs. $\ln\left(C_0Qt\right)$	a and q_{mdr}

Sources: Based on Lee et al., 2015; Biswas and Mishra, 2015.

molecules across the phase, which is another challenge in the use of such immobilized systems for large-scale applications. When loading biosorbent materials into the column, the packing density, volume, and strength significantly influence the performance of the process, as these parameters directly affect the column's operating pressure. If the packing density is too high, it may result in a drop in the pressure of the effluent runoff, affecting the overall efficiency of the process.

Non-ideal behavior of the bulk liquid flow on the column can result in channeling effects, as each sorbate molecule takes a tortuous pathway to reach the other end of the column and hence increases the time to elute out in the runoff. Also, it is obvious that diffusion at the interface is spontaneously mediated by the molecules; this channeling behavior of the sorbate molecules increases the retention time of the continuous flow system. As a consequence of this random molecule movement inside the column, the surface functional moieties of the sorbent at certain regions do not interact with the sorbate molecules, resulting in the formation of *stagnant* or *black zones*. Such non-ideal behavior of a continuous column can be detected by tracer studies using the residence time distribution (RTD) function and can be minimized.

Another challenge of using continuous fixed-bed studies is maintaining a uniform column temperature. The sorption process is dependent on the temperature, and controlling it over the entire bed length is challenging due to heat generated inside the column.

Multiple column bed operations connected in series or parallel are yet to be explored in wastewater treatment. In these columns, different kinds of biosorbents can be used to remove a wide range of contaminants or pollutants from the wastewater stream. In such cases, the challenges would be too significant to optimize the design parameters for each column; and maintaining uniform operating conditions for multiple columns connected in series or parallel is not feasible.

In the treatment of real industrial effluents, physicochemical properties such as turbidity, colloidal suspension, pH, biochemical oxygen demand (BOD), chemical oxygen demand (COD), and other ionic compounds can influence the behavior of the continuous flow studies. A comprehensive understanding of the biosorption properties of the continuous column system is required for application to industrial wastewater.

Conclusion

This chapter has presented a comprehensive report on continuous fixed-bed column studies for the biosorptive removal of various contaminants or pollutants from wastewater. The fundamentals of the biosorption process in the context of diffusive mass transfer were discussed, along with classes of biosorbents and biosorbate molecules. A wide range of operating modes for a typical biosorption process were systematically analyzed with their possible merits and potential limitations. Scholarly reports on dynamic bed biosorption of metals and other organic compounds were critically reviewed, and various influencing factors such as initial biosorbate concentration, bulk liquid flow rate, bed height, biosorbate solution pH, particle size distribution of biosorbents, temperature, etc., in determining the performance of the sorptive potential were discussed with possible technical insights. Also, numerical interpretations of the use of different models, including the Thomas model, Adams and Bohart model, Clark model, bed depth service time model, Yoon-Nelson model, Wolborska model, and modified dose-response model were made to understand the dynamics of the biosorbent column.

Research in fixed-bed column studies is still in the primitive stage as most work is focused on using a single-bed system with one type of biosorbent and treating synthetic wastewater. Multiple-column studies with more than one biosorbent targeting a set of sorbate molecules in a continuous mode of operation are the need of the hour for the effective removal of contaminants or pollutants from wastewater. Unfortunately, no such investigations were found in the literature. In addition, very few research works studied the regeneration or desorption potential of biosorbent columns. As the column's regeneration capability is a function of the overall economy of the process, the removal capacity from the bound surface of the biosorbents must be investigated further.

This chapter has discussed in detail various challenges in using biosorbents in continuous-flow studies, such as the particle size distribution of the biosorbents, carrier selection, diffusional resistance, pressure drop effects, channeling behavior, formation of stagnant or black zones, poor temperature control, etc. Investigating these challenges can be considered a research hotspot for future researchers.

References

Abdolali, A., Ngo, H.H., Guo, W. et al. (2017). Application of a breakthrough biosorbent for removing heavy metals from synthetic and real wastewaters in a lab-scale continuous fixed-bed column. *Bioresource Technology* 229: 78–87.

Afroze, S., Sen, T.K., and Ang, H.M. (2016). Adsorption performance of continuous fixed bed column for the removal of methylene blue (MB) dye using Eucalyptus sheathiana bark biomass. *Research on Chemical Intermediates* 42 (3): 2343–2364.

Ahmad, A.A. and Hameed, B.H. (2010). Fixed-bed adsorption of reactive azo dye onto granular activated carbon prepared from waste. *Journal of Hazardous Materials* 175 (1-3): 298–303. doi:10.1016/j.jhazmat.2009.10.003.

Aksu, Z. (2005). Application of biosorption for the removal of organic pollutants: a review. *Process Biochemistry* 40 (3): 997–1026.

Aksu, Z., Çağatay, S.S., and Gönen, F. (2007). Continuous fixed bed biosorption of reactive dyes by dried Rhizopus arrhizus: determination of column capacity. *Journal of Hazardous Materials* 143 (1–2): 362–371.

Amin, M.T., Alazba A.A., and Shafiq M. (2017). Batch and fixed-bed column studies for the biosorption of Cu(II) and Pb(II) by raw and treated date palm leaves and orange peel. *Global NEST Journal* 19 (3): 464–478.

Ang, T.N., Young, B.R., Taylor, M. et al. (2020). Breakthrough analysis of continuous fixed-bed adsorption of sevoflurane using activated carbons. *Chemosphere* 239: 124839.

Aranda-García, E. and Cristiani-Urbina, E. (2020). Hexavalent chromium removal and total chromium biosorption from aqueous solution by Quercus crassipes acorn shell in a continuous up-flow fixed-bed column: Influencing parameters, kinetics, and mechanism. *PloS One* 15 (1): e0227953.H

Arunachalam, T., Karpagasundaram, M., and Rajarathinam, N. (2017). Ultrasound assisted green synthesis of cerium oxide nanoparticles using Prosopis juliflora leaf extract and their structural, optical and antibacterial properties. *Materials Science-Poland* 35 (4): 791–798.

Barquilha, C.E.R., Cossich, E.S., Tavares, C.R.G. et al. (2019). Biosorption of nickel(II) and copper(II) ions from synthetic and real effluents by alginate-based biosorbent produced from seaweed Sargassum sp. *Environmental Science and Pollution Research* 26 (11): 11100–11112. doi:10.1007/s11356-019-04552-0.

Biswas, S. and Mishra, U. (2015). Continuous fixed-bed column study and adsorption modeling: removal of lead ion from aqueous solution by charcoal originated from chemical carbonization of rubber wood sawdust. *Journal of Chemistry and Chemical Engineering.* doi:10.1155/2015/907379.

Brady, J.M., Tobin, J.M., and Roux, J.-C. (1999). Continuous fixed bed biosorption of Cu2 ions: application of a simple two parameter mathematical model. *Journal of Chemical Technology & Biotechnology* 74 (1): 71–77.

Chang, S., Fu, H., Wu, X. et al. (2018). Batch and fixed-bed column studies for selective removal of cesium ions by compressible Prussian blue/polyurethane sponge. *RSC Advances* 8 (64): 36459–36467.

Chowdhury, S. and Saha, P.D. (2013). Adsorption of malachite green from aqueous solution by naoh-modified rice husk: Fixed-bed column studies. *Environmental Progress & Sustainable Energy* 32 (3): 633–639.

Dai, Q.H., Bian, X.Y., Li, R. et al. (2019). Biosorption of lead(II) from aqueous solution by lactic acid bacteria. *Water Science and Technology* 79 (4): 627–634. doi:10.2166/wst.2019.082.

Das, A., Bhowal, A., and Datta, S. (2008). Continuous biosorption in rotating packed-bed contactor. *Industrial & Engineering Chemistry Research* 47 (12): 4230–4235. doi:10.1021/ie070679g.

Dhankhar, R. and Hooda, A. (2011). Fungal biosorption – an alternative to meet the challenges of heavy metal pollution in aqueous solutions. *Environmental Technology* 32 (5): 467–491. doi:10.1080/09593330.2011.572922.

Dorado, A.D., Gamisans, X., Valderrama, C. et al. (2014). Cr(III) removal from aqueous solutions: a straightforward model approaching of the adsorption in a fixed-bed column. *Journal of Environmental Science And Health. Part, A., Toxic/Hazardous Substances & Environmental Engineering* 49 (2): 179–186.

Henriques, C.A., da Costa, A.C.A., dos Reis, M.M. et al. (2011). Batch and fixed-bed column biosorption of manganese ion by Sargassum filipendula. *Electronic Journal of Biotechnology* 14 (5). doi:10.2225/vol14-issue5-fulltext-8.

Duga, N.D.F., Imperial, P.E.A., Soriano, A.N., and Nieva, A.D. (2016). Packed bed biosorption of lead and copper ions using sugarcane bagasse. *ASEAN Journal of Chemical Engineering* 16 (1): 23–37. doi:10.22146/ajche.49671.

Cheraghi, E., Ameri, E., and Moheb, A. (2016). Continuous biosorption of Cd(II) ions from aqueous solutions by sesame waste: thermodynamics and fixed-bed column studies. *Desalination and Water Treatment* 57 (15): 6936–6949.

Fat'hi, M.R., Asfaram, A., Hadipour, A. et al. (2014). Kinetics and thermodynamic studies for removal of acid blue 129 from aqueous solution by almond shell. *Journal of Environmental Health Science and Engineering* 12 (1): 62.

Fernández-González, R., Martín-Lara, M.Á., Blázquez, G. et al. (2019). Recovering metals from aqueous solutions by biosorption onto hydrolyzed olive cake. *Water* 11 (12): 2519.

Fourest, E., Canal, C., and Roux, J.-C. (1994). Improvement of heavy metal biosorption by mycelial dead biomasses (Rhizopus arrhizus, Mucor mieheiandPenicillium chrysogenum): pH control and cationic activation. *FEMS Microbiology Reviews* 14 (4): 325–332. doi:10.1111/j.1574-6976.1994.tb00106.x.

Gadd, G.M. (2009). Biosorption: Critical review of scientific rationale, environmental importance and significance for pollution treatment. *Journal of Chemical Technology and Biotechnology* 84 (1): 13–28.

Gadd, G.M. and Fomina, M. (2011). Uranium and Fungi. *Geomicrobiology Journal* 28 (5–6): 471–482. doi:10.1080/01490451.2010.508019.

Ghasemi, M. Keshtkar, A. Dabbagh R., and Jaber Safdari S. (2011). Biosorption of uranium in a continuous flow packed bed column using *cystoseira indica* biomass. *Journal of Environmental Health Science and Engineering* 8 (1): 65–74.

Gupta, V.K. and Ali, I. (2013). Water treatment for inorganic pollutants by adsorption technology. *Environmental Water*: 29–91. doi:10.1016/b978-0-444-59399-3.00002-7.

Hardiljeet, K.B., Meera, J., and O'Carroll, M.D. (2011). Kinetics and thermodynamics of cadmium ion removal by adsorption onto nano zerovalent iron particles. *Journal of Hazardous Materials* 186 (1): 458–465. doi:10.1016/j.jhazmat.2010.11.029.

Hasanzadeh, M., Ansari, R., and Ostovar, F. (2016). Synthesis and application of CeO2/sawdust nanocomposite for removal of As(III) ions from aqueous solutions using a fixed bed column system. *Global NEST Journal* 19 (1): 7–16.

Hu, X., Cao, J., Yang, H. et al. (2020). Pb2 biosorption from aqueous solutions by live and dead biosorbents of the hydrocarbon-degrading strain Rhodococcus sp. HX-2. *PLOS ONE* 15 (1): e0226557. doi:10.1371/journal.pone.0226557.

Karthika, T., Thirunavukkarasu, A., and Ramesh, S. (2010). Biosorption of copper from aqueous solutions using Tridax Procumbens. *Recent Research in Science and Technology* 2 (3): 86–91.

Khosravi, V., Ardejani, F.D., Aryafar, A. et al. (2015). Biosorption of metal ions from sarcheshmeh acid mine drainage by immobilized bacillus thuringiensis in a fixed-bed column. *World Academy of Science, Engineering and Technology International Journal of Chemical and Molecular Engineering* 9 (11).

Kotrba, M. (2011). Microbial biosorption of metals – general introduction. In: *Microbial Biosorption of Metals* (ed. Kotrba, M. Mackova, and T. Macek), 1–6. Springer, Dordrecht.

Lee, C.-G., Kim, J.-H., Kang, J.-K. et al. (2015). Comparative analysis of fixed-bed sorption models using phosphate breakthrough curves in slag filter media. *Desalination and Water Treatment* 55 (7): 1795–1805. doi:10.1080/19443994.2014.930698.

Lee, H.S. and Suh, J.H. (2000). Continuous biosorption of heavy metal ions by Ca-loadedlaminaria japonica in fixed bed column. *The Korean Journal of Chemical Engineering* 17 (4): 477–479.

Li, P.-S. and Tao, H.-C. (2015). Cell surface engineering of microorganisms towards adsorption of heavy metals. *Critical Reviews in Microbiology* 41 (2): 140–149.

López-Cervantes, J., Sánchez-Machado, D.I., Sánchez-Duarte, R.G. et al. (2018). Study of a fixed-bed column in the adsorption of an azo dye from an aqueous medium using a chitosan–glutaraldehyde biosorbent. *Adsorption Science & Technology* 36 (1-2): 215–232.

Maria, M.E. and Mansur, M.B. (2017). Mathematical modeling of manganese adsorption onto bone char in a continuous fixed bed column incorporating backmixing and shriking core approaches. *Brazilian Journal of Chemical Engineering* 34 (3): 901–909.

Michalak, I., Chojnacka, K., and Witek-Krowiak, A. (2013). State of the art for the biosorption process—a review. *Applied Biochemistry and Biotechnology* 170 (6): 1389–1416.

Moyo, M., Pakade, V.E., and Modise, S.J. (2017). Biosorption of lead(II) by chemically modified Mangifera indica seed shells: Adsorbent preparation, characterization and performance assessment. *Process Safety and Environmental Protection* 111: 40–51.

Nassar, M.M., Awida, K.T., Ebrahiem, E.E. et al. (2003). Fixed-bed adsorption for the removal of iron and manganese onto palm fruit bunch and maize cob. *Adsorption Science & Technology* 21 (2): 161–175.

Nithya, R., Sivasankari, C., Thirunavukkarasu, A. et al. (2018). Novel adsorbent prepared from bio-hydrometallurgical leachate from waste printed circuit board used for the removal of methylene blue from aqueous solution. *Microchemical Journal* 142: 321–328.

Nithya, R., Sivasankari, C., and Thirunavukkarasu, A. (2021a). Electronic waste generation, regulation and metal recovery: a review. *Environmental Chemistry Letters* 19 (2): 1347–1368.

Nithya, R., Thirunavukkarasu, A., Sathya, A.B. et al. (2021b). Magnetic materials and magnetic separation of dyes from aqueous solutions: a review. *Environmental Chemistry Letters* 19 (2): 1275–1294. doi:10.1007/s10311-020-01149-9.

Park, D., Yun, Y.-S., and Park, J.M. (2010). The past, present, and future trends of biosorption. *Biotechnology and Bioprocess Engineering* 15 (1): 86–102. doi:10.1007/s12257-009-0199-4.

Patel, H. (2020). Batch and continuous fixed bed adsorption of heavy metals removal using activated charcoal from neem (Azadirachta indica) leaf powder. *Scientific Reports* 10 (1). doi:10.1038/s41598-020-72583-6.

Rajarathinam, N., Arunachalam, T., Raja, S. et al. (2020). Fenalan yellow G adsorption using surface-functionalized green nanoceria: An insight into mechanism and statistical modelling. *Environmental Research* 181: 108920. doi:10.1016/j.envres.2019.108920.

Renu, A., Singh, K. et al. (2020). Continuous fixed-bed adsorption of heavy metals using biodegradable adsorbent: Modeling and experimental study. *Journal of Environmental Engineering* 146 (2): 04019110.

Riazi, M., Keshtkar, A.R., and Moosavian, M.A. (2014). Batch and continuous fixed-bed column biosorption of thorium(IV) from aqueous solutions: equilibrium and dynamic modeling. *Journal of Radioanalytical and Nuclear Chemistry* 301 (2): 493–503.

Saha, P.D., Chakraborty, S., and Chowdhury, S. (2012). Batch and continuous (fixed-bed column) biosorption of crystal violet by Artocarpus heterophyllus (jackfruit) leaf powder. *Colloids and Surfaces, B., Biointerfaces* 92: 262–270.

Shafeeyan, M.S., Wan Daud, W.M.A., and Shamiri, A. (2014). A review of mathematical modeling of fixed-bed columns for carbon dioxide adsorption. *Chemical Engineering Research and Design* 92 (5): 961–988.

Sheng, L., Zhang, Y., Tang, F. et al. (2018). Mesoporous/microporous silica materials: Preparation from natural sands and highly efficient fixed-bed adsorption of methylene blue in wastewater. *Microporous and Mesoporous Materials: The Official Journal of the International Zeolite Association* 257: 9–18.

Sivashankar, R., Thirunavukkarasu, A., Nithya, R. et al. (2020). Sequestration of methylene blue dye from aqueous solution by magnetic biocomposite: Three level Box–Behnken experimental design optimization and kinetic studies. *Separation Science and Technology* 55 (10): 1752–1765.

Teutscherova, N., Houška, J., Navas, M. et al. (2018). Leaching of ammonium and nitrate from Acrisol and Calcisol amended with holm oak biochar: A column study. *Geoderma* 323: 136–145.

Thirunavukkarasu, A., Muthukumaran, K., and Nithya, R. (2018). Adsorption of acid yellow 36 onto green nanoceria and amine functionalized green nanoceria: Comparative studies on kinetics, isotherm, thermodynamics, and diffusion analysis. *Journal of the Taiwan Institute of Chemical Engineers* 93: 211–225. doi:10.1016/j.jtice.2018.07.006.

Thirunavukkarasu, A. and Nithya, R. (2019). Response surface optimization of Cu (II) biosorption onto Candida tropicalis immobilized strontium alginate beads by Box-Behnken experimental design. *Journal of Environment and Biotechnology Research* 8 (2): 14–21.

Thirunavukkarasu, A. and Nithya, R. (2020a). Adsorption of acid orange 7 using green synthesized CaO/CeO2 composite: An insight into kinetics, equilibrium, thermodynamics, mass transfer and statistical models. *Journal of the Taiwan Institute of Chemical Engineers.* doi.org/10.1016/j.jtice.2020.04.007.

Thirunavukkarasu, A. and Nithya, R. (2011). Response surface optimization of critical extraction parameters for anthocyanin from Solanum melongena. *Journal of Bioprocessing & Biotechniques* 1 (1). doi:10.4172/2155-9821.1000103.

Thirunavukkarasu, A. and Nithya, R. (2021). *The Role of Nanomaterials in Wastewater Treatment – Sustainable Bioprocessing for a Clean and Green Environment.* CRC Press.

Thirunavukkarasu, A., Nithya, R., and Sivashankar, R. (2020). A review on the role of nanomaterials in the removal of organic pollutants from wastewater. *Reviews in Environmental Science and Biotechnology* 19 (4): 751–778.

Thomas, W.J., John Thomas, W., and Crittenden, B. (1998). Processes and cycles. *Adsorption Technology & Design*: 96–134. doi:10.1016/b978-075061959-2/50006-9.

Vera, L.M., Bermejo, D., Uguña, M.F. et al. (2018). Fixed bed column modeling of lead(II) and cadmium(II) ions biosorption on sugarcane bagasse. *Environmental Engineering Research* 24 (1): 31–37. doi:10.4491/eer.2018.042.

Vijayaraghavan, K. and Yun, Y.-S. (2008). Bacterial biosorbents and biosorption. *Biotechnol. Adv.* 26: 266–291.

Vilvanathan, S. and Shanthakumar, S. (2017). Continuous biosorption of nickel from aqueous solution using Chrysanthemum indicum derived biochar in a fixed-bed column. *Water Science and Technology: A Journal of the International Association on Water Pollution Research* 76 (7–8): 1895–1906.

Yunnen, C., Ye, W., Chen, L. et al. (2017). Continuous fixed-bed column study and adsorption modeling: Removal of arsenate and arsenite in aqueous solution by organic modified spent grains. *Polish Journal of Environmental Studies* 26 (4): 1847–1854.

Zhang, L., Chen, Y., and Dai, S. (2011). Continuous packed bed biosorption of reactive red on pretreated Penicilium sp. In: *2011 International Conference on Consumer Electronics, Communications and Networks (CECNet)*, 2070–2073.

Zhong, Q.-Q., Xu, X., Shen, L. et al. (2018). Breakthrough curve analysis of enteromorpha prolifera packed fixed-bed column for the biosorption. *Environmental Engineering Science* 35 (8): 864–874.

9

Applications of Electrospun Membranes Immobilized with Biosorbents for the Removal of Contaminants

Noel Jacob Kaleekkal[1], Maheswari Purushothaman[2], and G Nandu[1]

[1] *Membrane Separation Group, Department of Chemical Engineering, National Institute of Technology Calicut, Kozhikode, Kerala, India*
[2] *Department of Chemistry, SRM Valliammai Engineering College, SRM Nagar, Kattankulathur, Tamilnadu, India*

Introduction

Water stress is a severe problem affecting more than 40% of the world's population. The population explosion and rampant industrialization, coupled with climate change, have contributed to this water stress. Conserving and augmenting existing water resources is the major challenge that must be addressed to overcome the predicted 40% water shortage (by 2030) (United Nations, 2021). Water security is maintained by water resource management, including planning, developing, and managing available resources (Khulbe and Matsuura, 2018). The program's biggest challenge is to secure adequate resource provision and mitigate the contaminants present in water sources. Various toxic pollutants such as heavy metals (Ersahin et al., 2012), hydrocarbons (Tan et al., 2015; Tempelman et al., 2019), pesticides (Mukherjee et al., 2020), drugs (Rivera-Utrilla et al., 2013; Marjani et al., 2020; Nakhjiri and Heydarinasab, 2019), dyes (Gugliuzza and Drioli, 2013), organic materials (Marjani et al., 2020), parasites (Naserirad et al., 2016), and microorganisms (Nasirabadi et al., 2016; Ghadiri et al., 2013) are the primary sources of contaminants in water from industrial activities.

Toxic heavy metals such as chromium (Cr), mercury (Hg), arsenic (As), cadmium (Cd), lead (Pb), zinc (Zn), and copper (Cu) in the water stream pose a severe threat to human health and agriculture as they are not biodegradable and can accumulate in the food chain (Bobade and Eshtiaghi, 2015; Al-Rashdi et al., 2011). Heavy metal accumulation in fish, oysters, mussels, sediments, and other components of aquatic ecosystems have been reported worldwide (Amin et al., 2014; Falconer and Humpage, 2005). The UN reports that thousands of children die every day due to the consumption of contaminated water and related diseases. Various water treatment technologies such as physicochemical methods, ozonation, advanced oxidation, ion exchange, chemical precipitation, and electrochemical technologies are widely explored for water recovery/treatment (Chand and Agarwal, 1994;

Biosorption for Wastewater Contaminants, First Edition.
Edited by Rangabhashiyam Selvasembian and Pardeep Singh.

Kadirvelu et al., 2003; Kurniawan et al., 2006; Pal, 2018). These techniques' limitations include high operating costs, incomplete pollutant removal, high energy/pressure requirements, and the creation of secondary pollutants. Biosorption is a promising alternative method for removing dyes and other toxic organic compounds from various industrial effluents (Das et al., 2008; Chojnacka, 2010).

Biosorption and Nanofibers

Biosorption is an eco-friendly process that uses organic/organic-derived materials as adsorbents. It is a simple physicochemical process that involves binding contaminants onto the functional groups present on the surface of the biosorbent. Natural materials such as microorganisms, agricultural waste, industrial byproducts, and biopolymers are potential biosorbents (Annadurai et al., 2002; Vijayaraghavan et al., 2013; Jan et al., 2015; Ang et al., 2018) studied. Industrial waste is also explored for its effectiveness as a biosorbent due to its availability and low cost. For example, agricultural waste contains a high percentage of cellulose and lignin that possess polar functional groups such as amino, carbonyl, alcohol, phenol, and ether groups, with a high binding potential to metal ions. The availability of free electrons is the main reason to form a complex with heavy metals in wastewater treatment. Due to its unique chemical composition (presence of hemicellulose, lipids, lignin, monosaccharides, and starch with different groups) and their availability, agricultural waste seems to be a viable option for heavy metal removal (Reddy and Ramana, 2011; Chao and Chang, 2012; Gaur et al., 2014; Fosso-Kankeu and Mulaba-Bafubiandi, 2014). Wheat bran (Kaya et al., 2014), *Oryza sativa* husk (Renu et al., 2017), rose biomass (Afroze and Sen, 2018), and sugarcane bagasse (Canettieri et al., 2018) have been investigated for their potential as biosorbents to remove heavy metals.

Bioaccumulation is an active process in which the contaminants intracellularly accumulate on the biosorbent (microorganisms). It takes place in two ways: (i) adsorption of contaminants onto cells (fast reaction), and (ii) transportation of contaminants inside cells by active transport (slow reaction). It is an irreversible, complex process that depends on the metabolism of the cell (Chojnacka, 2010). Biosorption is advantageous and simple process, requires no additional chemicals, is cost-effective, forms very little sludge, is high efficiency, has excellent regeneration ability, and causes no increase in chemical oxygen demand (COD).

Electrospinning has emerged as an attractive technique to produce polymer fibers by melting or solution spinning; the fibers range from a few hundred nanometers to micron size. These fibers have received significant attention for various applications: energy storage, catalyst, biosensors, protective clothing, tissue engineering, etc. (Wang et al., 2002; Lu et al., 2014; Borbás et al., 2016; Thakkar and Misra, 2017). Electrospun nanofibrous membranes possess a large specific surface area, ultra-high porosity, and good flexibility and can be easily modified to incorporate biosorbents.

Therefore, this chapter summarizes the currently available approaches for fabricating biosorption immobilized electrospun polymer fibrous membranes and discusses their application in wastewater treatment.

Electrospinning

Electrospinning is a facile technique used to manufacture nanofibers that can be formed into a non-woven mat. More than 100 polymers (synthetic and natural) have been utilized to prepare electrospun mats either by melting or spinning (Haider et al., 2018). This process has numerous advantages over other fiber manufacturing processes, including cost-effectiveness, simplicity, efficiency, and ease of control of the fiber structure and diameter (nm to μm). Typically, the electrospinning setup consists of four major components: (i) charged electrodes connected to a DC power supply, (ii) a syringe containing the polymer solution attached to a metallic needle, (iii) a high-precision syringe pump, and (iv) a grounded collector (drum, plate, or other morphology).

The polymer solution is moved to the needle's tip using the pump (desired flow rate), and a high voltage is introduced into the polymer solution. This leads to an instability of charges on the formed droplet, and the droplet gets an induced charge (Suja et al., 2017). The repulsion between the charges and the affinity to the grounded collector aid in overcoming the inherent surface tension, leading to the formation of a Taylor cone. A further increase in the voltage difference elongates the liquid jet, forming fibers that collect on the grounded collector. During this process, due to the jet's charge difference, it undergoes bending instability, causing a whip-like motion before reaching the collector. This motion leads to further entanglement and the production of thinner fibers due to the evaporation of the solvent. The dry electrospun fibers are collected on the support and can be further treated before use. Figure 9.1 shows a schematic diagram of the electrospinning process.

Membranes incorporating biosorbents such as chitosan and phosphorylated cellulose (Brandes et al., 2019), citric acid included cellulose (Zhang et al., 2020), and polycaprolactone (PCL) encapsulated fungal melamine (Tran-Ly et al., 2020) have been successfully prepared by the electrospinning approach. This technique is cost-effective and can be scaled up to produce fibers with the desired morphology and porosity and a larger surface area. These nanofibrous mats are excellent substrates for immobilizing microorganisms.

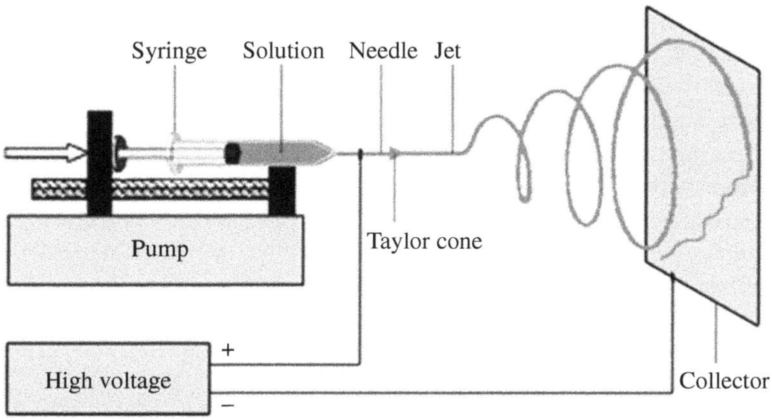

Figure 9.1 Schematic diagram of the electrospinning setup. *Source:* Ziabari et al., 2008.

They are widely studied for removing dyes and other metals for water remediation (Zamel et al., 2019).

The electrospun membranes must be optimized in terms of fiber diameter, membrane porosity, free volume, etc., to suit the incorporation of the biosorbents. The electrospinning technique is versatile for producing solid polymeric nanofibers and introducing desired secondary structures to the nanofibers – porous, hollow, or core-sheath morphology. These nanofibers can be modified to impart a surface charge or functional groups during or after the electrospinning process (Xue et al., 2017).

Factors Influencing Electrospun Fibers

Cellulose acetate (CA), a derivative of naturally available cellulose, is widely considered for developing electrospun nanofibrous membranes as it is biodegradable, biocompatible, and practically insoluble in water and has excellent mechanical/chemical stability and low cost (Khoshnevisan et al., 2018). CA with molecular weight ~100,000 Da forms uninterrupted uniform fibers when the polymer composition is between 9 and 15 wt.%. Solvents commonly used to dissolve CA are dimethylformamide and its mixture with acetone. At lower concentrations, the formation of the continuous jet was interrupted, whereas higher concentrations introduced beads due to the higher viscosity and entanglement of the polymer chains. At optimum concentrations, the tip-to-collector distance and applied voltage did not influence fiber production (Angel et al., 2020).

Chitosan is a polysaccharide produced by the thermochemical deacetylation of chitin and is non-toxic, biocompatible, and biodegradable. All of these properties make chitosan a suitable candidate for preparing nanofibrous biosorbents. Amino acid residues impart a positive charge to the polyelectrolyte in an acidic pH, improving its solubility in aqueous solutions. Its viscous nature and the 3D hydrogen network restrict the elongation of the polymer jet. The voltage applied is essential to overcome the solution surface tension, and an increase in voltage can lead to thinner fibers being produced. A low feed flow rate and minimum tip-to-collector distance are preferred for complete evaporation and solidifying the fibers before they are collected. The molecular weight of the chitosan influences solution viscosity, surface tension, and dielectric strength (Kalantari et al., 2019).

Alginate is a marine algae-derived anionic polysaccharide that is inert and biocompatible (due to the lack of bioactive groups). Pure alginate cannot be formed into electrospun nanofibrous mats using aqueous or organic solvents due to poor electrical conductivity, poor entanglement, and high surface tension (due to intramolecular H-bonding). The use of surfactants or suitable carrier polymers (polyethylene oxide or polyvinyl alcohol) is necessary to ensure the spinning of the alginate polymer. Chemical modification of the alginate structure is also an approach to improve polymer spinnability (Taemeh et al., 2020).

The solution parameters and process parameters that affect the fibers' dimension, morphology, and nature are concisely tabulated in Table 9.1.

Advantage of Electrospinning

Incorporating bacteria or other microorganisms into the electrospun membrane matrix can shield them from the harsh environment and other oxidative stresses they

Table 9.1 Factors affecting the morphology and structure of electrospun nanofibrous membranes.

Electrospinning parameter	Effect on morphology and structure of the membranes/fibers
Solution parameters	
Viscosity (polymer concentration)	Optimal range. Low viscosity leads to electrospraying, and too-high viscosity leads to the drying of solution at the tip (non-uniform fibers with beads). Within an optimal range, an increase in viscosity/concentration increases the fiber diameter.
Molecular weight	Higher-molecular-weight polymers can form microribbons even at low concentrations. Promotes stretching and entanglement of the fibers. Bead-free at a higher molecular weight.
Solution conductivity	Increasing conductivity aids in reducing fiber diameter and may produce a broader distribution of fibers. Ionic salts are added to improve solution conductivity. Altering pH can change solution conductivity
Surface tension	No relationship can be predicted. Generally, lower surface tension is preferred as it decreases the resistance of the solution. Use of surfactants or change of solvent is preferred.
Solvent volatility	Greater solvent volatility produces pores and microvoids on the fibers, further increasing the surface area.
Process parameters	
Applied voltage	The effect of voltage difference is ambiguous. Greater applied voltage is shown to yield thinner fibers. The range of operation has been reported between 10 and 35 kV in most cases. Higher ΔV can lead to bead formation.
Flow rate	High flow rates are unsuitable as they cause bead formation. Lower flow rates yield narrower fibers.
Needle diameter	Particles can clog the outlet if the diameter is small. Small diameter promotes slower initiation of the charged jet (due to higher surface tension forces), leading to thinner fibers.
Distance between needle tip and collector	The reported distance is between 13 and 30 cm. At an optimum distance, the collected fiber is completely dried. At distances that are too short or too far, beads frequently form.
Collector	Metal flat-plate or rotating drums are widely used. Rotating drums can help to produced aligned fibers.
Environmental parameters	
Temperature	A high temperature favors the formation of thin fibers due to the reduction in viscosity: 23–32°C.
Humidity	High relative humidity produces pores on fibers due to microphase separation. Low humidity produces smooth fibers. Relative humidity is maintained between 25 and 50%.
Pressure	Low pressure (below atmospheric pressure) leads to polymer solution oozing from the needle, leading to instability during fiber formation. Rapid discharge of electrical charges at very low pressure prevents electrospinning from initiating.

are subjected to during biosorption. Moreover, microorganisms can naturally form biofilms on hydrophobic substrates to enhance cell viability, adhesion, and biochemical activity (Sarioglu et al., 2016). Nano-sized biomass-derived materials such as polysaccharides, proteins, and organic acids have exhibited great promise as cost-effective green biosorbents to remove heavy metals, organic contaminants, etc. However, in powder form, these biosorbents pose a threat as secondary pollutants, and the energy involved in their recovery is enormous. Electrospun membranes address this issue as the biosorbents are immobilized without much loss in their adsorption capacity (Mahmoodi et al., 2020).

Zeolite Imidazole Framework-8 (ZIF-8) incorporated into chitosan (CS) / polyvinyl alcohol (PVA) electrospun fiber can adsorb dyes utilizing the electrostatic interaction. These membranes have increased biocompatibility, reduced secondary pollutants, and augmented capture efficiency due to the large surface area. MWCNT-laccase introduced into electrospun membranes was far to free laccase. These membranes had excellent stability and enhanced storage and could withstand harsher operating conditions (Dai et al., 2016). Thiol-containing cellulose nanofibers were anchored onto the PAN support by thermal cross-linking to prepare a robust biosorbent that had increased adsorption capacity and greater efficiency and could be regenerated in EDTA solution (Yang et al., 2014).

Electrospun Biosorbent Membranes

Chitin is a naturally abundant polysaccharide material derived from the shells and exoskeletons of crustaceans. It cannot be processed and is generally deacetylated to form chitosan, which contains many amino acid residues on its polymer backbone that can adsorb toxic metals from wastewater (Kai et al., 2014).

Naturally derived chitosan has poor solubility in conventional polar solvents, and the solution is generally highly viscous. The fibers produced are rigid and possess poor mechanical strength. Various polymers such as polyethylene oxide (PEO), polyvinyl alcohol (PVA), cellulose, polycaprolactone, and nylon-6 have been incorporated into the dope solution to improve the general processibility of the produced nanofibers (Aliabadi et al., 2014). The –OH and $-NH_2$ groups are responsible for the electrostatic attraction or ion exchange or chelation mechanisms that permit the uptake of heavy metal ions (Esmaeili and Aghababai Beni, 2018).

Cellulose is another abundantly available, eco-friendly biopolymer that can be converted to cellulose acetate, which can be employed to fabricate electrospun membranes. These membranes possess a large number of –OH groups that can act as the sites for adsorption of oppositely charged moieties, making them a suitable biosorbent (Zhuang et al., 2021).

Gums are naturally occurring hydrocolloids that are complex heterogeneous polysaccharides linked by glycosidic bonding. Gums such as gum acacia, gum tragacanth, gum karaya, and kondagogu gum have been produced as electrospun membranes and used to remove dyes, heavy metals, and trace metals from dilute solutions (Padil et al., 2015; Padil et al., 2019).

Immobilized Membranes for Heavy Metal Removal

Tannin-immobilized gelatin/PVA nanofiber was used to recover uranium (VI) ions from simulated seawater. The maximum adsorption capacity was 170 mg/g and 1.4 μg/g at the optimal pH of 5.5 from an initial solution containing 80 mg/L and 3 μg/L, respectively. The fibers possessed a kelp-like structure with good floatability and avoided agglomeration in test conditions. The SEM images with bayberry tannic and uranium adsorbed fibers are shown in Figure 9.2 (Meng et al., 2019).

Porous ultrafine fibers (d = 0.94 ± 0.36 μm) possessing a ribbon-like structure exhibited a greater removal (88%) of copper (II) compared to non-porous ultrafine fibers (70.8%). The increased specific surface area of 56.57 m^2/g and the negative ΔG° make the adsorption process feasible. The positive value of ΔS° indicated that randomness increased at the solid/solution interface, with some structural changes in both the adsorbates and adsorbents during the adsorption process (Liu et al., 2015). Arsenic (V) is highly toxic and can be adsorbed onto porous, chitosan-based electrospun nanofibers. The maximum adsorption was at pH 4.5, which produced more protonated sites that could interact with negatively charged arsenic (V). The ionic strength effect study indicated the adsorbed ions formed an outer-sphere surface complex, and the adsorption efficiency decreased in the presence of

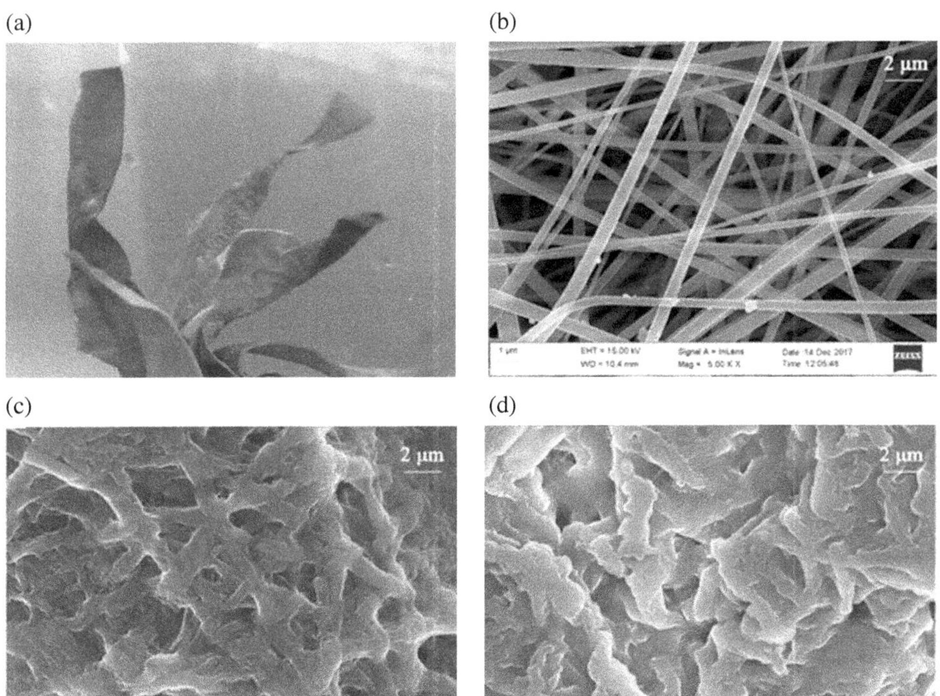

Figure 9.2 (a) Digital photo of gelatin/PVA composite nanofiber band loaded with bayberry tannin (GPNB-BT); (b) scanning electron microscope images of GPNB; (c, d) GPNB-BT before and after adsorption of uranium. *Source:* Meng 2019, p. 9-17 / with permission of Elsevier.

competing anions (SO_4^{2-}, PO_4^{3-}) (Min et al., 2015). Adding iron to the chitosan electrospun nanofibrous membranes improved the adsorption capacity for As (V) even at neutral pH, and the adsorption followed pseudo-second-order kinetics. The maximum adsorption capacity was reported to be 11.2 mg/g (pH 7.2) within 100 minutes, and the membrane could be easily regenerated with a mild alkaline wash. Interestingly, these membranes retained more than 98% efficiency even after 10 consecutive cycles (Min et al., 2016).

Morganella morganii STB5, hexavalent chromium reducing bacteria, was immobilized on electrospun polystyrene and polysulfone to remediate chromium-containing wastewater. The more hydrophobic PS membrane exhibited better bacterial attachment and biofilm growth, leading to better performance (72 h). Both membranes could be stored for a sufficient duration and repeatedly reused to remove chromium (Sarioglu et al., 2016). Fungal melanin is extracted from *Armillaria cepistipes* (Empa 655) and acts as a promising biosorbent for removing toxic heavy metals. The PCL/melamine or PU/melamine fibers were distinctly more porous, as evidenced from SEM images, and the polymers did not significantly influence the adsorption of heavy metals onto the active melamine sites. The removal was in the order of $Pb^{2+} > Cr^{3+} > Ni^{2+} > Cd^{2+} > Zn^{2+} > Ca^{2+}$ (very low adsorption of calcium ions) (Tran-Ly et al., 2020).

Citric acid-modified sustainable, low-cost cellulose nanofibrous membranes were suitable for removing Cr (VI) from an aqueous solution. The adsorption was favored at low pH (2) due to the formation of a more significant number of protonated sites. However, as the pH increased to 6.5, the membranes were negatively charged, and electrostatic repulsion (CrO_4^{2-}) hindered the adsorption (Zhang et al., 2020). Figure 9.3 shows the general mechanisms for the adsorption of heavy metals on the surface of immobilized biosorbents (Kumar and Hansda, 2015; Tsezos et al., 2006).

Lysinibacillus sphaericus CBAM5 immobilized PCL microfibrous mats were used to concentrate gold from electronic processing wastewater. These immobilized bacteria adsorbed and accumulated more than 93% of the gold within two hours (Páez-Vélez et al., 2020). A layer-by-layer approach was used to immobilize a cheap fermentation byproduct, *Saccharomyces Cerevisiae* (by electrospraying), onto poly (ε-caprolactone)/chitosan/rectorite electrospun nanofibrous membrane. The Langmuir isotherm model explained the Pb(II) biosorption onto the sandwiched membrane (Figure 9.4). A high biosorption capacity of 238 mg/g was reported even after three adsorption-desorption cycles (Xin et al., 2017).

The adsorptive membrane prepared by PVA and polyacrylic acid fabricated by electrospinning exhibited an adsorption capacity of 288 mg-PB(II)/g-membrane for an initial concentration of 1 mg/L. The nanofiber membrane exhibited removal efficiency due to the combined effect of adsorption and filtration, especially in treating tap water containing Pb(II) at near-neutral pHs. Extended X-ray absorption fine structure (EXAFS) spectroscopy determined the Pb-Pb shell in the membranes' structure after filtration of tap water at pH 7, making this technology commercially attractive (Zhang et al., 2019). The cellulose was obtained from agave bagasse biowaste derived from tequila production, and polycaprolactone was used to prepare the electrospun nanofibrous membranes. Filtration with the PCL/CNF membranes showed 100% turbidity removal, 100% conductivity, and heavy metal removal on the order of 75–99% for iron and chromium (Hinestroza et al., 2020).

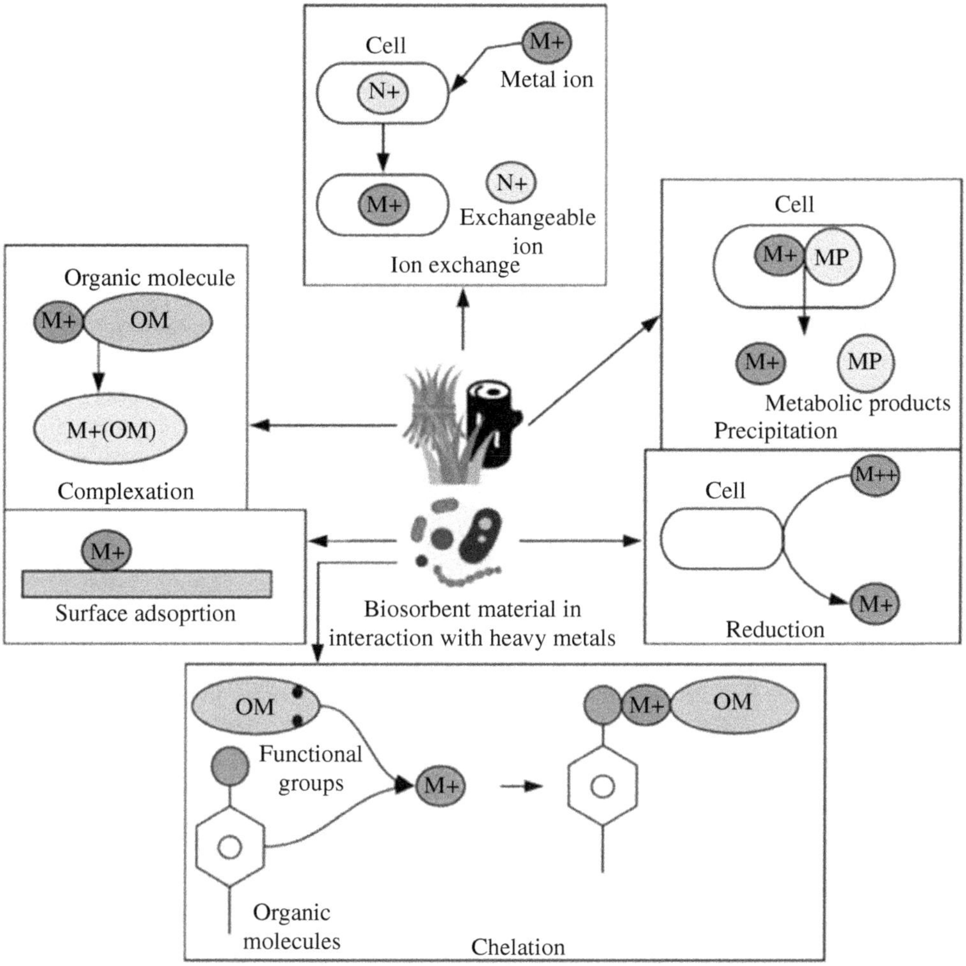

Figure 9.3 General mechanisms of heavy metal removal using biosorbents. *Source:* Calderón et al., 2020.

Figure 9.4 Schematic diagram illustrating the fabrication of sandwich-structured *S. cerevisiae* loaded nanofibrous mats by alternately electrospinning and electrospraying. *Source:* Xin 2017, p. 365-372 / with permission of Elsevier.

Immobilized Membranes for Dye Removal

Dyes are colored organic pigments used in industries such as textiles, cosmetics, printing, leather, etc., to impart color to various products. Of the 1.6 million ton of dyes used worldwide annually, more than 10–15% are discarded in wastewater. The presence of these dyes increases the COD, colors the water, and deteriorates the water quality. Therefore, dye removal is a vital challenge (Varghese et al., 2019).

Fibers derived from agro-waste (lemongrass leaf) were combined with cellulose acetate to prepare electrospun fibers to remove crystal violet dye. The electrostatic interactions between the positively charged quaternary ammonium and hydroxyl groups are responsible for strong adsorption. H-bonding and ion-dipole forces also contribute to binding stability. An adsorption capacity of 33.47 mg/g was reported at a pH of 5 over an equilibrium time of 4 hours (initial dye concentration = 60mg/L) (Putri et al., 2020). *Clavibacter michiganensis* (bacteria) immobilized on PCL or polylactic acid (PLA) nanofibrous polymeric webs was explored for the removal of reactive dye Setazol blue BRF-X. These immobilized-bacteria webs prepared by Sarioglu et al. are potentially reusable, and the kinetic plots revealed biological remediation rather than simple adsorption. The treatment of contaminated wastewater could be improved by increasing the number of immobilized bacteria (Sarioglu et al., 2017a). *Pseudomonas aeruginosa* encapsulated into a PVA or PEO electrospun matrix retained their viability for months at 4 ^{0}C. These membranes efficiently removed methylene blue dye (initial concentration of 10–15 mg/L)(Sarioglu et al., 2017b).

A pearl necklace-like ZIF-8@chitosan/PVA electrospun nanofiber was synthesized and investigated for its efficiency in removing malachite green (MG) from wastewater. Increasing the cycles of 2-methylimidazole coating produced a distinct rhombic dodecahedral morphology on the fibers (Figure 9.5). These membranes displayed excellent chemical stability and impressive adsorption capacity (1000 mg/g), and the adsorption kinetics fit the pseudo-second-order model (Mahmoodi et al., 2020).

Polysulfone nanofibrous webs were used to immobilize toxin-resistant *Lysinibacillus* sp. NOSK isolated from soil. Bioremoval (live immobilized-bacteria membranes) and biosorption (dead immobilized-bacteria membranes) efficiencies were investigated for the simultaneous removal of reactive black 5 (RB5) and Cr(VI). These nanofibrous mats had several advantages, such as smaller growth space, ease of operation, and reusability. Reusability test results indicate that the immobilized mats can be reused for at least seven cycles with 28.1 and 66.7% removal efficiencies for RB5 and Cr(VI), respectively (Keskin et al., 2015).

Immobilized Membranes for Removal of Organic Contaminants

Laccase is a multi-copper polyphenol oxidase that can oxidize a broad range of aromatic organic compounds via the single-electron oxidation mechanism. Laccase and poly(D,L-lactide-*co*-glycolide) (PDLGA) were electrospun together to form an immobilized biosorbent that removed toxic bisphenol A from aqueous systems. The addition of multi-walled CNTs to the fibers enhanced the tensile strength and improved the membranes' operating range (more comprehensive temperature and pH). The optimized nanofibers' activity reached 85% of free laccase, demonstrating its excellent practical potential for treating bisphenol A (BPA) wastewater (Dai et al., 2016).

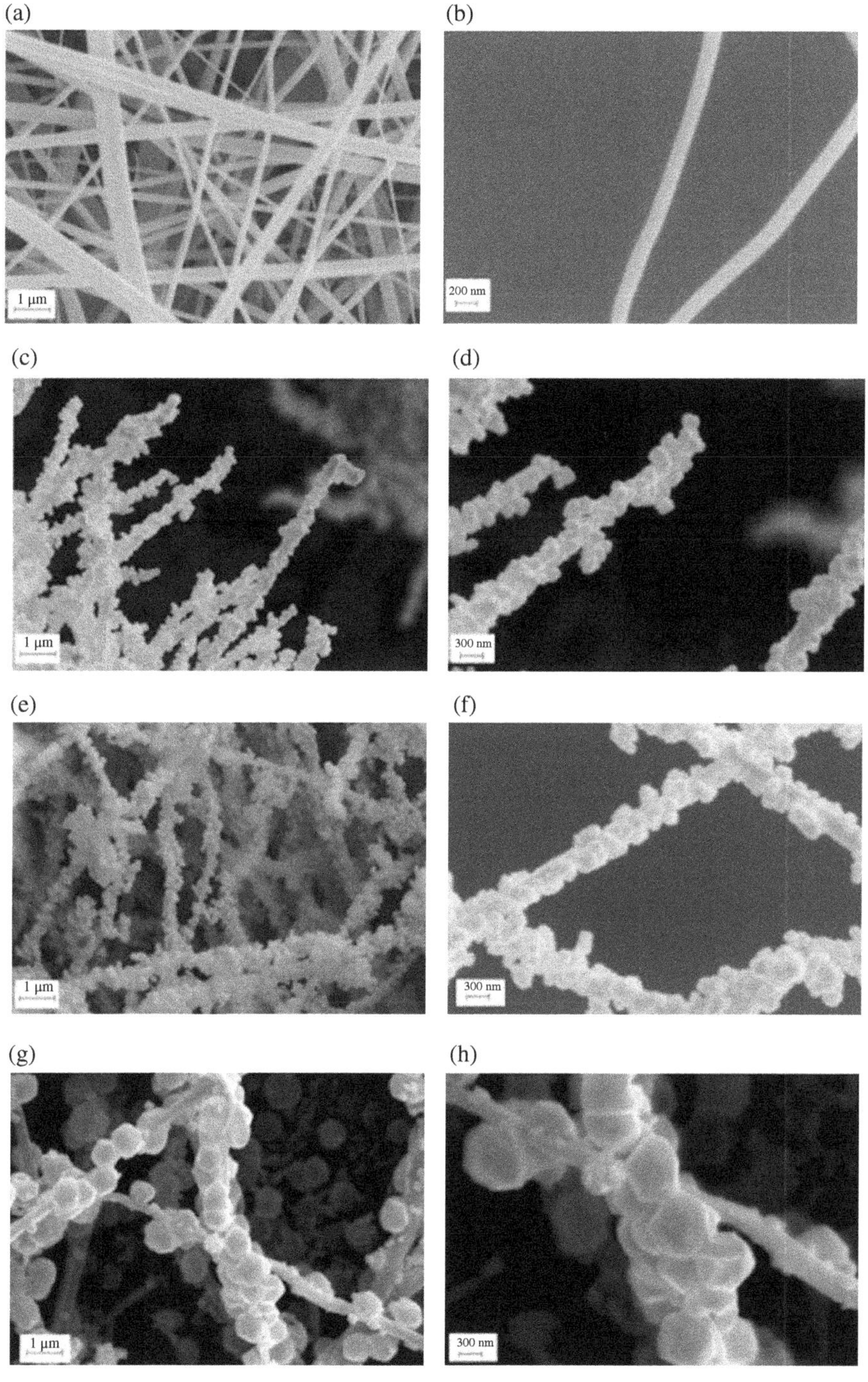

Figure 9.5 (a, b) Scanning electron microscope images of synthesized nanomaterials; (c, d) chitosan (CS) and polyvinyl alcohol (PVA)-electrospun nanofiber (ENF); (e, f) ZIF-8@CS and PVA-ENF (one coating of ZIF-8); (g, h) ZIF-8@CS and PVA-ENF (two coatings of ZIF-8. *Source:* Mahmoodi 2020, p. 227 / with permission of Elsevier.

A poly(vinyl alcohol)/alginate/chitosan (PVA/Alg/CS) composite nanofiber immobilized with zinc oxide was prepared via the electrospinning technique by Elkady et al. Multilayer adsorption of phenol onto the fibers was indicated from the experimental studies. The adsorption fit well with pseudo-order and intra-particle diffusion kinetic models. The thermodynamic studies showed an increase in randomness at the solid interface during the adsorption process indicated by the positive $\Delta S°$ value (Elkady et al., 2020).

A new approach to treat oil spills used hydrocarbon-degrading gammaproteobacteria (marine) and actinobacteria (soil) immobilized on a biodegradable electrospun support (PLA and polycaprolactone). These electrospun biosorbents removed 100% of oil and were degraded (66%) within 10 days. The biofilm-associated cells were more resilient and withstood lower temperatures (Catania et al., 2020).

Conclusion

Electrospinning nanofibrous mats and membranes is emerging as a popular technology in a large number of applications due to the membranes' exceptionally high surface area, > 80% porosity, greater availability of specific sites for adsorption, ability to incorporate a host of biosorbents, and ability to retain the viability and activity of microbial cells/derivatives. However, specific challenges still need to be addressed before these technologies can be commercialized. (i) Scaling up the production of nanofiber membrane from lab scale (a few cm^2) to a large scale (hundreds of m^2) is a significant hurdle. The current needle-based electrospinning techniques are time-consuming and low yield. (ii) Adding biosorbents can interrupt the fiber uniformity (diameter); hence the influence of individual biosorbents derived from various sources must be investigated. (iii) Low mechanical strength and preparing site-specific adsorption are other significant challenges. Low mechanical strength can limit the regeneration and reuse of the biosorbent immobilized electrospun membranes. Wastewater contains multiple charged species (pollutants), and selectively retaining them on the biosorbent has to be investigated in greater detail.

Overall, the immobilization of biosorbents on electrospun membranes is a facile technique and aids in the biosorbents' recovery and reuse. The future direction of research focuses on improving the number of times electrospun membranes can be reused. Another aspect being considered is the biodegradability of the electrospun membranes and biosorbents. Biosorbent immobilized membranes can be studied in greater detail for removing various emerging contaminants.

References

Afroze, S. and Sen, T.K. (2018). A review on heavy metal ions and dye adsorption from water by agricultural solid waste adsorbents. *Water, Air, & Soil Pollution* 229 (7). doi:10.1007/s11270-018-3869-z.

Aliabadi, M., Irani, M., Ismaeili, J. et al. (2014). Design and evaluation of chitosan/hydroxyapatite composite nanofiber membrane for the removal of heavy metal ions from aqueous solution. *Journal of the Taiwan Institute of Chemical Engineers* 45 (2): 518–526. doi:10.1016/j.jtice.2013.04.016.

Al-Rashdi, B., Somerfield, C., and Hilal, N. (2011). Heavy metals removal using adsorption and nanofiltration techniques. *Separation & Purification Reviews* 40 (3): 209–259. doi:10.1080/15422119.2011.558165.

Amin, M.T., Alazba, A.A., and Manzoor, U. (2014). A review of removal of pollutants from water/wastewater using different types of nanomaterials. *Advances in Materials Science and Engineering* 2014: 1–24. doi:10.1155/2014/825910.

Ang, K.L., Li, D., and Nikoloski, A.N. (2018). The effectiveness of ion exchange resins in separating uranium and thorium from rare earth elements in acidic aqueous sulfate media. Part 2. Chelating resins. *MineralsEengineering* 123: 8–15.

Angel, N., Guo, L., Yan, F. et al. (2020). Effect of processing parameters on the electrospinning of cellulose acetate studied by response surface methodology. *Journal of Agriculture and Food Research* 2: 100015. doi:10.1016/j.jafr.2019.100015.

Annadurai, G., Juang, R., and Lee, D. (2002). Use of cellulose-based wastes for adsorption of dyes from aqueous solutions. *Journal of Hazardous Materials* 92 (3): 263–274. doi:10.1016/s0304-3894(02)00017-1.

Bobade, V. and Eshtiaghi, N. (2015). Heavy metals removal from wastewater by adsorption process: a review. *Asia Pacific Confederation of Chemical Engineering Congress* 2015.

Borbás, E., Sinkó, B., Tsinman, O. et al. (2016). Investigation and mathematical description of the real driving force of passive transport of drug molecules from supersaturated solutions. *Molecular Pharmaceutics* 13 (11): 3816–3826. doi:10.1021/acs.molpharmaceut.6b00613.

Brandes, R., Belosinschi, D., Brouillette, F. et al. (2019). A new electrospun chitosan/phosphorylated nanocellulose biosorbent for the removal of cadmium ions from aqueous solutions. *Journal of Environmental Chemical Engineering* 7 (6): 103477. doi:10.1016/j.jece.2019.103477.

Calderón, O.A.R., Abdeldayem, O.M., Pugazhendhi, A., and Rene, E.R. (2020). Current updates and perspectives of biosorption technology: an alternative for the removal of heavy metals from wastewater. *Current Pollution Reports* 6 (1): 8–27.

Canettieri, E.V., da Silva, V.P., Neto, T.G.S. et al. (2018). Physicochemical and thermal characteristics of sugarcane straw and its cellulignin. *Journal of the Brazilian Society of Mechanical Sciences and Engineering* 40 (9): 416.

Catania, V., Lopresti, F., Cappello, S. et al. (2020). Innovative, ecofriendly biosorbent-biodegrading biofilms for bioremediation of oil-contaminated water. *New Biotechnology* 58: 25–31. doi:10.1016/j.nbt.2020.04.001.

Chand, S. and Agarwal, V.K. (1994). Removal of hexavalent chromium from wastewater by adsorption. *Indian J Environ. Health* 36: 151–8.

Chao, H.-P. and Chang, C.-C. (2012). Adsorption of copper(II), cadmium(II), nickel(II) and lead(II) from aqueous solution using biosorbents. *Adsorption* 18 (5–6): 395–401. doi:10.1007/s10450-012-9418-y.

Chojnacka, K. (2010). Biosorption and bioaccumulation – the prospects for practical applications. *Environment International* 36 (3): 299–307. doi:10.1016/j.envint.2009.12.001.

Dai, Y., Yao, J., Song, Y. et al. (2016). Enhanced performance of immobilized laccase in electrospun fibrous membranes by carbon nanotubes modification and its application for bisphenol A removal from water. *Journal of Hazardous Materials* 317: 485–493. doi:10.1016/j.jhazmat.2016.06.017.

Das, N., Vimala, R., and Karthika, P. (2008). Biosorption of heavy metals – an overview. *Indian Journal of Biotechnology* 7: 159–69.

Elkady, M., Salama, E., Amer, W.A. et al. (2020). Novel eco-friendly electrospun nanomagnetic zinc oxide hybridized PVA/alginate/chitosan nanofibers for enhanced phenol decontamination. *Environmental Science and Pollution Research International* 27 (34): 43077–43092.

Ersahin, M.E., Ozgun, H., Dereli, R.K. et al. (2012). A review on dynamic membrane filtration: Materials, applications and future perspectives. *Bioresource Technology* 122: 196–206. doi:10.1016/j.biortech.2012.03.086.

Esmaeili, A. and Aghababai Beni, A. (2018). Optimization and design of a continuous biosorption process using brown algae and chitosan/PVA nano-fiber membrane for removal of nickel by a new biosorbent. *International Journal of Environmental Science and Technology* 15 (4): 765–778.

Falconer, I. and Humpage, A. (2005). Health risk assessment of cyanobacterial (blue-green algal) toxins in drinking water. *International Journal of Environmental Research and Public Health* 2 (1): 43–50. doi:10.3390/ijerph2005010043.

Fosso-Kankeu, E. and Mulaba-Bafubiandi, A.F. (2014). Review of challenges in the escalation of metal-biosorbing processes for wastewater treatment: Applied and commercialized technologies. *African Journal of Biotechnology* 13 (17): 1756–1771. doi:10.5897/ajb2013.13311.

Gaur, N., Flora, G., Yadav, M. et al. (2014). A review with recent advancements on bioremediation-based abolition of heavy metals. *Environ. Sci.: Processes Impacts.* 16 (2): 180–193. doi:10.1039/c3em00491k.

Ghadiri, M., Fakhri, S., and Shirazian, S. (2013). Modeling and CFD simulation of water desalination using nanoporous membrane contactors. *Industrial & Engineering Chemistry Research* 52 (9): 3490–3498. doi:10.1021/ie400188q.

Gugliuzza, A. and Drioli, E. (2013). A review on membrane engineering for innovation in wearable fabrics and protective textiles. *Journal of Membrane Science* 446: 350–375. doi:10.1016/j.memsci.2013.07.014.

Haider, A., Haider, S., and Kang, I.-K. (2018). A comprehensive review summarizing the effect of electrospinning parameters and potential applications of nanofibers in biomedical and biotechnology. *Arabian Journal of Chemistry* 11 (8): 1165–1188. doi:10.1016/j.arabjc.2015.11.015.

Hinestroza, H.P., Urena-Saborio, H., Zurita, F. et al. (2020). Nanocellulose and polycaprolactone nanospun composite membranes and their potential for the removal of pollutants from water. *Molecules* 25 (3): 683. doi:10.3390/molecules25030683.

Jan, K., Riar, C.S., and Saxena, D.C. (2015). Engineering and functional properties of biodegradable pellets developed from various agro-industrial wastes using extrusion technology. *Journal of Food Science and Technology* 52 (12): 7625–7639. doi:10.1007/s13197-015-1938-5.

Kadirvelu, K., Kavipriya, M., Karthika, C. et al. (2003). Utilization of various agricultural wastes for activated carbon preparation and application for the removal of dyes and metal ions from aqueous solutions. *Bioresource Technology* 87 (1): 129–132. doi:10.1016/s0960-8524(02)00201-8.

Kai, D., Liow, S.S., and Loh, X.J. (2014). Biodegradable polymers for electrospinning: Towards biomedical applications. *Materials Science and Engineering: C* 45: 659–670. doi:10.1016/j.msec.2014.04.051.

Kalantari, K., Afifi, A.M., Jahangirian, H. et al. (2019). Biomedical applications of chitosan electrospun nanofibers as a green polymer – Review. *Carbohydrate Polymers* 207: 588–600. doi:10.1016/j.carbpol.2018.12.011.

Kaya, K., Pehlivan, E., Schmidt, C. et al. (2014). Use of modified wheat bran for the removal of chromium(VI) from aqueous solutions. *Food Chemistry* 158: 112–117. doi:10.1016/j.foodchem.2014.02.107.

Keskin, N.O.S., Celebioglu, A., Sarioglu, O.F. et al. (2015). Removal of a reactive dye and hexavalent chromium by a reusable bacteria attached electrospun nanofibrous web. *RSC Advances* 5 (106): 86867–86874. doi:10.1039/c5ra15601g.

Khoshnevisan, K., Maleki, H., Samadian, H. et al. (2018). Cellulose acetate electrospun nanofibers for drug delivery systems: Applications and recent advances. *Carbohydrate Polymers* 198: 131–141. doi:10.1016/j.carbpol.2018.06.072.

Khulbe, K.C. and Matsuura, T. (2018). Removal of heavy metals and pollutants by membrane adsorption techniques. *Applied Water Science.* 8 (1). doi:10.1007/s13201-018-0661-6.

Kumar, V. and Hansda, A. (2015). Biosorption of copper by bacterial adsorbents: a review. *Research Journal of Environmental Toxicology.* 9 (2): 45–58. doi:10.3923/rjet.2015.45.58.

Kurniawan, T.A., Chan, G.Y.S., Lo, W.-H. et al. (2006). Physico-chemical treatment techniques for wastewater laden with heavy metals. *Chemical Engineering Journal* 118 (1–2): 83–98. doi:10.1016/j.cej.2006.01.015.

Liu, R.-L., Tang, C.-Y., Zhao, J.-Y. et al. (2015). Electrospun membranes of nanoporous structure cellulose acetate and its adsorptive behaviors using Copper(II) as models. *Desalination and Water Treatment* 56 (7): 1768–1775. doi:10.1080/19443994.2014.958108.

Lu, W., Sun, J., and Jiang, X. (2014). Recent advances in electrospinning technology and biomedical applications of electrospun fibers. *Journal of Materials Chemistry B* 2 (17): 2369. doi:10.1039/c3tb21478h.

Mahmoodi, N.M., Oveisi, M., Taghizadeh, A. et al. (2020). Synthesis of pearl necklace-like ZIF-8@chitosan/PVA nanofiber with synergistic effect for recycling aqueous dye removal. *Carbohydrate Polymers* 227: 115364. doi:10.1016/j.carbpol.2019.115364.

Marjani, A., Nakhjiri, A.T., Adimi, M. et al. (2020). Effect of graphene oxide on modifying polyethersulfone membrane performance and its application in wastewater treatment. *Scientific Reports* 10 (1). doi:10.1038/s41598-020-58472-y.

Meng, J., Lin, X., Zhou, J. et al. (2019). Preparation of tannin-immobilized gelatin/PVA nanofiber band for extraction of uranium (VI) from simulated seawater. *Ecotoxicology and Environmental Safety* 170: 9–17.

Min, L.-L., Yuan, Z.-H., Zhong, L.-B. et al. (2015). Preparation of chitosan based electrospun nanofiber membrane and its adsorptive removal of arsenate from aqueous solution. *Chemical Engineering Journal* 267: 132–141. doi:10.1016/j.cej.2014.12.024.

Min, L.-L., Zhong, L.-B., Zheng, Y.-M. et al. (2016). Functionalized chitosan electrospun nanofiber for effective removal of trace arsenate from water. *Scientific reports*, 6: 32480.

Mukherjee, A., Mehta, R., Saha, S. et al. (2020). Removal of multiple pesticide residues from water by low-pressure thin-film composite membrane. *Applied Water Science* 10 (12): 244. doi:10.1007/s13201-020-01315-y.

Nakhjiri, A.T. and Heydarinasab, A. (2019). CFD Analysis of CO2 sequestration applying different absorbents inside the microporous PVDF hollow fiber membrane contactor. *Periodica Polytechnica Chemical Engineering* 64 (1): 135–145. doi:10.3311/ppch.13172.

Naserirad, S., Shirazi, M.M.A., Kargari, A. et al. (2016). Application of membrane separation technology in downstream processing of bacillus thuringiensis biopesticide: a review. *Journal of Membrane Science and Research* 2 (2). doi:10.22079/jmsr.2016.19154.

Nasirabadi, P.S., Saljoughi, E., and Mousavi, S.M. (2016). Membrane processes used for removal of pharmaceuticals, hormones, endocrine disruptors and their metabolites from wastewaters: a review. *Desalination and Water Treatment* 57 (51): 24146–24175. doi:10.1080/19443994.2016.1140081.

Padil, V.V.T., Senan, C., and Černík, M. (2019) "Green" polymeric electrospun fibers based on tree-gum hydrocolloids. In: *Materials for Biomedical Engineering* 127–172. doi:10.1016/b978-0-12-816872-1.00005-4.

Padil, V.V.T., Stuchlík, M., and Černík, M. (2015). Plasma modified nanofibres based on gum kondagogu and their use for collection of nanoparticulate silver, gold and platinum. *Carbohydrate Polymers* 121: 468–476. doi:10.1016/j.carbpol.2014.11.074.

Páez-Vélez, C., Castro-Mayorga, J.L., and Dussán, J. (2020). Effective gold biosorption by electrospun and electrosprayed bio-composites with immobilized Lysinibacillus sphaericus CBAM5. *Nanomaterials* 10 (3): 408. doi:10.3390/nano10030408.

Pal, P. (2018). Treatment and disposal of pharmaceutical wastewater: toward the sustainable strategy. *Separation and Purification Reviews* 47 (3): 179–198.

Putri, K.N.A., Keereerak, A., and Chinpa, W. (2020). Novel cellulose-based biosorbent from lemongrass leaf combined with cellulose acetate for adsorption of crystal violet. *International Journal of Biological Macromolecules* 156: 762–772. doi:10.1016/j.ijbiomac.2020.04.100.

Reddy, D.H.K., Ramana, D.K.V. et al. (2011). Biosorption of Ni(II) from aqueous phase by Moringa oleifera bark, a low cost biosorbent. *Desalination* 268 (1–3): 150–157. doi:10.1016/j.desal.2010.10.011.

Renu, M.A., Singh, K., Upadhyaya, S. et al. (2017). Removal of heavy metals from wastewater using modified agricultural adsorbents. *Materials Today: Proceedings* 4 (9): 10534–10538. doi:10.1016/j.matpr.2017.06.415.

Rivera-Utrilla, J., Sánchez-Polo, M., Ferro-García, M.Á. et al. (2013). Pharmaceuticals as emerging contaminants and their removal from water. *A review. Chemosphere* 93 (7): 1268–1287. doi:10.1016/j.chemosphere.2013.07.059.

Sarioglu, O.F., Celebioglu, A., Tekinay, T. et al. (2016). Bacteria-immobilized electrospun fibrous polymeric webs for hexavalent chromium remediation in water. *International Journal of Environmental Science and Technology* 13 (8): 2057–2066. doi:10.1007/s13762-016-1033-0.

Sarioglu, O.F., Keskin, N.O.S., Celebioglu, A. et al. (2017a). Bacteria immobilized electrospun polycaprolactone and polylactic acid fibrous webs for remediation of textile dyes in water. *Chemosphere* 184: 393–399. doi:10.1016/j.chemosphere.2017.06.020.

Sarioglu, O.F., Keskin, N.O.S., Celebioglu, A. et al. (2017b). Bacteria encapsulated electrospun nanofibrous webs for remediation of methylene blue dye in water. *Colloids and Surfaces B: Biointerfaces* 152: 245–251. doi:10.1016/j.colsurfb.2017.01.034.

Suja, P.S., Reshmi, C.R., Sagitha, P. et al. (2017). Electrospun nanofibrous membranes for water purification. *Polymer Reviews* 57 (3): 467–504. doi:10.1080/15583724.2017.1309664.

Taemeh, M.A., Shiravandi, A., Korayem, M.A. et al. (2020). Fabrication challenges and trends in biomedical applications of alginate electrospun nanofibers. *Carbohydrate Polymers* 228: 115419. doi:10.1016/j.carbpol.2019.115419.

Tan, K.B., Vakili, M., Horri, B.A. et al. (2015). Adsorption of dyes by nanomaterials: Recent developments and adsorption mechanisms. *Separation and Purification Technology* 150: 229–242. doi:10.1016/j.seppur.2015.07.009.

Tempelman, K., Casanova, S., and Benes, N.E. (2019). The effect of hydrocarbon pollution on polysulfone-based membranes in aqueous separations. *Separation and Purification Technology* 224: 348–355. doi:10.1016/j.seppur.2019.05.013.

Thakkar, S. and Misra, M. (2017). Electrospun polymeric nanofibers: New horizons in drug delivery. *European Journal of Pharmaceutical Sciences* 107: 148–167.

Tran-Ly, A.N., Ribera, J., Schwarze, F.W.M. et al. (2020). Fungal melanin-based electrospun membranes for heavy metal detoxification of water. *Sustainable Materials and Technologies* 23: e00146. doi:10.1016/j.susmat.2019.e00146.

Tsezos, M., Remoundaki, E., and Hatzikioseyian, A. (2006). Biosorption-principles and applications for metal immobilization from wastewater streams. EU-Asia Workshop on Clean Production and Nanotechnologies.

United Nations (UN). (2021). UN world water development report 2021. https://www.unwater.org/publications/un-world-water-development-report-2021.

Varghese, A.G., Paul, S.A., and Latha, M.S. (2019). Remediation of heavy metals and dyes from wastewater using cellulose-based adsorbents. *Environmental Chemistry Letters* 17 (2): 867–877.

Vijayaraghavan, J., Sardhar Basha, S.J., and Jegan, J. (2013). A review on efficacious methods to decolorize reactive azo dye. *Journal of Urban and Environmental Engineering*: 30–47. doi:10.4090/juee.2013.v7n1.030047.

Wang, X., Drew, C., Lee, S.-H. et al. (2002). Electrospun nanofibrous membranes for highly sensitive optical sensors. *Nano Letters* 2 (11): 1273–1275. doi:10.1021/nl020216u.

Xin, S., Zeng, Z., Zhou, X. et al. (2017). Recyclable Saccharomyces cerevisiae loaded nanofibrous mats with sandwich structure constructing via bio-electrospraying for heavy metal removal. *Journal of Hazardous Materials* 324: 365–372. doi:10.1016/j.jhazmat.2016.10.070.

Xue, J., Xie, J., Liu, W. et al. (2017). Electrospun nanofibers: new concepts, materials, and applications. *Accounts of Chemical Research* 50 (8): 1976–1987. doi:10.1021/acs.accounts.7b00218.

Yang, R., Aubrecht, K.B., Ma, H. et al. (2014). Thiol-modified cellulose nanofibrous composite membranes for chromium (VI) and lead (II) adsorption. *Polymer* 55 (5): 1167–1176. doi:10.1016/j.polymer.2014.01.043.

Zamel, D., Hassanin, A.H., Ellethy, R. et al. (2019). Novel bacteria-immobilized cellulose acetate/poly(ethylene oxide) nanofibrous membrane for wastewater treatment. *Scientific reports* 9 (1): 18994.

Zhang, D., Xu, W., Cai, J. et al. (2020). Citric acid-incorporated cellulose nanofibrous mats as food materials-based biosorbent for removal of hexavalent chromium from aqueous solutions. *International Journal of Biological Macromolecules* 149: 459–466. doi:10.1016/j.ijbiomac.2020.01.199.

Zhang, S., Shi, Q., Christodoulatos, C. et al. (2019). Adsorptive filtration of lead by electrospun PVA/PAA nanofiber membranes in a fixed-bed column. *Chemical Engineering Journal* 370: 1262–1273. doi:10.1016/j.cej.2019.03.294.

Zhuang, S., Zhu, K., and Wang, J. (2021). Fibrous chitosan/cellulose composite as an efficient adsorbent for Co(II) removal. *Journal of Cleaner Production* 285: 124911. doi:10.1016/j.jclepro.2020.124911.

Ziabari, M., Mottaghitalab, V., and Haghi, A.K. (2008). Evaluation of electrospun nanofiber pore structure parameters. *Korean Journal of Chemical Engineering* 25 (4): 923–932. doi:10.1007/s11814-008-0151-x.

10

Biosorption of Precious Metals from Wastewater

Amit Kumar Tiwari[1], Jay Mant Jha[2], and Dan Bahadur Pal[1]

[1] *Department of Chemical Engineering, Birla Institute of Technology, Mesra, Ranchi, Jharkhand, India*
[2] *Department of Chemical Engineering, Maulana Azad National Institute of Technology, Bhopal, Madhya Pradesh, India*

Introduction

Palladium, silver, gold, platinum, etc., are costly metals that are used extensively in various industries as catalysts. Commercial ventures such as the chemical industry, agricultural industry, and pharmaceutical industries commonly utilize these metals due to their special physicochemical competence (Ramesh et al., 2008). These materials also have historical importance because of their economical uses as currency and are good investments as commodities.

According to ISO 4217, valuable materials such as silver (Ag), platinum (Pt), palladium (Pd), gold (Au), etc., are recognized forms of currency worldwide. Due to the limited availability of these precious metals, extracting and removing them from effluents and other industrial residues is essential. As per Jacobsen (2005), various industrial cycling techniques (by application of heat and leaching processes) have been broadly utilized to regain these valued metal components from wastewater.

Due to increased demand for precious metals, especially platinum, gold, and palladium, the extraction and refining of precious metals is increasing worldwide. As we know, conventional techniques are not economically suitable for this purpose; therefore, a technological transformation toward organic extraction of metal is needed for extraction and recovery. These biological techniques are useful for recovering metal from the increased volumes of wastewater produced by such industries. In the most cases, these industrial effluents contain a substantial quantity of valuable metals. Conventional processes (such as extraction (using solvents), chemical precipitation and ion exchange, etc.) to remove low concentrations of dissolved metallic materials from wastewater, have various disadvantages such as fragmentary removal of metals, high cost, the need for huge amounts of reagents and energy, and production of residual toxic precipitates and other unusable waste that requires proper disposal facilities (Goksungur et al., 2005). Wastewaters from industries such as platinum group metal (PGM) refineries have a high concentration of

Biosorption for Wastewater Contaminants, First Edition.
Edited by Rangabhashiyam Selvasembian and Pardeep Singh.

chloride with low pH; PGMs in these wastewaters are found as anionic chloro-complexes with complex chemical nature. Their constitution depends on factors such as concentration of Cl, hydrogen ion concentration and intensity of heat present in aqueous solution, ion strength, and age of the aqueous medium. The formation of metal complexes by PGMs is based on the chemical composition of the aqueous solution. The composition may affect the mechanism of adsorption and the suitability of the metal for sorption sites on the biomass (Chassary et al., 2005).

The most important PGM chloro-complexes depend on the total chloride concentration (Harris 1993). Adsorption using biological materials can be applied to removing metals or metalloids, composite materials, and metallic particles from an aqueous solution (Gadd, 1992). The biosorption mechanisms are based on physicochemical interactions among the metallic ions and the active components on the cell surface, such as ion exchange, electrostatic interaction, and binding (Ozer et al., 2004). The active elements such as hydroxyl, amine, carboxylate, and phosphoryl groups are mostly decoyed in inter-activities within the compounds (polysaccharides, lipids, and proteins) in the cell wall (Dziwulska et al., 2004).

Biosorption is a comparatively new method for removing heavy and costly metals from aqueous solutions, and it has been shown to be suitable for removing metallic pollutants. The primary benefits of biosorption are its usefulness and low cost for reducing heavy metals. This process is more suitable for treatment to dilute heavy metal wastewater. Various types of inexpensive plant waste materials, e.g. potato peels (Amana et al., 2008), wood scobs (Kaczala et al., 2009), pulse husks (Saeed et al., 2005), eggshells (Park et al., 2007), seed shells (Amuda et al., 2009), coffee beans (Oliveira et al., 2008), pectin from beets (Beta vulgaris) (Mata et al., 2009), and citrus peels (Schiewer et al., 2009) have been broadly tested as biosorbents for removing heavy metals. Large amounts of other waste materials (chemical waste) are generated that are suitable agents for the precipitation and reduction of metals.

A great deal of research is studying biosorption of metals using algal species as biomass. These biosorbents are characterized by their best properties (such as low-cost materials with rapid adsorption) and have been applied as good sources. Unfortunately, these experiments and studies are still in the conceptual model and hypothetical or experiential stage. In addition, the segregation of biomass or biosorbents may be tricky after completing the process (adsorption).

Therefore, the development of a suitable, low-priced, eco-friendly process is required to recover precious metallic elements from industrial effluents. Biosorption may be considered a promising technique for recovering these metals because various types of dead microbial biomass have good efficiency to attach to metals in industrial waste (Dobson and Burgess, 2007). Microbial biomasses are comparatively inexpensive and readily available in large amounts. Due to the inherent value of these metals and their harmfulness to human, animal, and aquatic life, finding an appropriate way to extract and remove them from bodies of water bodies and industrial effluents is vital (Gadd et al., 2009). Therefore, different improved procedures have been reported for removing valuable metals from water (Table 10.1).

Of these, some methods are not suitable because they consume large amounts of chemicals, are expensive, and require special care (Spitzer et al., 2004).Some are not selected because of their poor efficiency at low ion concentrations (Xie et al., 2014), and a few

Table 10.1 Improved procedures for removal of costly metals.

Method	References
Extraction	(Bulgariu et al., 2011)
Membrane separation electrolysis	(Kargari et al., 2004)
Electrochemical processes and adsorption	(Soleimani et al., 2008)

methods have problems such as a lack of selectivity and the risk of producing other chemical pollutants during the process. People are working to develop efficient new techniques to solve these problems through research and experiments (Soleimani et al., 2008). Biosorption is a recent development as a subclass of adsorption in which biomaterials and microorganisms are used (Das et al., 2010); resources are readily available, and operational costs are low. Water quality is a basic requirement because humans use water for many purposes; and metallic contamination in water can cause health problems. The maximum permitted limit of silver (Ag) in potable water is 0.1 ppm (WHO, 2011).Various conventional methods (Table 10.2) are used to extract and remove silver from industrial effluent. However, these methods also have many drawbacks, such as lower efficiency and high energy requirements to collect relatively small amounts of metal ions from dilute aqueous solutions (Çoruh et al., 2010). Given this situation, biosorption is an effective, low-cost, eco-friendly technique for handling industrial effluents containing a low concentration of silver.

A number of biomaterials have been found suitable and employed as biosorbents to remove silver (Table 10.3).Algae are considered the most suitable potential biosorbents

Table 10.2 Conventional methods for recovery of precious metals.

Method
Ion exchange
Chemical precipitation
Ultra-filtration
Electrolytic process
Adsorption and membrane processes

Table 10.3 Biomaterials employed as biosorbents.

Biomaterials	References
Yeast (waste)	Zhao et al., 2015
Yeast (fresh)	Chen et al., 2014
Bacterial spp.	Zinicovscaia et al., 2017
Molds and yeast	Das et al., 2010
Algal spp.	Cechinel et al., 2016

because they are renewable, low-cost materials and have a high capacity for biosorption of metal. In the last two decades, a great deal of work has been done to realize gold from effluents, but now researchers are showing increased attention in recovering other costly metals such as silver, platinum, palladium, etc.

Outline of Treatment Methods

Various processes and methods are used to reclaim precious metals such as gold, platinum, and silver (as shown in Table 10.4).In the last few years, considerable attention has been given to research based on biosorption using different biosorbents, because it is a less costly and organic process that has been proven more effective than traditional processes to displace and eliminate metals. It can be commercially used to recover metals such as gold, silver, platinum, etc.

Biosorbents

Biosorbents of Gold

Common uses of gold are in the manufacture of electronic items, jewelry, and drugs for arthritis (Shaw, 1999). Due to the increased demand for gold, there is a growing need to recover it from industrial effluents that contain high concentrations of gold (e.g. effluents from refineries and the mining industry). The recovery of gold has been scrutinized by

Table 10.4 Processes and techniques for the improvement of precious metals

Precious metals	Processes / techniques	References
Gold	Zinc dust cementation C adsorption Electro-deposition Solvent extraction Ion exchange	Wan and Miller, 1990 Xu et al., 1995 Wan and Miller, 1986 Mooiman and Miller, 1991 Gomes et al., 2001
Silver	Precipitation Ion exchange Reductive exchange Electrolytic process	Cotton and Ikinson, 1972
Platinum group	Solvent extraction Ion exchange Membrane process Adsorption	Barnes and Edwards, 1982 Els et al., 2000 Takahiko et al., 1996 Liu et al., 2000 Veglio and Beolchini, 1997

researchers using biosorbents such as fungi (Gomes et al., 1998), algae (Kuyucak and Volesky, 1989), and yeast (Karamuchka and Gadd, 1999). Activated carbon has also been used to absorb gold particles from wastewater.

Gold is required for manufacturing durable electronic devices using advance integrated circuits (AICs). The total demand for gold is increasing continuously in manufacturing AICs (bonding wires and coatings for contacts) and connectors. Globally, around 250 tons/year of gold are required in the electronics industry, which is about 8% of total valuable metals utilization (Barbieri et al., 2010). These metals are required to provide good chemical stability and conduction to electronic devices; therefore, it is the main reason for utilizing these important metals in the electronics manufacturing sector. According to Natarajan and Ting (2014), electronic waste contains concentrations of gold up to 10 times greater (approximately 10–10,000 grams/ton) than waste from gold ore.

In the last few decades, it has been realized that biotechnology could be one of the best ways to recover metal from primary waste and secondary waste resources. Bioleaching helps recover metals by using microbial abilities to transform metals into dissoluble, extractable, and recoverable components (Liu et al., 2008). As microbiological biomass can combine with metal ions outside the plasma membrane of its cells, the biomass can act as a metal sink. These biomasses can also accumulate pigments and extracellular and intracellular polysaccharides on or around cells. Hydroxyls, carboxyl, and amines are the functional groups and can attach to metal ions during biosorption (Choi and Yun, 2006). A suitable combination of microbes and chemicals can make the recovery of metals more productive (Singh and Pant, 2016). Ammoniacal thiosulfate (AT) is a well-recognized leaching agent (chemical) for gold, but as it is a metallo-thiosulphate complex, it is unstable and only exists when Cu is present. It also shows lixiviation with other metals such as Fe and Cu present in waste (Grosse et al., 2003).

Biosorbents of Silver

For many years, the photographic and imaging industries have used silver extensively in the preparation of various photography tools. Silver is also a known environmental pollutant because it is released into the environment during industrial applications, which leads to contamination of the water ecosystem (especially for water organisms) (Pedroso et al., 2007). Most of the silver is recovered globally from wastes such as photographic films, X-ray films, and jewelry industries (Ajiwe and Anyadiegwu, 2000).

Due to the promising properties of silver nanoparticles (AgNPs), silver has been used as a coating material in recent decades. The main properties of AgNPs are that they provide a large surface area so that nanostructures can be applied in fields such as medicine, biotechnology, food technology, electronics, environmental bioremediation, biosensor manufacturing, agriculture, and pharmaceuticals. AgNPs are used extensively to produce electrical and electronics items, healthcare products, food items, and textiles (Calderon-Jimenez et al., 2017), which can be estimated by the numbers of suppliers, quantity supplied worldwide, and by its global utilization. According to Syafiuddin et al. (2017), due to the excellent execution of AgNPs in electronic items and electrical appliances, demand will continue to grow through 2022 in these fields. Fortunately, among the 2000 nanomaterials produced globally, around 20% (435) are from silver-based nano-products; overall manufacturing of nano-products is around 2,90,299–3,81,018 kg/yr (Pulit-Prociak and Banach, 2016). The use of AgNPs is not restricted to industrial products (for commercial uses); they can be used in environmental fields (for bioremediation) (Yadav et al., 2017) and treatment of wastewater (Kumar et al., 2018).

A huge quantity of AgNPs are needed to fulfill the market demand; therefore, they must be synthesized artificially on a commercial level with physicochemical techniques (γ-ray radiation) (Flores-Rojas et al., 2018), chemical reduction (Khan et al., 2017a), electrochemical processes (Nasretdinova et al., 2015), etc. However, these methods produce toxicants that may be hazardous to animals and the environment. These methods are also consumes a great deal of energy and are costly. To resolve these problems, extensive attempts have been made in the last few years to encourage the commercial synthesis of AgNPs by successful and financially viable means. In the last five years, many biological (biogenic) studies have been carried out to solve this issue (Rafique et al., 2017).

Biosorbents of PGMs (Palladium and Platinum)

PGMs such as palladium and platinum are high-priced products that are used extensively in various industries due to their physicochemical behavior. To fully utilize these resources, it is necessary to recover the maximum PGMs effectively from natural ores and industrial waste. Palladium is primarily used in cars and trucks as catalysts for emission control, in electronic parts, as an alloy for dental treatment, and as a catalyst in various chemical processes (Crundwell et al., 2011). Palladium is required in huge quantities to produce end-use products, so palladium recycling is constantly increasing. To save our natural mineral resources and also meet growing consumer demand, we must recover palladium.

Various wet methods (chemical) such as crystallization, hydrolysis, solvent extraction, precipitation, metal reduction, and adsorption are used to extract and purify PGMs from liquids (effluents) (Crundwell et al., 2011). Apart from conservative recycling techniques, which are commonly applied to recycle costly metals, biological and organic techniques can also offer effective and eco-friendly options for recycling these metals from wastes even at ambient temperatures. Microbial species are utilized to extract/remove aqueous Pd ions (II) from weak aqueous solutions through their cells at ambient temperature. Different groups or species of microorganisms are used to shrink Pd (II) ions to hard Pd (0) at neutral pH in anaerobic conditions in various studies; this process is known as bio-reductive deposition or bio-mineralization. For these processes, scientists suggest various types and species of microorganisms, including the following:

- Sulfate-reducing bacteria
  - *Desulfovibrio desulfuricans* (Yong et al., 2002)
  - *D. fructosivorans* (Mikheenko et al., 2008)
- Bacteria for reducing metal ions
  - *Shewanella oneidensis* (Ogi et al., 2011)
  - *S. algae* (Tamaoki et al., 2010)
  - *S. loihica* (Wang et al., 2018)
  - *Geobacter sulfurreducens* (Tuo et al., 2013)
- Facultative anaerobic bacterium (gram-negative)
  - *Escherichia coli* (Courtney et al., 2016)
- Facultative anaerobic bacterium (gram-positive)
  - *Staphylococcus sciuri* (Sobjerg et al., 2011)

Factors Affecting Biosorption

The biosorption process involves microorganisms and the environmental conditions required for the growth and activity of those microorganisms. The physicochemical properties of the raw materials are also responsible for the proper growth and development of microbes. Microbes are sensitive to pH, acidity, temperature, moisture, etc., and these factors are essential for achieving the complete biosorption of any material. The primary factors that can affect biosorption include:

- pH of the mixture
- Operational temperatures
- Dosage of biomass
- Ionic potency
- Initial concentration of solute
- Rate and period of agitation

pH of the mixture

The pH is a vital and variable factor that can influence the formation of new forms of metal in aqueous solution during complexation, redox-reaction, hydrolysis, etc., especially during recovery of metals (Esposito et al., 2002). According to the studies of Fiol et al. (2006), pH can influence the dissociation state of the binding site; it can also change the entire alchemy of the targeted metal, aqueous solution, legands, and redox-potentials.

Gold is commonly found in the anionic form in aqueous solution or industrial effluents. Sorption of gold depends on the aqueous pH; the maximum value of biosorption can be achieved with an immobilized persimmon process using tannins if the pH of aqueous solution is 3.0 (Sakaguchi et al., 1995). As suggested by Pethkar and Paknikar(1998), around 80% recovery of gold can be achieved using fungal species (*Cladosporium cladosporoides*) with acidic conditions between pH 1.0 and 5.0. When an inactivated alfalfa biomass was used for biosorption, the uptake of gold at pH 2.0 was slightly higher (Gardea-Torresdey et al., 1999). The consequences of different pH values on exaltation of gold were also studied associated with new and distinct gold species and fungal biosorbents. It was found that pH 3.9 was suitable for maximum gold uptake based on the above-mentioned observations. This pH influences the binding sites' dissociation and also affects the chemistry of the solution and metals (Fiol et al., 2006).

The pH of the substrate is an important factor and is responsible for the effectiveness of the sorption process; at a certain pH, only a few selected metals can be effectively absorbed. Apart from other factors, the availability of the metal and effects of the metal pH such as precipitation and transformation can affect the biosorption process (Esposito et al., 2002). Successful biosorption of the base metal cations is extremely dependent on the pH and usually requires a pH range of 3.0 to 7.0.

Operational Temperatures

Temperature can affect biosorption and efficiency significantly, and the operational temperature of the aqueous solution or mixture is an important factor in biosorption (Panda et al., 2006).However, researchers produced several models claiming that an increase in temperature reduced biosorption efficiency, which is a clear indication of an exothermic process. In few studies, it is also reported that low temperatures stimulate the sorption of metals on binding sites during the biosorption process.

Aksu et al. (1992) reported that the biosorption process and its efficiency are not affected by temperature in the range between 20 and 35 °C. Similarly, Bhainsa and D'Souza (1999) also stated that many reactions are exothermic or self-governing (Sag and Kutsal, 2000). Ishikawa et al. (2002) observed a considerable effect of temperature on gold exaltation at elevated temperature (65 °C). Enzymatic sorption using immobilized tannin was carried out by Wang et al. (2005), who studied the effect of temperature (30–50 °C) on the sorption process of platinum. The maximum exaltation (Qmax) increased with increased temperature: at 50 °C, the Qmax of Pt (II) was 0.599 mmol/g and Pd (II) was 1.29 mmol/g. A study by Ma et al. (2006), for the sorption of platinum and palladium, noticed that higher adsorption was achieved at higher temperatures.

Dosage of Biomass

Biosorbent dosages have a strong influence on the degree of biosorption. A high dose of biosorbent in biomasses usually increases the amount of solute biosorbent. It is due to the enlarged surface area of the biosorbent, which ameliorates the number of binding sites (Esposito et al., 2001). But the amount of biosorbed solute/unit weight of biosorbent decreases with a greater biosorbent dosage. This may be due to the complex activity of various factors with each other. An insufficient amount of solute may also be an important factor at elevated sorbent dosages, which can completely cover the exchangeable sites on the biosorbent, resulting in low solute exaltation (Tangaromsuk et al., 2002).

Gadd et al. (1988) suggested that interference between binding sites cannot be ruled out due to increased biosorbent dosage because this promotes low specific uptake. The effects of immobilized bayberry tannin (collagen fiber membrane) dosage were also studied by Ma et al. (2006) on the adsorption extent of platinum (IV) and Pd (II). The quantity of membrane affects the adsorption extent of the two metal ions. At 293 K, the adsorption extents of Pt (IV) and Pd (II) by using 0.100 g membrane were 68.4 and 44.2%, respectively, while the results were increased to 97.4 and 95.0%, respectively, with 0.400 g membrane.

Ionic Potency

According to Borrok and Fein (2005), the biosorption process may be influenced by another crucial parameter: ionic strength. Ionic potency affects the capacity of the biomass surface for adsorption of solute. Patterns such as ion competition and changes in metal activity are the best examples of the effect of ionic strength. When two different phases –biomass surface and solute– are combined in an aqueous medium, they are bound to be wrapped by an electrical double layer due to electrostatic interaction. Consequently, an increase in ionic strength results in decreased adsorption (Donmez and Aksu, 2002).

A few inorganic ions (such as Cl) may create specific metal ion complexes and affect the entire sorption (Borowitzka, 1988). According to Godlewska-Zylkiewicz (2003), in environments with high chloride, the existence of a large number of Cl ions creates complications: a strong rivalry arises between the anions of the chloride and the species, which can restrict metal sorption in effluents. In the case of biosorption of Pt and Pd, the same results were observed by Dziwulska et al. (2004).

Initial Concentration of the Solute

The initial concentration of metal ions is an essential factor that affects the biosorption process. It is reported that a higher concentration results in greater solute uptake (Binupriya et al., 2007) due to a low concentration of solute at the initial stage. If the ratio of the initial moles of solute is low compared to the available surface area, more sites are available for sorption. But at a higher concentration, fewer sites are available for sorption compared to the solute concentration; therefore, the removal of solute strongly depends on the initial concentration of solute.

Rate and Period of Agitation

The efficiency of the biosorption process may be affected by the diffusion of external film. Using accurate agitation, the mass-transfer rate and resistance can be regulated. If the agitation speed increases, the dispersion rate of the solute also increases due to greater turbulence and a reduction in the thickness of the boundary layer of the liquid (Evans et al., 2002). In this situation, the value of the exterior dispersion coefficient increases (Shen and Duvnjak, 2005). Finally, if the agitation rate is greater, the thickness of the boundary layer is less, which usually enhances the rate of diffusion of solute through the boundary layer.

Gold recovery efficiency was investigated by Tasdelen et al.(2009) as a function of agitation rate; the study was carried out between 20 to 140 rpm. The study duration was limited to 30 min; during the investigation, it was noticed that agitation of 120–140 rpm was optimal, and a greater shaking rate resulted in an increase in gold recovery efficiency for samples containing varying amounts of gold.

Biosorption Equilibrium Models

Different models of biosorption have been presented by various researchers, which can be used to transfer technology from lab scale to commercial scale. These lab experiments may be useful in understanding the mechanism of a process, analysis of investigational data, imagining process conditions, and process optimization (Limousin et al., 2007).Biosorption of valuable metals from industrial effluents and wastewater is a form of solid-liquid absorption. Primary testing of solid and liquid absorption process is primarily based on (i) equilibrium batch sorption (EBS) and (ii) dynamic continuous flow sorption (DCFS). Gadd et al. (1998) suggested that the equilibrium of the biosorption process can be achieved by fitting the investigation data with models to assess the isotherm adsorption equilibrium.

Models of adsorption isotherms presented by various researchers are as follows:

- Freundlich model (Freundlich, 1906)
- Langmuir model (Langmuir, 1918)
- Sips model (Sips, 1948)

- Toth model (Toth, 1971)
- Radke-Prausnitz model (Radke and Prausnitz, 1972)

These models are used to quantify and study similarities between the performances of various biosorbents. Langmuir isotherms and Freundlich isotherms are the most important and widely used isotherms to study adsorption. A Langmuir isotherm is based on a surface having binding sites, equivalent sorption energies, and no interaction between sorbed species. In comparison, the Freundlich isotherm is based on an empirical equation, exponential distribution of sorption sites, energies, etc.

Desorption and Recovery

The release of an adsorbed substance from a surface is necessary when biomass production is a costly venture. It is necessary to reduce the process and product costs and process dependency; this may be possible by uninterrupted delivery of the biosorbent through desorption. An ideal desorption process depends on the choice of suitable elutants and biosorbents and the mechanism of biosorption. A good elutant should have the following properties:

- Should not be harmful to the biomasses
- Low-cost
- Eco-friendly
- Effective

Some researchers have conducted detailed experiments to identify suitable elutants for the maximum recovery of valuable materials such as platinum, gold, silver, etc. Ichikawa et al. (2002) studied the release of Au (gold) using sodium cyanide and NaOH. A 0.1 MNaOH solution was sufficient to desorb more than 95% of the Au (I) sorbed by eggshells (membrane); on the other hand, a 0.1 M sodium cyanide solution was enough to desorb greater than 92% of the Au (III). It is reported that the ESM column can be applied to recover around 98% of gold. Kuyucak and Volesky (1989) reported that gold was removed from marine algae (*Saragasum natans*) using a combination of 0.1 M thiourea and a 0.02 M solution of ferric ammonium sulfate.

Continuous Biosorption

At present, continuous biosorption studies are gaining importance in evaluating the feasibility and technicality of the biosorption process for real applications. For economical and convenient biosorption processes, packed-bed columns (PBCs) have been certified as the most useful and efficient of the column configurations (Chu, 2004). PBCs have numerous technical attributes, such as high yield and ease of scaling up the biosorption process (Aksu, 2005). Fluidized reactors and continuous stirred-tank reactors (column reactors) are almost never recommended for biosorption (Solisio et al., 2000). If the biosorbent is used in powder form, then continuous reactors such as stirred-tank reactors are helpful in the biosorption process (Cossich et al., 2004)); but this approach involves high fixed and operational costs (Volesky, 1987). Fluidized bed columns, which can work endlessly, require a high flow rate to distribute biosorbent particles symmetrically in the suspension (Muraleedharan et al., 1991).

Utilization of Industrial Discharge/Wastes for Biosorption

Biosorption is a proven process for the extraction and collection of valuable metals from contaminated water. But its performance in actual aqueous environments is a serious problem. Few investigations have been carried out to study the compatibility of biosorbents to collect these high-priced metals from industrial waste. *Cladosporium cladosporioides* biomass beads were tested for their applicability as fungal biosorbents on Au sorption; studies were conducted by Pethkar and Paknikar (1998) using electroplating industrial effluent containing 46 mg gold concentration per gram of waste. Gold uptake by the biosorbent beads was around 55% in batch experiments. The observed gold loading capacity was around 36 mg/gram. The recovery of Ag was reported by Ajiwe and Anyadiegwu (2000); in the study, samples were taken from the X-ray film industry. The solution was prepared by dissolving the Ag mixture with concentrated HNO_3. Soleimani and Kaghazchi (2008) used activated shells of apricot stones to remove Au from electroplating wastewater. The findings suggest that this could be an effective adsorbent due to its physical, chemical, and adsorptive properties.

Conclusions

Due to the availability of plenty of biomaterials and the easy application process for metal recovery, biosorption processes to recover precious metal from industrial effluents have achieved great popularity in recent years. Compared to other conservative methods, biosorption processes offer benefits such as low cost, minimum chemical use, less biological waste, high competence in detoxifying effluents, etc. From research conducted over the past few decades, an improved understanding of valuable metal extraction and removal through biosorption using specific biosorbents has been developed. New, cost-effective, inexpensive biomaterials can be used to develop low-cost biosorption techniques for this purpose, and metal-containing industrial effluents can be utilized.

Even with the remarkable increase in published work, biosorption has very limited application in a commercial context. As biomasses, microorganisms, and biosorbents have been identified to recover metals from industrial wastes, there is a need for their commercial application. By commercializing this cost-effective, eco-friendly process, we can reduce processing costs and save our environment; these may be the primary attractions of the metal biosorption process.

References

Ajiwe, V.I.E. and Anyadiegwu, I.E. (2000). Recovery of silver from industrial wastes, cassava solution effects. *Separation and Purification Technology* 18: 89–92.

Aksu, Z. (2005). Application of biosorption for the removal of organic pollutants: a review. *Process Biochemistry* 40 (3-4): 997–1026. doi:10.1016/j.procbio.2004.04.008.

Aksu, Z. et al. (1992. The biosorption of copper (II) by C. vulgaris and Zramigera. *Environmental Technology* 13: 579–586.

Aman, T., Kazi, A.A., Sabri, M.U. et al. (2008). Potato peels as solid waste for the removal of heavy metal copper(II) from waste water/industrial effluent. *Colloids and Surfaces B: Biointerfaces*, 63 (1): 116–121. doi:10.1016/j.colsurfb.2007.11.013.

Amudaa O.S., F.E. Adelowoa, and M.O. Ologunde. (2009). Kinetics and equilibrium studies of adsorption of chromium (VI) ion from industrial wastewater using Chrysophyllum albidum (Sapotaceae) seed shells. *Colloid Surf* 68: 184–192.

Barbieri, L., Giovanardi, R., Lancellotti, I., and Michelazzi, M. (2010). A new environmentally friendly process for the recovery of gold from electronic waste. *Environ. Chem. Lett.* 8 (2): 171–178.

Barnes, J.E. and Edwards, J.D. (1982). Solvent extraction at Inco's Action precious metal refinery. *Chemistry and Industry* 6, 151.

Bhainsa, K.C. and D' Souza, S.F. (1999). Biosorption of uranium (VI) by Aspergillus fumigatus. *Biotechnology Techniques* 13: 695–699.

Binupriya, A.R., Sathishkumar, M., Kavitha, D. et al. (2007). Experimental and isothermal studies on sorption of Congo red by modified mycelial biomass of wood rotting fungus. *Clean* 35: 143–150.

Borrok, D. and Fein, J.B. (2005). The impact of ionic strength on the adsorption of protons, Pb, Cd and Sr onto the surface of gram negative bacteria; testing non electrostatic, diffuse, and triple layer models. *Journal of Colloid and Interface Science* 286: 110–126.

Bulgariu, L. and Bulgariu, D. (2011). Extraction of gold(III) from chloride media in aqueous polyethylene glycol-based two-phase system, *Sep. Purif. Technol.* 80: 620–625.

Calderon-Jimenez, B., Johnson, M.E., Montoro Bustos, A.R. et al. (2017). Silver nanoparticles: technological advances, societal impacts, and metrological challenges. *Front. Chem.* 5 (6).

Cechinel, M.A.P., Mayer, D.A., Pozdniakova, T.A. et al. (2016). Removal of metal ions from a petrochemical wastewater using brown macro-algae as natural cation exchangers. *Chem. Eng. J.* 286: 1–15

Chen, C., Wen, D., and Wang, J. (2014). Cellular surface characteristics of Saccharomyces cerevisiae before and after Ag(I) biosorption. *Bioresour. Technol.* 156: 380–383.

Choi, S.B. and Yun, Y.S. (2006). Biosorption of cadmium by various types of dried sludge: an equilibrium study and investigation of mechanisms. *J. Hazard. Mater.* 138 (2): 378–383.

Chu, K.H. (2004). Improved fixed bed models for metal biosorption. *Chemical Engineering Journal* 97, 233–239.

Çoruh, S., Senel, G., and Ergun, O.N. (2010). A comparison of the properties of natural clinoptilolites and their ion-exchange capacities for silver removal. *J. Hazard Mater.* 180, 486–492.

Cossich, E.S., da Silva, E.A., Tavares, C.R.G. et al. (2004). Biosorption of Chromium (III) by biomass of seaweed Sargassum species in a fixed bed column. *Adsorption* 10: 129–138.

Cotton, E.A. and Ikinson, G.W. (1972). *Advanced Inorganic Chemistry*, 3e., 721–725. Wiley.

Courtney, J., Deplanche, K., Rees, N.V., and Macaskie, L.E. (2016). Biomanufacture of nano-Pd (0) by Escherichia coli and electrochemical activity of bio-Pd(0) made at the expense of H2 and formate as electron donors. *Biotechnol. Lett.* 38: 1903–1910.

Das, N. (2010). Recovery of precious metals through biosorption-a review. *Hydrometallurgy* 103: 180–189.

Das, D., Das, N., and Mathew, L. (2010). Kinetics, equilibrium and thermodynamic studies on biosorption of Ag(I) from aqueous solution by macrofungus Pleurotus platypus. *J. Hazard Mater.* 184: 765–774.

Dobson, R.S. and Burgess, J.E. (2007). Biological treatment of precious metal refinery wastewater: a review. *Minerals Engineering* 20: 519–532.

Donmez, G. and Aksu, Z. (2002). Removal of Cr(VI) from saline wastewaters by Dunaliella species. *Process Biochemistry* 38: 751–762.

Dziwulska, U., Bajguz, A., and Godlewska-Zylkiewicz, B. (2004). The use of algae Chlorella vulgaris immobilized on Cellex-T support for separation/ preconcentration of trace amounts of platinum and palladium before GFAAS determination. *Analytical Lett* 37: 2189–2203.

Esposito, A., Pagnanelli, F., Lodi, A. et al. (2001). Biosorption of heavy metals by Sphaerotilus natans; an equilibrium study at different pH and biomass concentrations. *Hydrometallurgy* 60: 129–141.

Esposito, A., Pagnanelli, F., and Vegliò, F. (2002). pH-related equilibria models for biosorption in single metal systems. *Chemical Engineering Science* 57 (3): 307–313. doi:10.1016/ S0009-2509(01)00399-2.

Evans, J.R., Davids, W.G., MacRae, J.D., and Amirbahman, A. (2002). Kinetics of cadmium uptake by chitosan-based crab shells. *Water Research* 36: 3219–3226.

Fiol, N., Villaescusa, I., Martinez, M. et al. (2006). Sorption of Pb (II), Ni(II), Cu (II) and Cd (II) from aqueous solution by olive stone waste. *Separation and Purification Technology* 50: 132–140.

Flores-Rojas, G.G., Lopez-Saucedo, F., and Bucio, E. (2020). *Gamma-irradiation applied in the synthesis of metallic and organic nanoparticles: A short review. Radiation Physics and Chemistry*, 169.

Freundlich, H.M.F. (1906). Uber die adsorption in Losungen. *Zeitschrift für Physikalische Chemie* 57: 385–470.

Gadd, G.M. and White, C. (1992). Removal of thorium from simulated acid process streams by fungal biomass - potential for thorium desorption and reuse of biomass and desorbent. *J Chem Technol Biotechnol* 55: 39–44.

Gardea-Torresdey, J.L., Tiemann, K.J., Gamez, G. et al. (1999). Recovery of gold (III) by alfalfa biomass and binding characterization using x-ray microfluorescence. *Advances in Environmental Research* 3: 83–93.

Godlewska-Zylkiewicz, B. (2003). Biosorption of platinum and palladium for their separation / pre-concentration prior to graphite furnace atomic absorption spectrometric determination. *Spectrochimica Acta Part B: Atomic Spectroscopy Letters* 37: 2189–2203.

Goksungur, Y., Üren, S., and Güvenç, U. (2005). Biosorption of cadmium and lead ions by ethanol treated waste baker's yeast biomass. *Bioresour Technol* 96: 103–109.

Gomes, C.P., Almeida, M.F., and Loureiro, J.M. (2001). Gold recovery with ion exchange used resins. *Separation and Purification Technology* 24 (1–2), 35–57.

Gomes, N.C.M., Camargos, E.R.S., Dias, J.C.T., and Linardi, V.R. (1997). Gold and silver accumulation by Aspergillus niger from cyanide-containing solution obtained from the gold mining industry. *World Jl of Microbi and Biotech* 14 (1):149–149.

Grosse, A.C., Dicinoski, G.W., Shaw, M.J., and Haddad, P.R. (2003). Leaching and recovery of gold using ammoniacal thiosulfate leach liquors (a review). *Hydrometallurgy* 69 (1): 1–21.

Harris, G.B. (1993). A review of precious metals refining In: *Precious metals* (ed. R. Mischra), 351–374. IPMI.

Ishikawa, S.I., Suyama, K., Arihara, K., and Itoh, M. (2002). Uptake and recovery of gold ions from electroplating wastes using eggshell membrane. *Bior Tech* 81: 201–206).
Jacobsen, R.T. (2005). Removing contaminants from spent catalysts. *Chem Engg Prog* 101: 20–23.
Kaczala, F., Marques, M., and Hogland, W. (2009). Lead and vanadium removal from a real industrial wastewater by gravitational settling/sedimentation and sorption onto Pinus sylvestris sawdust. *Bioresource Technology* 100 (1): 235–243.
Karamuchka, V. and Gadd, G.M. (1999). Interaction of Saccharomyces cerevisiae with gold: toxicity and accumulation. *BioMetals* 12: 289–294.
Kargari A., Kaghazchi, T., Sohrabi, M. (2004). Soleimani, Batch extraction of gold(III) ions from aqueous solutions using emulsion liquid membrane via facilitated carrier transport, *J. Membr. Sci.* 233: 1–10.
Khan, Z., Hussain, J.I., Hashmi, A.A., and Al-Thabaiti, S.A. (2017a). Preparation and characterizations of silver nanoparticles using aniline. *Arab. J. Chem.* 10: S1506–S1511.
Kumar, S.S.D., Rajendran, N.K., Houreld, N.N., and Abrahamse, H. (2018). Recent advances on silver nanoparticle and biopolymer-based biomaterials for wound healing applications. *Int. J. Biol. Macromol.* 115: 165–175.
Kuyucak, N. and Volesky, B. (1989). Accumulation of gold by algal biosorbent. *Biorecovery* 1: 189–204.
Langmuir, I. (1918). The adsorption of gases on plane surfaces of glass, mica and platinum. *Journal of the American Chemical Society* 40 (9): 1361–1403.
Limousin, G., Gaudet, J.P., Charlet, L. et al. (2007). Isotherms: a review on physical bases, modeling and measurement. *Applied Geochemistry* 22: 249–275.
Liu, Y.G., Zhou, M., Zeng, G.M. et al. (2008). Bioleaching of heavy metals from mine tailings by indigenous sulfur-oxidizing bacteria: effects of substrate concentration. *Bioresour. Technol.* 99 (10): 4124–4129.
Liu, Y.Y., Fu, J.K., and Zhou, Z.H. (2000). A study of Pt4+ adsorption and its reduction by Bacillus megaterium. *Chem Res in Chin Un* 16 (3): 246.
Ma, H.W., Liao, X.P., Liu, X., and Shi, B. (2006). Recovery of platinum (IV) and palladium(II) by bayberry tannin immobilized collagen fiber membrane from water solution. *Journal of Membrane Science* 278: 373–380.
Mao, J., Won, S.W., Vijayaraghavan, K., and Yun, Y.S. (2009). Surface modification of Corynebacterium glutamicum for enhanced Reactive Red 4 biosorption. *Bior Tech* 100: 1463–1466.
Mata, Y.N., Blázquez, M.L., Ballester, A. et al. (2009). Sugar-beet pulp pectin gels as biosorbent for heavy metals: preparation and determination of biosorption and desorption characteristics. *Chem. Eng. J.* 150: 289–301.
Mikheenko, I.P., Rousset, M., Dementin, S., and Macaskie, L.E. (2008). Bioaccumulation of palladium by Desulfovibrio fructosivorans wild-type and hydrogenase-deficient strains. *Appl. Environ. Microbiol.* 74: 6144–6146.
Mooiman, M.B., Miller, J.D. (1991. The chemistry of gold solvent extraction from alkaline cyanide solution by solvating extractants. *Hydrometallurgy* 27, 29–46.
Muraleedharan, T.R., Iyengar, L., and Venkobachar, C. (1991). Biosorption: An attractive alternative for metal removal and recovery. *Current Science* 61 (6): 379–385.
Nasretdinova, G.R., Fazleeva, R.R., Mukhitova, R.K. et al. (2015). Electrochemical synthesis of silver nanoparticles in solution. *Electrochem. Commun.* 50: 69–72.
Natarajan, G. and Ting, Y.P. (2014). Pretreatment of e-waste and mutation of alkali-tolerant cyanogenic bacteria promote gold biorecovery. *Bior Techn* 152: 80–85.

Ogi, T., Honda, R., Tamaoki, K. et al. (2011). Direct room-temperature synthesis of a highly dispersed Pd nanoparticle catalyst and its electrical properties in a fuel cell. *Powder Technol* 205: 143–148.

Oliveira L., Franca, A.S., Alves, T.M., and, Rocha, S.D.F. (2008). Evaluation of untreated coffee husks as potential biosorbents for treatment of dye contaminated waters. *J. Hazard. Mater.* 155: 507–512.

Ozer, A., Ozer, D., and Ekíz, H.Í. (2004). The equilibrium and kinetic modeling of the biosorption of copper(II) ions on Cladophora crispata. *Adsorption* 10: 317–326.

Panda, G.C., Dasa, S.K., Chatterjee, S. et al. (2006). Adsorption of cadmium on husk of Lathyrus sativus: physicochemical study. *Colloids and Surfaces B: Biointerfaces* 50: 49–54.

Pant, D., Singh, P., and Upreti, M.K. (2014). Metal leaching from cathode ray tube waste using combination of Serratiaplymuthica and EDTA. *Hydrometallurgy* 146: 89–95.

Park, H.J., Jeong, S.W., Yang, J.K. et al. (2007). Removal of heavy metals using waste eggshell. *Journal of environmental sciences* 19 (12): 1436–1441.

Pedroso, M.S., Pinho, G.L.L., Rodrigues, S.C., and Bianchini, A. (2007). Mechanism of acute silver toxicity in the euryhaline copepod Acartia tonsa. *Aquatic Toxicology* 82: 173–180.

Pethkar, A.V. and Paknikar, K.M. (1998. Recovery of gold from solutions using Cladosporium cladosporioides biomass beads. *J of Biotech* 63: 121–136.

Pulit-Prociak, J. and Banach, M. (2016). Silver nanoparticlesa material of the future. . .? *Open Chem.* 14: 76–91.

Radke, C.J. and Prausnitz, J.M. (1972. Adsorption of organic solutes from dilute aqueous solutions on activated carbon. *Ind and Engg Chem Fund* 11: 445–450.

Rafique, M., Sadaf, I., Rafique, M.S., and Tahir, M.B. (2017). A review on green synthesis of silver nanoparticles and their applications. *Artif. Cells. Nanomed. Biotechnol.* 45: 1272–1291.

Ramesh, A., Hasegawa, H., Sugimoto, W. et al. (2008). Adsorption of gold (III), platinum(IV) and palladium(II) onto glycine modified crosslinked chitosan resin. *Biores Tech* 99: 3801–3809.

Saeed, A., Iqbal, M., and Akhtar, M.W. (2005). Removal and recovery of lead (II) from single and multiple (Cd, Ni, Cu, Zn) solutions by crop milling waste (black gram husk). *J Hazard Mater* 117: 65–73.

Sag, Y. and Kutsal, T. (2000). Determination of the biosorption heats of heavy metal ions on Zoogloea ramigera and Rhizopus arrhizus. 6: 145–151.

Schiewer, S. and Pati, S.B. (2008). Modeling the effect of pH on biosorption of heavy metals by citrus peels. *J. Hazard. Mater.* 157: 8–17.

Shaw, C.F. (1999). Gold-based therapeutic agents. *Chemical Reviews* 99: 2589–2600.

Shen, J. and Duvnjak, Z. (2005). Adsorption kinetics of cupric and cadmium ions on corncob particles. *Process Biochemistry* 40: 3446–3454.

Sips, R. (1948). Structure of a catalyst surface. *Journal of Chemical Physics* 16, 490–495.

Sobjerg, L.S., Lindhardt, A.T., Skrydstrup, T. et al. (2011). Size control and catalytic activity of bio-supported palladium nanoparticles. *Colloids Surf. B* 85: 373–378.

Soleimani M. and Kaghazchi, T. (2008). Adsorption of gold ions from industrial wastewater using activated carbon derived from hard shell of apricot stones-an agricultural waste, *Bioresour. Technol.* 99: 5374–5383.

Soleimani, M. and Kaghazchi, T. (2008). Activated hard shell of apricot stones: a promising adsorbent. *Chin J of Chem Engg* 16 (1): 112–118.

Solisio, C., Lodi, A., and Borghi, M.D. (2000). The effect of acid pretreatment on the biosorption of chromium(III) by Sphaerotilus natans from industrial wastewater. *Water Res* 34: 3171–3178.

Spitzer M. and Bertazzoli, R. (2004). Selective electrochemical recovery of gold and silver from cyanide aqueous effluents using titanium and vitreous carbon cathodes, *Hydrometallurgy* 74: 233–242.

Syafiuddin, A., Salmiati, M.R.S., Kueh, A.B.H., and Hadibarata, T. (2017). A review of silver nanoparticles: Research trends, global consumption, synthesis, properties, and future challenges. *Nur. J. Chin. Chem. Soc.* 64: 732–756.

Takahiko, K., Masahiro, G., and Fumiyuki, N. (1996). Separation of platinum and palladium by liquid surfactant membranes utilizing a novel bi-functional surfactant. *J of Mem Sc* 120: 77.

Tamaoki, K., Saito, N., Ogi, T. et al. (2010). Microbial reduction and recovery of palladium using metal ion-reducing bacterium Shewanella algae. *Kagaku Kogaku Ronbunshu* 36: 288–292.

Tangaromsuk, J., Pokethitiyook, P., Kruatrachue, M., and Upatham, E.S. (2002). Cadmium biosorption by Sphingomonas paucimobilis biomass. *Biore Tech* 85: 103–105.

Toth, J. (1971). State equations of the solid gas interface layer. *Acta Chimica Academiae Hungaricae* 69: 311–317.

Tuo, Y., Liu, G., Zhou, J. et al. (2013). Microbial formation of palladium nanoparticles by Geobacter sulfurreducens for chromate reduction. *Bioresour. Technol.* 133: 606–611.

Veglio, F. and Beolchini, F. (1997). Removal of metals by biosorption: A review. *Hydrometallurgy* 44: 301–316.

Volesky, B. (1987). Biosorbents for metal recovery. *Trends in Biotechnology* 5: 96–101.

Wan, R.Y. and Miller, J.D. (1986). Solvation extraction and electrode position of gold from cyanide solutions. *Journal of Metals* 35–40.

Wan, R.Y. and Miller, J.D. (1990). Research and development activities for the recovery of gold from alkaline cyanide solutions. *Min Proc and Ext Metall Rev* 6: 143–190.

Wang, W., Zhang, B., Liu, Q.L. et al. (2018). Biosynthesis of palladium nanoparticles using Shewanella loihica PV-4 for excellent catalytic reduction of chromium(VI). *Environ. Sci. Nano* 5: 730–739.

Willst, A.J., Atkinson, K., Will B.A. et al. (2000). The adsorption of precious metals and base metals on a quaternary ammonium group ion exchange resin. *Minerals Engineering* 13: 401.

Xie F., Zhang, T.A., Dreisinger, D., and Doyle, F., (2014). A critical review on solvent extraction of rare earths from aqueous solutions, *Miner. Eng.* 56: 10–28.

Xu, Q., Meng, X., and Han, K.N. (1995). The electrochemical behavior of the dissolution of gold–silver alloys in cyanide solutions. SME.

Yadav, K.K., Singh, J.K., Gupta, N., and Kumar, V.J. (2017). A review of nanobioremediation technologies for environmental cleanup: a novel biological approach. *Mater. Environ. Sci.* 8: 740–757.

Yong, P., Rowson, N.A., Farr, J.P.G. et al. (2002). Bioreduction and biocrystallization of palladium by Desulfovibrio desulfuricans NCIMB 8307. *Biotechnol. Bioeng.* 80: 369–379.

Zhao, Y., Wang, D., Xie, H. et al. (2015). Adsorption of Ag (I) from aqueous solution by waste yeast: kinetic, equilibrium and mechanism studies. *Bioproc. Biosyst. Eng.* 38: 69–77.

Zinicovscaia, I., Cepoi, L., Chiriac, T. et al. (2017). Application of Arthrospira (Spirulina) platensis biomass for silver removal from aqueous solutions. *Int. J. Phytoremediation* 19: 1053–1058.

11

Biosorption as a Strategy for the Recovery of Rare Earth Elements

João Pedro Neves Goldeinstein[1] *and João Paulo Bassin*[1,2]

[1] *Civil Engineering Program, COPPE, Federal University of Rio de Janeiro, Rio de Janeiro, Brazil*
[2] *Chemical Engineering Program, COPPE, Federal University of Rio de Janeiro, Rio de Janeiro, Brazil*

Rare Earth Elements (REEs)

Modern society increasingly demands for rare earth elements (REEs), leading to the challenging task of exploring and efficiently recovering them from previous usages. First, it is necessary to understand what these resources are and why they are so relevant nowadays. According to the International Union of Pure and Applied Chemistry (IUPAC), REEs comprise 17 elements: the lanthanoids (also known as lanthanides), yttrium, and scandium. These last two metals are placed among the REEs because they are commonly found in the same ore deposits as the lanthanoids and exhibit similar chemical properties. It is important to state that, since promethium, one of the lanthanides, has a short half-life and is extremely hard to find, and scandium behaves differently in nature than the other elements for practical purposes, REEs are an even smaller group (Wall, 2013).

This category was named *rare* to describe the difficulty that early chemists faced in separating each metal from its stable form as natural oxides once their behavior showed a strong chemical coherence. The first rare earth was discovered by Arrhenius in 1787 near Stockholm (Sweden) and was a mixture of REEs. They show a wide range of technological applications, being incorporated in high-temperature superconductors, magnets, catalysts, and others. These substances were also extensively used as cathode ray tubes for televisions and computers in the past.

Later, these elements were divided into three smaller groups: light, middle, and heavy REEs. Following the periodic table displacement, the light REEs are La, Ce, Pr, and Nd (atomic numbers 57–60), the middle REEs go from Sm to Gd (atomic numbers 61–63), and the heavy REEs are the rest (atomic numbers 64–71) and yttrium (atomic number 39). In contrast, scandium (atomic number 21) does not have enough properties to be classified as heavy or light. These metals have peculiar physical and chemical properties among the lanthanoids: as the atomic number increases, their atomic size smoothly decreases. Such a phenomenon is also known as *lanthanides contraction*. Table 11.1 shows the REEs' chemical characteristics.

Biosorption for Wastewater Contaminants, First Edition.
Edited by Rangabhashiyam Selvasembian and Pardeep Singh.

Table 11.1 Chemical characteristics of the rare earth elements.

Rare earth element	Chemical symbol	Atomic number	Atomic weight	Density at 25 °C (kg/m^3)	Melting point (°C)	Color of +3 cation in solution
Scandium	Sc	21	44.96	2992	1541	Colorless
Yttrium	Y	39	88.91	4475	1522	Colorless
Lanthanum	La	57	138.91	6174	918	Colorless
Cerium	Ce	58	140.12	6711	798	Colorless
Praseodymium	Pr	59	140.91	6779	931	Yellow-green
Neodymium	Nd	60	144.24	7000	1021	Rose
Promethium	Pm	61	144.91	7220	1042	Pink
Samarium	Sm	62	150.36	7536	1074	Yellow
Europium	Eu	63	151.96	5248	822	Colorless
Gadolinium	Gd	64	157.25	7870	1313	Colorless
Terbium	Tb	65	158.93	8267	1356	Pale pink
Dysprosium	Dy	66	162.50	8531	1412	Pale yellow-green
Holmium	Ho	67	164.93	8797	1474	Yellow
Erbium	Er	68	167.26	9044	1529	Pink
Thulium	Tm	69	168.93	9325	1545	Greenish tint
Ytterbium	Yb	70	173.04	6966	819	Colorless
Lutetium	Lm	71	174.97	9842	1663	Colorless

Despite their names, rare earth elements are plentiful in the planet's crust – they are much more abundant than precious metals such as silver, gold, and platinum, with the exception of promethium (Wall, 2013). Usually, REEs are found in ore deposits in a +3 valence state, as metal oxides. Due to the considerable attempts to establish a sustainable, low-carbon, circular economy, since REEs are used in green technologies, these resources are being increasingly consumed. It is important to mention that despite the necessity for these metals in a futuristic economy, their recovery is a much better approach than opening new mining sites.

Figure 11.1 shows some of the applications of REEs: compact fluorescent lights, batteries for electric vehicles, wires, aluminum alloys, super-magnets, lasers, wind turbines, and other high-tech machinery and equipment that requires specific components made of such elements. Regardless of the variety of minerals containing REEs, mining involves significant risks and negative environmental impacts. During operation, along with the exorbitant energy expenditure, mine tailings and wastewaters carry significant amounts of REEs and other contaminants that adversely affect the watershed – the natural paths toward the shores – and can pollute the oceans and accumulate throughout the food chain. Furthermore, after mine closure, old sites become a permanent loss of rich ecosystems and landscapes.

Currently, the leading countries exploring deposits of REEs are the United States, Australia, and China, with the latter representing more than 90% of worldwide production. However, Brazil and Germany have considerable reserves of these metals and are constantly part of the negotiations in this mining sector. At this point, proper recycling of these elements is uncommon. With that in mind, recovering valuable elements such as REEs has been a promising strategy to avoid environmental degradation, save time, and sustain a cleaner and healthier economy. The opportunities for recovery come from pre-consumer manufacturing scraps/residues, end-of-life products, and industrial waste (Wang et al., 2017). For that, many technologies such as biosorption, ion-exchange processes, solvent extraction, adsorption, and precipitation are being applied (Ramasamy et al., 2018).

This chapter outlines the main methods to recover these elements and focuses on a literature review of the biosorption approach of REE recovery. After acknowledging the primary strategies to extract these resources from waste, the chapter presents some of the

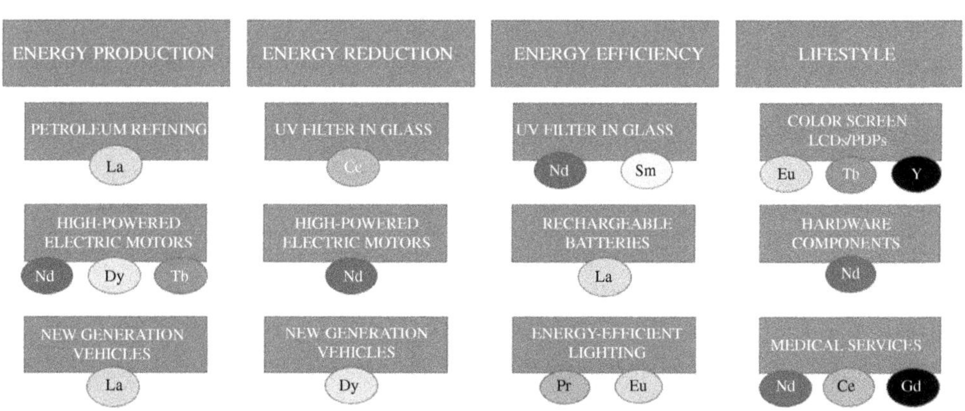

Figure 11.1 Applications of REEs. *Source:* Adapted from Wang et al., 2017.

most recent case studies and laboratory experiments on this topic. Finally, the chapter clarifies the real achievements and bottlenecks of today's research on the subject.

Methods to Recover Rare Earth Elements

REEs have individual properties that make their separation from minerals advantageous, and the high-purity demand stimulates the development of new technologies. The first trials to recover these substances began in the late 1960s when scientists used ion-exchange or solvent-extraction technologies to separate REEs from industrial leachates.

Solvent Extraction

In solvent extraction, compounds are separated based on their solubilities, and it is important to mention that this does not destroy the substance of interest and instead provides valuable benefits. This method requires mixers, settlers, or separatory funnels filled with the solvent and the metal solution, and the system configuration depends on the intended level of purity of the REEs. The solvent, which can be cation exchangers (acidic extractants), solvation extractants (neutral extractants), or anion exchangers (basic extractants), is also a cost variable in the operation. Among them, D2EHPA (di-(2-ethylhexyl)phosphoric acid), HEHEHP (2-ethylhexyl phosphoric acid mono-2-ethylhexyl ester), Versatic 10, TBP (tributylphosphate), and Aliquat 336 are commercially used for REE recovery industry; the names and chemical formulas are listed in Table 11.2.

Usually, this technology has three steps: extraction, washing, and re-extraction. First, as shown in Figure 11.2, the solvent and aqueous solution are gathered, both immiscible phases, and a mass transfer of rare earths (solute) occurs until equilibrium is reached. Later, the organic phase (extract), less dense and made of organic complexes, is separated from the aqueous phase (raffinate) and washed to remove impurities that are co-extracted. In the end, the re-extraction step takes place, where a reverse extraction occurs, migrating the element of interest back to a new aqueous phase, now ready for proper recovery. To become an economically viable method, the solvent is regenerated in an acidic or basic environment, and this process usually happens during re-extraction.

Solvent-extraction for REE recovery is the most-used technology, but it has several disadvantages such as extreme pH conditions, generation of undesirable byproducts,

Table 11.2 Commercial and full names of solvents used for REE recovery.

Usual name or brand	Full name or abbreviation	Chemical formula
D2EHPA	di-(2-ethylhexyl)phosphoric acid	$C_{16}H_{35}O_4P$
HEHEHP	2-ethylhexyl phosphoric acid mono-2-ethylhexyl ester	$C_{16}H_{35}O_3P$
Versatic 10	Neodecanoic acid	$C_{10}H_{20}O_2$
TBP	Tributylphosphate	$C_{12}H_{27}O_4P$
Aliquat 366	Methyltrioctylammonium chloride	$C_{25}H_{54}ClN$

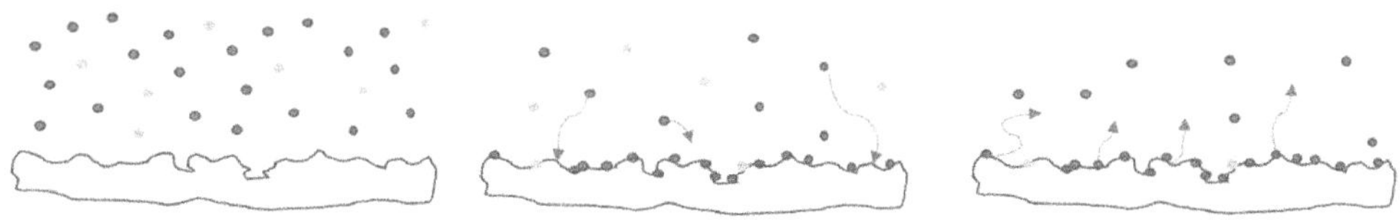

Figure 11.2 Scheme of adsorption and desorption phenomena. *Source:* Adapted from Mesgari-Shadi et al., 2018.

environmental pollution, hazards to operator health, toxic wastewater, solvent loss, and high operational costs.

Ion Exchange

Unlike solvent extraction, ion exchange does not involve liquid waste, phase separations, or solvent loss. This technique is based on chelating resins with selectivity properties. The separation takes place from strong bonding ions between REEs and the resin surface. Free ions from the solid material (exchanger) are replaced by ions with a similar charge from the solution containing the REE (medium). The cation exchangers allow the passage of cations but reject anions, while the opposite occurs with anion exchangers. This selectivity can be a challenging method to exclusively recover the elements of interest since the medium usually presents other substances with the same electric charge as the resin (Hassanvand et al., 2017).

The ion exchangers can be organic or inorganic, natural or synthetic. Among the organic alternatives, plant or animal cells can act as exchangers due to their carboxylic groups, whereas many clays can act as inorganic extracts due to their granulometry and chemical properties. Most of these systems operate cyclically, where the liquid solution containing rare earths flows through column-shaped resins until saturation, also called *exhaustion*. At that moment, the material needs backwashing with a specific cleaning solution capable of regenerating the media (Hassanvand et al., 2017).

The application of chelating resins for the separation of REEs is recent, and they are commonly *cation exchangers*, although anion exchangers can also be found but are not well understood. Some commercially available resins are Dowex AG 50 W-X8, Dowex AG 50WX12, Bio Rad AG 50-X8, Ostion LGKS 0800, sulfonated polystyrene, and bonded-phase silica and Amberlite IR-120 (Khawassek et al., 2019).

Adsorption

Adsorption is considered a separation method for recovering REEs and is based on the adhesion of molecules from two different materials due to surface forces. When the interactions are based on Van der Waals forces, it is called *physical adsorption;* and when they rely on chemical forces, it is called *chemisorption* (Chiou, 2003; Berger and Bhown, 2011).

The system consists of a solution called the *adsorbate*, and the solid material, which uptakes the chemicals, referred to as the *absorbent*. Normally, the absorbent has a high surface area or porosity in order to optimize the adsorption process. The adsorption equilibrium is mainly controlled by the properties of the adsorbent and the adsorbate, the composition of the solution, and adsorption and desorption kinetics, and its behavior can be explained by isotherms (Langmuir and others) (Raghav and Kumar, 2018).

Since REEs can be weathered as ions, they travel with water and can be easily absorbed. In the light of that, REEs started to be studied for adsorption recovery using two kinds of adsorbent: raw and modified. Among them are nano carbon shells, modified clays, nanoparticles, hydrogels, sediments, microcapsules, and more (Anastopoulos et al., 2016).

Last but not least, any metal of interest that is adsorbed, such as the REEs, needs to be detached from the adsorbent's surface. When they are not strongly bonded, distilled water can be a simple solution to desorb them for use (Anastopoulos et al., 2016). Figure 11.2 shows a scheme of the sequenced steps of the adsorption method.

Chemical Precipitation

One of the most traditional methods for recovering metals is chemical precipitation, and it was first discovered by Liesegang in the nineteenth century. This technique consists of converting a substance into an insoluble compound, which makes it possible to extract the element from a medium. The recently formed compound is called a *precipitate*, and the solution left behind is called a *supernatant*.

When chemical precipitation is used for recovering REEs, some impurities usually precipitate with them by coprecipitation or adsorption. Different agents (precipitants) are used to obtain these elements from a solution at a wide range of pH values. Since carbonates are readily available at low cost, they are used to precipitate REEs at high pHs, but other widely used agents such as phosphates, sulfates, nitrates, chlorides, fluorides, and oxalates show better results when the pH is lower than 7. According to the work of Han et al. (2020), nitrates, chlorides, and sulfates show the best recovery results, but they state the importance of controlling pH as one of the significant variables.

As seen previously, solutions with phosphoric acids also provide ion-exchange processes with the REEs; therefore, applying this kind of chemical provides two simultaneous methods for recovering rare earths.

Biosorption

Finally, biosorption is one of the most cost-effective biotechnologies for recovering valuable metals. The *bio* prefix denotes the dependency of biological entities called *biomaterials*, also known as *biosorbents*, including algae, fungi, bacteria, biopolymers, industrial and agricultural wastes, etc., with binding properties that adhere to metals, allowing their separation from aqueous solution. Since biomass is inexpensive and widely available, biosorption for REE recovery is a promising alternative. Besides its relatively low cost, biosorption is defined as a green technology since living, dead, or inactive microbial communities do not represent any environmental risks. Inactive or dead cells are advantageous since living cells have growth limitations and the REE solution may be harmful to organisms. In addition, they can be easily stored. Another positive aspect of the biosorption process is that it can be used *in situ* with proper design (Das and Das, 2013).

Many researchers consider this type of extraction a subdivision of the adsorption process because the elements of interest chemically engage with a solid phase, which, in this case, must be a wisely chosen biomaterial. It is a physicochemical process independent of any metabolization process happening in the cells' interior since it depends entirely on surfaces. The removal efficiency of the metal from the waste stream requires previous knowledge of how it interacts with each kind of microbial biomass, including environmental variables such as pH, temperature, contact time, biosorbent dosage, and metal concentration (Das and Das, 2013). The literature also shows that biosorption sometimes follows Langmuir, Freundlich, or Brunauer-Emmett-Teller (BET) isotherms. While the Langmuir approach relies on single-layer sorption, Freundlich and BET use multi-layer-based adsorption (Andres, Texier, and Le Cloirec, 2003).

All kinds of biomaterials interact with at least one organic or inorganic compound. For example, there are gram-positive and gram-negative bacteria provide different sites for metal biosorption. In gram-positive bacteria, carboxylic groups are the primary binding sites for metal cations, while in gram-negative species, the phosphate groups are responsible (Fomina and Gadd, 2014). Furthermore, in some other bacteria, the metal-binding components are polysaccharides, lipids, proteins, sulfate groups, peptidoglycan, and proteinaceous S-layer. The varieties of components acting as binding sites go far beyond this list, depending specifically on the organism involved in the biosorption process (Breuker et al., 2020).

The biosorption process consists of two steps. First, the metal ions from the solution are adsorbed when they interact with functional groups on the living cell surfaces due to the active sites on the organism. Later, once adsorbed, these metals can cross the cell wall and reach the cell membrane and cytoplasm. The composition of the cell wall varies depending on the microbe. For example, there have been reports of successful interactions between lanthanides and well-known brown algae named *Sargassum*. This seaweed has more than 70% of its titrated sites composed of alginate carboxylic groups. Moreover, it also presents sulfonic acid that can improve biosorption. In the case of microalgae, amino, carboxyl, phosphate, and sulfate represent the major active sites for biosorption and have been confirmed to interact with rare earths such as neodymium. Unlike other biosorbents, algae generally do not produce toxic substances, making them an environmentally better alternative (Gad, 2016).

Alternatively, bacteria, such as *Pseudomonas sp.*, are commonly made of peptidoglycans which can sequester metal ions. In microalgae, amino, carboxyl, phosphate, and sulfate represent the major active sites and have been confirmed to interact with rare earths such as neodymium. On the other hand, in filamentous fungi, amino or non-amino polysaccharides can retain REEs. Thereby, structure, functional groups, and surface area can be considered the significant variables among organisms when choosing a proper biosorbent (Das and Das, 2013).

After biosorption, desorption is a significant aspect of the method. Its success relies on a proper selection of an elutant, which depends on the biosorbent itself. The elutant must be eco-friendly, low-cost, and incapable of damaging the biomass. Extensive experiments have been conducted to choose the best elutant. Some studies performed by Das and Das (2013) show the application of nitric acid or chlorine acid solutions immediately after adsorption of REEs.

In addition, to obtain a feasible biosorption process at full scale, scientists have found continuous and regenerable systems of utmost importance. Among the various attempts, they have used packed-bed columns or, rarely, fluidized and continuous-stirred tank reactors. This last configuration is usually based on powder biosorbent and has high investment

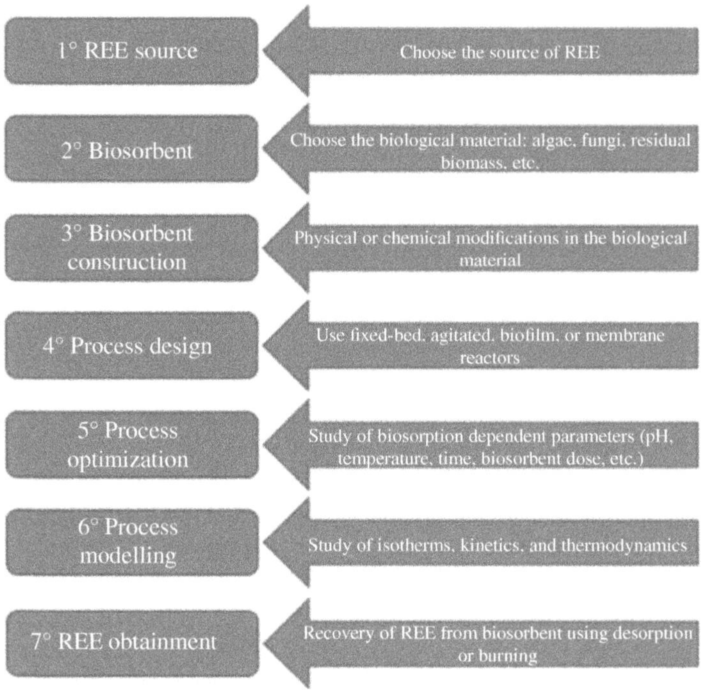

Figure 11.3 Flowchart strategy for REE biosorption recovery. *Source:* Adapted from Giese, 2020.

and operational costs. On the other hand, column-shaped packed-bed reactors are more reasonable due to their higher efficiency. In the end, biosorption technology seeks a successful recovery of metals so they can be reintroduced into the markets, assembled with developing strategies to regenerate biomass (Das and Das, 2013).

Concomitantly, chemical modification of biomaterials has been conducted to enhance binding sites, hence improving biosorption. Alkaline and acid pretreatment, as well as polymerization, are being used to achieve a more efficient method. Along with the enhancement of the process, the inactivation of biomass has been conducted by heating, drying, or freezing techniques because they do not damage the adsorption capacity of cells (Giese, 2020). In light of all he strategies for biosorption, a summarizing flowchart is shown in Figure 11.3.

With that in mind, the following section presents a collection of scientific studies that demonstrate how the adsorption process efficiency varies according to the combination of the rare earth and adopted biosorbent. It must be considered that high extraction levels mean betterresults since recovery for further use is the primary goal of biosorption.

Biosorption Approach for Recovering Rare Earth Elements

A pioneering study was conducted by Korenevsky, Sorokin, and Karavaiko (1997) using batch cultures of *Candida utilis* (fungi) for REE recovery. They acknowledged the importance of pH control to biosorb these elements. No sorption of scandium, lanthanum, samarium, and yttrium was found at pH values below 5. However, at pHs above 7, hydroxides of REEs started to deposit on cell surfaces. The pattern of deposits was strongly correlated with the cytoplasmic membrane of the microbial cells.

The influence of higher pHs was also observed by Hosomomi et al. (2013) when *Escherichia coli* was used to adsorb REEs. This bacterium showed greater selectivity for heavier REEs, specifically scandium. During the tests, 99% of Nd, Dy, and Lu was removed after only 5 minutes using *Escherichia coli*, and equilibrium was achieved in 30 minutes. Despite the bacteria strongly adsorbed Y, Nd, Dy, and Lu at low pH, the increasing pH dissociated carboxyl and phosphate groups from the microbe and positively charged the REEs, promoting bindings with the negative sites of cell walls. The authors confirmed that very acidic solutions pronated these functional groups and that pH above 2 allows the uptake of REEs by not only phosphate groups but also carboxyl groups, potentially removing the elements from the medium.

On Japan's coast, Sakamoto, Kano, and Imaizumi (2008) sampled and analyzed five living and dried brown algae species (*Dictyota dichotoma, Ecklonia stolonifera, Sargassum hemiphyllum, Sargassum honeri,* and *Undaria pinnatifida*) in order to determine the levels of lanthanides biosorbed. The authors also studied the accumulation of REEs among the *Sargassum hemiphyllum* organs. They found that light REEs were generally detected, except in *Dictyota dichotoma*. The concentrations of La, Yb, and Eu were higher in *Sargassum hemiphyllum* than in *Undaria pinnatifida*. Finally, the concentration of rare earths (La, Yb, and Eu) in organs were in the following descending order: "main branch," "leaf," "vesicle." The authors suggest that this distribution strongly depends on the element's chemical behavior.

Another interesting study using *Sargassum* was carried out in Egypt to evaluate the capacity of this brown algae to capture REEs from El Atshan Trachytic samples (Eastern desert of Egypt). After the samples had been in contact with *Sargassum,* the author found that Ho, Er, and Eu concentrations decreased. The main conclusion from the data was that light REEs levels decrease with the increase in their atomic number upon biosorption treatment. On the other hand, heavy REEs showed an increase after treatment, but the researcher did not investigate this phenomenon (Gad, 2016).

In a different approach, Wang et al. (2010) showed that *Buccinum Tenuissimum* shell biomass could be a promising biomaterial for recovering rare earths and is satisfactorily described by Langmuir and Freundlich isotherms. In a pH range of 3 to 7 and 100 $\mu g/dm^3$ of standard solution using 3.0 g/dm^3 biomass dosage, 98% of the REEs were removed within a 7 hour contact time. When the authors fixed the pH at 5, 95% of the elements were biosorbed in 10 minutes, and 99% in 30 minutes, showing no appreciable change after that. Varying the biomass concentration, they found that the optimal dosage was 2.0 g/dm^3 of biomass under the same conditions. During their analysis, the researchers changed the specific surface of the biomass through heat treatment, however, this approach showed a decreased factor by less than one-eighth, which was unfavorable compared to the original state of the shells.

Botalov et al. (2017) studied recovering REEs using single- and multi-component solutions containing more than 2 mg/L of Y, Mo, La, Ce, Nd, Dy, and other metals such as Al, Ca, and Fe. The authors explained that Al, Fe, and Ca were used to investigate the competition between REEs and these metal ions for active sites of the green algae's surface during biosorption. The designed batch system worked at pH 1.5, 3, and 5 with *Chlorella vulgaris* at dosages of 0.5, 1, and 3 g/L. They achieved the aimed concentration at pH 1.5; however, they observed that higher pH caused precipitation of rare earths, Al, and Fe. At pH 5, Sc and Mo had almost completely disappeared. Varying pH concentrations from 1.5 to 5, reductions of 31, 24, 25, and 25% were obtained for Y, La, Ce, and Nd, respectively. In addition, the removals of Sc, Y, Mo, La, Ce, Nd, V, and Dy were higher at 3 g/L microalgae dosage; and, in the same conditions, the recoveries of Y, La, Ce, Nd, and Dy were 67, 75, 70, 75, and 77%, respectively.

Another assessment related to biosorption of REEs using single- and multi-solute solutions was conducted by Barros et al. (2019). The division of single- and multi-component solutions was adopted to assess the xenobiotic effects that each metal can cause in bacteria. A porous mineral zeolite was used with and without a *Bacillus cereus* biofilm to adsorb REEs and then look for successful desorption. An initial concentration of 20 mg/L for La, Ce, Y, Pr, Tb was used in column setups. From the assays, it was observed that more than 90% of lanthanum and cerium were removed from the medium. However, when suspended *Bacillus* was used, the adsorption decreased to 70 and 80%, respectively. The bacteria showed a change from gram-positive staining to gram-negative staining when in contact with the metal ions, presenting higher stress in Ce and Pr solutions, and a thinner and fragile cell structure was observed through the microscope. In all conditions, Eu, Y, and Tb showed a smaller reduction in concentration, even after 72 hours, and the researchers concluded that these three metals have a lower affinity for the zeolite than the others. They also confirmed that the biofilms reduced the surface area of the zeolite, which explains the lower uptake of the REEs when biomass was present. In the end, more than 60% of Eu, Tb, and Y were recovered, while La, Ce, and Pr had lower desorption levels, amounting to 20, 30, and 40%, respectively, regardless of the conditions, due to their higher affinity for zeolite.

Walnut shells have been used as a biosorbent for removing pharmaceuticals, heavy metals, and pesticides. Since this is an abundant type of agricultural waste, the use of such material would be cost-effective if it could recover REEs from solution. Based on that, Gallardo et al. (2020) carried out a study with these shells, previously dried at 105 ºC for 24 hours, to evaluate biosorption and desorption along with strategic modifications. They prepared a liquid phase containing La^{3+}, Sm^{3+}, Eu^{3+}, Gd^{3+}, and Nd^{3+}, varied pH values from 3 to 7, and used shells with two particle sizes (75 and 2000 μm) during biosorption. Most REEs were biosorbed at pH 4; only lanthanum showed better results at pH 5.5. With a size of 75 μm, the adsorption levels of Eu, Gd, Sm, and La were 88, 84, 93, and 85%, respectively; and with a size of 2000 μm, they showed 89, 97, 99, and 85% adsorption. When they modeled the process using the Langmuir approach, they found a coefficient of 0.999, indicating that it was very suitable for data. The authors concluded that a raw material such as walnut shells had excellent results for biosorption of REEs regardless of particle size.

Despite the disadvantage of using living biomass, due to growth limitation, as mentioned earlier, Breuker et al. (2020) achieved better performance using living *Bacillus subtilis* and *Pichia sp.* rather than dead. Considering that metal bioleaching processes usually occur in acidic solutions ($pH < 3$) and may contain many REEs, the researchers aimed to biosorb metal ions with different organisms in acidic environments. First, *Bacillus subtilis* was mixed with a prepared solution of La, Nd, Dy, and Yb, each element with a concentration of 15 μmol/L. From this test, *Bacillus subtilis* indicated a strong affinity for Yb, but more than 96% of the REEs were removed from the aqueous solution, regardless of the amount of biomass. In accordance with previous studies, and as expected from what could be observed so far, by increasing REE/biomass ratio, biosorption was relatively compromised. Additionally, gram-negative bacteria *Leisingera methylohalidivorans* exhibited great results for all REE removals, while *Phaeobacter inhibens* seemed more reasonable for heavy REE removal. Finally, *Catenulostroma chromoblastomyces* and *Pichia sp.* showed a preference for middle REEs, especially *Pichia sp.*, which demonstrated a potential capacity for anthropogenic gadolinium recovery from wastewaters.

In view of the studies carried out on the topic, it can be seen that many parameters are involved in the biosorption process of rare earths. Worldwide collaborators have contributed to this field, pointing out important aspects such as pH, chemical and physical modifications of sorbents, REE/biomass ratio, contact time, and different affinities among REEs and biomasses.

Final Considerations

The increasing demand for rare elements is not negligible, given the expansion of high-tech industries associated with a circular and greener economy. The exploration of new ore deposits through mining goes along with permanent negative impacts on the environment, notwithstanding the unsustainable mindset associated with it. This paradigm urges changes in the way society handles limited resources.

The use of biosorption for recovering REEs has already been proved to be a reliable and promising strategy. Many methodologies were presented in the chapter, showing an extensive, unfinished, and trustworthy list of biomaterials. Agricultural waste, such as walnut shells, and cost-effective microorganisms, such as algae, fungi, and bacteria, have shown biosorption and desorption characteristics based on binding interactions among functional groups (sulfate anion, phosphate, amino, carboxyl, peptidoglycan, polysaccharides, alginate, etc.) on cell walls with physicochemical properties of rare earths dissolved in a liquid phase.

REEs may cause stressed environments for living organisms that act as biosorbents. Nevertheless, researchers have found strategies such as alkaline or acid pretreatment, drying, freezing, or heating the biomass to enhance biosorption capacity by changes in the active sites and specific surface.

Finally, based on the positive aspects and successful results presented in the literature review, biosorption can be more advantageous as a recovery method of REEs than competing processes for the same purpose, such as ion exchange, conventional adsorption, solvent extraction, and chemical precipitation.

References

Anastopoulos, I., Bhatnagar, A., and Lima, E.C. (2016). Adsorption of rare earth metals: A review of recent literature. *Journal of Molecular Liquids* 221: 954–962.

Andres, Y., Texier, A.C., and Le Cloirec, P. (2003). Rare earth elements removal by microbial biosorption: a review. *Environmental Technology* 24 (11): 1367–1375.

Barros, Ó., Costa, L., Costa, F. et al. (2019). Recovery of rare earth elements from wastewater towards a circular economy. *Molecules* 24 (6): 1005.

Berger, A.H. and Bhown, A.S. (2011). Comparing physisorption and chemisorption solid sorbents for use separating CO2 from flue gas using temperature swing adsorption. *Energy Procedia* 4: 562–567.

Botalov, M., Bunkov, G., Rychkov, V. et al. (2017). Recovery of rare earth elements (REEs) from uranium containing solutions using biosorption. In: AIP Conference Proceedings 1886 (1): 020033. AIP Publishing LLC.

Breuker, A., Ritter, S.F., and Schippers, A. (2020). Biosorption of rare earth elements by different microorganisms in acidic solutions. *Metals* 10 (7): 954.

Chiou, C.T. (2003). Fundamentals of the adsorption theory. *Partition and Adsorption of Organic Contaminants in Environmental Systems*: 39–52.

Fomina, M. and Gadd, G.M. (2014). Biosorption: current perspectives on concept, definition and application. *Bioresource Technology* 160: 3–14.

Gad, S. (2016). Biosorption of rare earth elements using biomass of sargassum on El-Atshan Trachytic sill, Central Eastern Desert, Egypt. *Egyptian Journal of Petroleum 25* (4): 445–451.

Gallardo, K., Castillo, R., Mancilla, N., and Remonsellez, F. (2020). Biosorption of rare earth elements from aqueous solutions using walnut shell. *Frontiers in Chemical Engineering 2* (4).

Giese, E.C. (2020). Biosorption as green technology for the recovery and separation of rare earth elements. *World Journal of Microbiology and Biotechnology* 36 (4): 1–11.

Han, K.N. (2020). Characteristics of precipitation of rare earth elements with various precipitants. *Minerals* 10 (2, 178.

Hassanvand, A., Wei, K., Talebi, S. et al. (2017). The role of ion exchange membranes in membrane capacitive deionisation. *Membranes* 7 (3): 54.

Hosomomi, Y., Baba, Y., Kubota, F. et al. (2013). Biosorption of rare earth elements by E. coli. *Journal of Chemical Engineering of Japan*: 13we031.

Khawassek, Y.M., Eliwa, A.A., El Sayed, A.H. et al. (2019). Adsorption of rare earth elements by strong acid cation exchange resin thermodynamics, characteristics and kinetics. *SN Applied Sciences* 1 (1): 1–11.

Mesgari-Shadi, A., Sarrafzadeh, M.H., Divband, B. et al. (2018). Batch adsorption/desorption for purification of scfv antibodies using nanozeolite microspheres. *Microporous and Mesoporous Materials* 264: 167–175.

Raghav, S. and Kumar, D. (2018). Adsorption equilibrium, kinetics, and thermodynamic studies of fluoride adsorbed by tetrametallic oxide adsorbent. *Journal of Chemical and Engineering Data* 63 (5): 1682–1697.

Ramasamy, D.L., Puhakka, V., Repo, E. et al. (2018). Two-stage selective recovery process of scandium from the group of rare earth elements in aqueous systems using activated carbon and silica composites: Dual applications by tailoring the ligand grafting approach. *Chemical Engineering Journal* 341: 351–360.

Wall, F. (2014). Rare earth elements. *Critical Metals Handbook*, 312-339.

Wang, Y., Koto, Y., Sakamoto, N. et al. (2010). Biosorption of rare earth elements, thorium and uranium using Buccinum tenuissimum shell biomass. *Radioisotopes* 59 (9): 549–558.

Wang, K., Adidharma, H., Radosz, M. et al. (2017). Recovery of rare earth elements with ionic liquids. *Green Chemistry* 19 (19): 4469–4493.

Zhang, Z., Jia, Q., and Liao, W. (2015). Progress in the separation processes for rare earth resources. *Handbook on the Physics and Chemistry of Rare Earths* 48: 287–376.

12

Deployment of Used Biosorbents in Environmental Remediation: Prospects and Challenges

Shashikant Shivaji Vhatkar[1], *Guru Charan Sahu*[2], *and Ramesh Oraon*[1]

[1] *Department of Nanoscience and Technology, Central University of Jharkhand, Brambe, Ranchi, Jharkhand, India*
[2] *Department of Chemistry, Ranchi University, Ranchi, Jharkhand, India*

Introduction

Anthropogenic activities (mining, refining, etc.) and insensitive waste disposal have led to environmental pollution. The need to recycle waste is of global importance due to its advantages, including sustainability (Kovacs and Szemmelveisz, 2017; Zeng et al., 2019; Soliman and Moustafa, 2020). Recycling presents benefits including minimizing waste and reducing stress on environmental resources. For instance, the recovery of metals from waste (such as industrial effluents) could lessen their environmental impact by decreasing industrial demand for raw metals (Hoang et al., 2018; Kumar et al., 2021).

In another example, water contaminated with heavy metals affects the food chain and environment (Wang and Yang, 2016). Heavy metals are toxic pollutants that commonly find their way into waterways through anthropogenic activities (mining, pesticides, and so on). They are also not biodegradable (density > 5 g/cm^3). As a result, when they are introduced into water source (surface or groundwater) in excessive concentrations, they end up in primary consumers on the food chain (such as small animals), which cannot break them down (Kacholi and Sahu, 2018; Sang et al., 2018). With every step up the food chain, the quantity of heavy metals increases, thereby poisoning the entire ecosystem, including humans. Another common pollutant is dye, which has its source in textile industries. Like heavy metals, dyes are toxic and non-biodegradable and eventually end up in the food chain (Premkumar et al., 2018; Tkaczyk et al., 2020).

Several remediating physicochemical techniques are already in place: treating industrial wastewater chemically, metal leaching, filtration, and so on (Mohd Udaiyappan et al., 2017; Kinidi et al., 2018). These were later found to be less efficient. Additionally, they created extra waste, most of which was toxic byproducts. In recent years, adsorption has been recognized as an attractive alternative to filtration technology due to its simplicity and environmentally friendly approach (Maleki et al., 2016; Panda et al., 2017; Ahmad et al., 2019).

Biosorption for Wastewater Contaminants, First Edition.
Edited by Rangabhashiyam Selvasembian and Pardeep Singh.

Adsorption techniques are favored over filtration because of their low cost, processing approach, and environmental impact. Unlike absorption, adsorption is a surface phenomenon that involves the adhesion of a sorbate (such as contaminant molecules or atoms) through Van der Wall interactions (Crini et al., 2019). For instance, biosorbents such as rice husks, orange peels, seeds, and oyster shells adsorb pollutants and dyes from water with efficiency as high as 99% (Crini et al., 2019). In addition to their abundance and low cost, biosorbents are reusable, making them sustainable candidates for effective environmental remediation. As a result, biosorbents have been attracting the interest of researchers and academicians worldwide.

Biosorbents, which have an environmental origin, offer several additional benefits over conventional adsorbents in terms of biodegradability, cost, durability, and – most important – reusability. In addition, their large surface area plus the presence of superficial functional groups, together with the fact that they can be tailored, make biosorbents an efficient and ideal candidate for sustainable environmental remediation (Cho et al., 2016; Putro et al., 2017; Xie et al., 2020; Gül et al., 2020).

Because the resulting biosorption is advantageous, waste treatment through used biosorbents has attracted the interest of academicians worldwide. Effective reutilization of biosorbents would reduce the environmental burden and promote the conservation and protection of environmental resources.

Mechanism Studies

It is difficult for heavy metals to degrade environmentally; however, their oxidation state can be altered. The altered compounds are more stable and less toxic than the original compounds (Sobhanan et al., 2020). Several biosorption mechanisms are explained next.

Adsorption

Adsorption is a process in which an adsorbent draws adsorbate molecules through physicochemical interactions (such as redox) at the interfacial region (Burakov et al., 2018). These generally accumulate due to the adhesive action of the atoms, molecules, and ions of the adsorbate onto the adsorbent. On the other hand, in absorption, atoms are absorbed by the substrate chemical process. Absorption also involves a phase change, causing an irreversible reaction (Chen et al., 2018).

The term *sorption* is generally used to express both these processes (adsorption and absorption). In practice, adsorption is preferred over absorption due to its many advantages. For instance, biomaterials such as cellulose, lignin, etc., are used as biosorbents. These biosorbents exhibit adsorption when exposed to adsorbates such as contaminants (heavy metals, for example). Additionally, these biosorbents undergo desorption, so they can be reused (Kumar et al., 2017). Many other advantages are discussed in detail later in this chapter.

Ion-Exchange Resin

An ion-exchange resin is a hydrophilic matrix of beads consisting of cross-linked cation-exchanger (anionic) polymers. They are ion-functionalized for selective binding (such as heavy metal ions) in solution (Choi et al., 2020). In practice, strong acidic anion resins (positively charged) maintain their charge in a wide pH range and attract negatively charged ions, whereas cation resins (negatively charged) draw positively charged ions (Nekouei et al., 2019). Regeneration of resins is generally performed by flushing trapped ions with the help of a regenerant (for example, brine solution) (Amphlett et al., 2018).

Popular applications of ion-exchange resins involve the removal and recovery of heavy metal ions at wastewater treatment plants (Soyluoglu et al., 2020) and from industrial effluents (uranium, platinum group metals, lanthanides, and so on). They are also deployed in the semiconductor industry, at nuclear power plants, the pharmaceutical industry, and at research facilities to purify water and other solutions at ionic levels (Zagorodnyaya et al., 2017; Gedgagov et al., 2018; Bajwa et al., 2020).

Complexation

Complexation is a selective technique that employs chelating agents (such as EDTA) to obtain a stable metal complex through coordination (Ball et al., 2002; Han et al., 2006). These stable metal complexes stay passive in physicochemical activities such as sorption, precipitations, redox, etc. In practice, complexation is characterized either by donor-acceptor interactions or by charge transfer. The former facilitates stable bond formation through dipole interactions, whereas the latter involves electrostatic interactions resulting in strong inter-molecular bonding (Kim et al., 2002; Duarte et al., 2012; Zhang et al., 2016). For instance, a complexing agent (hydroxyisobutyric acid) has been used to treat industrial effluent containing a significant amount of rare earth metals. Although the complexation technique is expensive in terms of energy and capital and has minimal reproducibility along with stability issues, it has been deployed in several applications, such as treating industrial effluents.

Microprecipitation

Microprecipitation is a phenomenon at the surface of or within adsorbents (biosorbents) at the expense of physicochemical interactions (Wang et al., 2019a). Sequestration of heavy metals usually occurs at the sorption surface. With reference to biosorbents, the adsorbate may be a biomaterial (such as cellulose or lignin) or living or dead microbial mass. Living microbial biomass facilitates precipitation based on metabolic responses generated by cells, whereas the latter depends on the physicochemical interaction at the interface (Kuyucak and Volesky, 1988; Lee et al., 1997).

Pyrometallurgical Processes

These are high-temperatures refining processes facilitating the recovery of useful metals (iron, zinc, nickel, cobalt, etc.) from mineral ores and residues (Tuncuk et al., 2012). In general, thermal treatment results in the degradation of impurities, which results in the extraction of metals. Metallurgical processes such as calcination and smelting are generally categorized as pyrometallurgy (Shamsuddin 1986; Debnath et al., 2018). Although the process is energy-intensive, it has been the conventional means for metal extraction and recovery from mineral deposits and industrial effluents, respectively, for a very long time (Işıldar et al., 2018).

Hydrometallurgical Processes

Hydrometallurgy is a metal extraction process for less electropositive metals (such as gold, silver, etc.), which involves selective leaching, followed by purification techniques (like adsorption and solvent extraction) (Kinoshita et al., 2003; Erüst et al., 2013). The leaching process involves forming an aqueous solution of desired mineral ores or metal residues or even industrial effluents (Rocchetti et al., 2013). Dissolved metal is generally recovered in solid form for fabrication purposes (Jadhav and Hocheng, 2015).

Biosorption

Microbial biomass (bacteria, fungi, etc.) performs ex situ in the remediation of heavy metals because of its cell wall structure (Mcafee et al., 2001; Das and Das, 2013). This cell wall contains several functional groups, mainly hydroxyl (–OH), thiol (–SH), carboxy (–COOH), and sulphonic acid (–HCO_3). In addition, the affinity of microbial biomass to sorbate (heavy metal ions) due to its metabolism further promotes adsorption binding. The interaction inactivates heavy metal ions through precipitation, complexation, or ion exchange (Veglio and Beolchini, 1997). Biosorption offers several advantages with respect to efficiency (highest possible), cost of nutrients (not required), and – most important – restoration of biosorbents as well as recovery of metals from waste (Hima et al., 2007; Jadhav and Hocheng, 2015).

Bioaccumulation and Principles

Unlike biosorption, bioaccumulation occurs in living things (such as microbial mass: fungi, bacteria, etc.) directly through intracellular sequestration of contaminant (like heavy metal ions). The process depends on organisms' metabolism, thus limiting it to living beings – thus the name, *bioaccumulation* (see Figure 12.1) (Brady and Duncan, 1994; Lo et al., 2014).

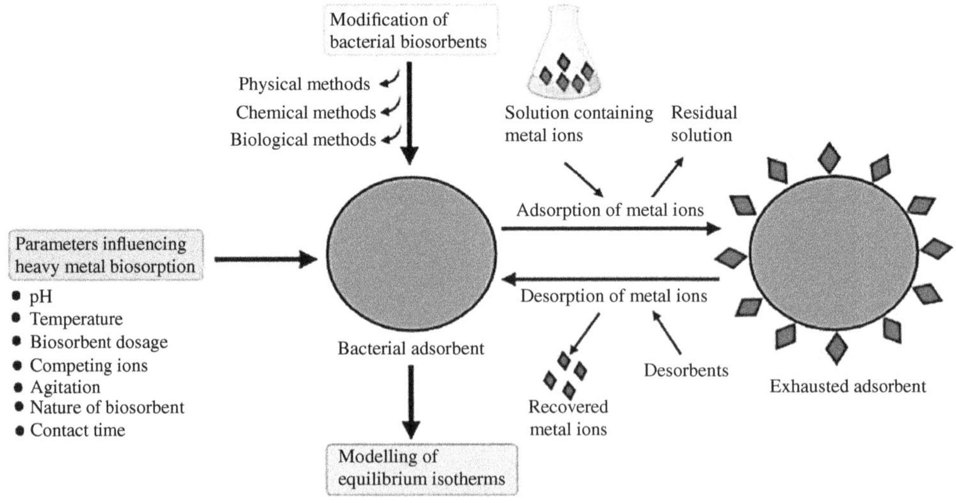

Figure 12.1 Schematic of biosorption and mechanisms for recovery and regeneration. *Source:* Priyadarshanee and Das, 2021

Bioaccumulation occurs in several stages with distinctive junctures, including metal-ligand interaction (functional groups), protein transport, intracellular binding, and accumulation. First, metal-ligand interaction promotes the formation of metal surface complexes (MSCs) (Olguín and Sánchez-Galván, 2012). These complexes are then transported into the cellular matrix by proteins through microbial ATP synthesis, resulting in intracellular metal-protein bindings such as Ppytochelatin (PC) and low-molecular-weight metallothionein (MT). The final sequestration occurs due to several biological processes (such as methylation). Additionally, the accumulated contaminants are trapped in a vacuole, thereby preventing harm to the organism (Mcafee et al., 2001; Chojnacka, 2010). Biosorption by various microbes is tabulated in Table 12.1. Table 12.2 lists several microbiomes for multiple intracellular sequestration.

Table 12.1 Biosorption of heavy metals by various microbial masses.

Microbe (binomial name)	Heavy metal	References
Aspergillus, Aspergillus niger, Aspergillus terreus	Pb (II)	(Sriharsha et al., 2017)
L. hexandra	Chromium (VI)	(Wang et al., 2020)
Bacillus thuringiensis var. thuringiensis	Chromium (VI)	(Şahin and Öztürk, 2005)
T. citronoviridae, F. solani	Pb, Cu	(Liaquat et al., 2020)
Pleurotus ostreatus	Cu(II), Cd(II), Cr(III)	(Kocaoba and Arısoy, 2011)
Mortierella	cobalt	(Pal et al., 2006)
Bacillus subtilis	Cd^{2+}	(Huang et al., 2020)
Geobacillus toebii sub.sp. decanicus, Geobacillus hermoleovorans sub.sp. stromboliensis	Cd^{2+}, Cu^{2+}, Ni^{2+}, Zn^{2+}, Mn^{2+}	(Özdemir et al., 2009)
Phomopsis sp. XP-8	Au^{3+}	(Xu et al., 2019)

Table 12.2 Intracellular sequestration of metals by various microbes.

Intracellular metal sequestration	Microbe/genome (binomial name)	References
Uranium	*R. arrhizus*	(Kuyucak and Volesky, 1988)
Uranium (VI), thorium (IV)	*A Pseudomonas*	(Sar et al., 2004)
Strontium ion (Sr(II)	*Bacillus pumilus*	(Dai et al., 2020)
^{137}Cs, ^{233}U, ^{239}Pu, ^{241}Am, ^{90}Sr, ^{237}Np	*Arthrospira platensis CNMN–CB-02*	(Zinicovscaia et al., 2020)
Zn	*Streptomyces K11*	(Sedlakova-Kadukova et al., 2019)
Uranium(VI)	*Bacillus mojavensis*	(Özdemir et al., 2017)
Pb(II)	*Penicillium funiculosum fungus*	(Mahmoud et al., 2012)

Table 12.3 Bioleaching of heavy metals with bioremediation agents.

Microbe/genome (bioremediation agents)	Heavy metal	References
Providencia sp.	Mn(II)	(Li et al., 2020a)
Acidithiobacillus ferrooxidans	Copper, iron	(Murugesan et al., 2020); (Chen et al., 2011)
Acidithiobacillus thiooxidans	Li, Co, Mn	(Naseri et al., 2019)
Talaromyces islandicus, Aspergillus terreus	Pb	(Sharma et al., 2020)

Biotransformation

Biotransformation involves redox-based enzyme-promoted transformations of contaminants (Jobby et al., 2018). In practice, these microorganism-based transformations enable precipitations of heavy metals followed by immobilization. These immobilized contaminants are insoluble or less soluble in water (Emenike et al., 2018).

Bioleaching

This is the process of converting toxic heavy metals to a non-toxic form through redox activity (Ishigaki et al., 2005; Schippers et al., 2013; Gomes et al., 2018; Kadivar et al., 2020). When the oxidation state is changed, metal ions precipitate. Once precipitated, they are insoluble in water and become immobilized. This process is an enzyme-catalyzed transformation. Basically, the microorganisms provide the reducing power as well as the oxidizing power (Yang et al., 2014; Gomes et al., 2020). Bioleaching by several microbes is summarized in Table 12.3.

Recovery of Metals through Used Biosorbents

Recovery of a Single Metal with Used Biosorbents

Several biosorbents have been reported in the treatment of effluents and are both economically and environmentally friendly. Biosorbents do not generate additional waste (also known as *secondary waste*) in the recovery process; nor is the microbial mass affected by the effluent composition (Mohammed and Chong, 2014; Ahmad and Haydar, 2016; E. et al., 2020). As a result, their recovery from secondary industry effluents becomes necessary and important. Among many hydrometallurgy methods (such as ion exchange and filtration), biosorption offers advantages that include low operating costs, enviro-friendly processes, and many more (Ayub et al., 2020). In addition, biosorption presents several advantages over traditional or conventional methodologies in terms of capital investment (installation, operation), energy, recovery efficiencies, and, more important, reusability (Jagaba et al., 2020).

Vanadium (V)

Vanadium is a ductile transition metal and is generally found as its most stable oxide (V) among other three oxide forms (II, III, IV). Its physical (stronger), chemical (catalyst), and optical properties make vanadium an essential element for a wide range of applications, including chemical industries (such as in the production of sulphuric acid), energy (batteries), machining tools (manufacturing high-speed cutting tools), aerospace applications (such as in the design of high-strength turbine blades), and electronics (supercapacitors) (Ajeya et al., 2020; Deliormanlı et al., 2020; Isacfranklin et al., 2020; Li et al., 2020b; Mangini et al., 2020). In addition, vanadium oxide is utilized for blocking infrared radiation.

Although vanadium is an important compound for industrial applications, treated and untreated effluents stress environmental resources due to the toxicity of all vanadium compounds. Conventional environmental remediation (such as filtration and other chemical processes) is undesirable and unsustainable for several reasons including high cost, generation of secondary waste, and energy-intensive processes. With advances in nanotechnology and materials science, biosorbents have gained interest due to their natural abundance, low cost, and ability to be reused. For instance, Ghanim et al. (2020) modified seaweed (*Ascophyllum nodosum*) hydro char with KOH and deployed it as a biosorbent to recover V(V) from alkaline residue leachates. The biosorbent was further optimized (pH 4, contact time 60 minutes, 293 K), thereby resulting in a maximum uptake of 12.3 mg/g.

Chromium (Cr)

Chromium is one of the most abundant transition metals in the earth's crust. The widespread presence of chromium in organic (chromium III) and inorganic systems (chromium VI) as a result of its varied stable oxidation states makes it an essential metal. In addition, chromium and its derivativeshave extensive industrial applications (chemicals, refractories, etc.) as well as biomedical applications (bio-imaging) (Zhang and Fan, 2020). However, this has also resulted in insensitive disposal of chromium (VI) into the environment through multiple means including industrial effluents (from metallurgy, tanning, and dyes) and incineration of electronic waste. As a result of the disposal of this industrial waste, these metals get into the soil and water, thereby entering the food chain.

Although the maximum uptake limit for hexavalent chromium is set at 0.5 mg/L, chromium (VI) concentrations several times higher have been found in waste. However, the many applications of chromium have also led to an increase in Cr (VI) demand due to the demand for electronic gadgets. As a result, the need to remediate and recover Cr (VI) through a green route is considered sustainable.

Li et al. (2008) developed functional magnetic beads as a biosorbent to recover chromium (VI) from wastewater through biosorption. The optimized (pH 1.0) biosorbent had a maximum sorption capacity of 6.73 mg/g. The sorption capacity was attributed to rapid separation and recovery as the result of incorporated magnetic nanoparticles. In another instance, olive stones were surface-modified using zero-valent iron and magnetite nanoparticles to recover Cr (VI) from synthetic solutions. After optimizing the sorption parameters (pH 3, contact time 120 minutes), the maximum adsorption was found to be 4.11 mg Cr/g. Furthermore, the recovery efficacy was more than 98% with a mixture of deionized water, oxalic acid, and NaOH (Vilardi et al., 2018). Recently, several biosorbents have been reported in this regard. For instance, an alginate-PEI membrane was deployed to recover Cr (VI) with a maximum adsorption capacity of 314 mg/g. This is attributed to ionic interactions between amine and carboxylic groups and crosslinked amine groups with glutaraldehyde (Mo et al., 2019).

Nickel (Ni)

Nickel is a lustrous, hard, ductile transition metal. In practice, it has been explored commercially for industrial applications such as electrochemistry, plastic manufacturing, energy devices, automobiles, textiles, dyes, and so on. As a result, these industries generate a huge quantity of heavy-metal-rich effluent, including nickel. Insensitive disposal into the environment causes the accumulation of nickel in the food chain, resulting in toxicity to terrestrial and aquatic life. Nickel toxicity leads to serious health concerns, including headache, nausea, breathing difficulties, and weakness.

To remediate this, industrial effluents have been conventionally treated with techniques including adsorption, ion-exchange resins, solvent extractions, and so on. Although these techniques help remediate the issue, they consume more resources in terms of energy and money. They also generate secondary waste, which causes additional environmental distress. Unlike these techniques, biosorbents treat effluent economically, require minimal energy expense, and can be reused for removing and recovering heavy metals. Zafar et al. (2007) obtained a biosorbent by treating biomass (rice bran) with H_3PO_4 for adsorbing nickel from an aqueous solution. This optimized (pH 6) biosorbent exhibited a maximum sorption activity of 102 mg/g.

Copper (Cu)

Cooper is a ductile, abundant (50 ppm), polycrystal metal. It can be extracted from a variety of minerals, such as chalcopyrite. The physical and chemical properties of copper make it an ideal candidate for several industrial applications. Consequently, these industries release treated, partially treated, or untreated effluents containing some percentage of lost metals, including copper. This unintended disposal results in severe environmental distress and serious effects in humans (cardiovascular problems, diabetes, etc.), animals, and plants.

Additionally, excessive absorption of copper by soil damages plants (such as leaf chlorosis) and retards normal growth. It also causes oxidative stress in reactive oxygen species, resulting in metabolic and intra-cellular damage. Conventional treatments like ion-exchange resins and precipitation have several disadvantages and are not economically viable for small-scale implementations. These shortcomings demand a technique that will de-stress the environment by remediating and providing recovery from industrial effluents and wastewater.

Biosorbents have been gaining worldwide attention because they can be tailored for enhanced copper binding. In addition, biosorbents allow low-cost, selective remediation and reusability with no secondary waste. For instance, Chua et al. (1998) obtained a biosorbent by using magnetite-immobilized cells of *Pseudomonas putida* 5-X to remove and recover copper from aqueous solution. The biosorbent exhibited a recovery efficiency of 95%. In another interesting report by Sinha et al. (2018), the authors developed a multi-stage metal recovery process involving biomass and electrochemical treatment to recover functioning copper from waste printed circuit boards. This biosorbent was obtained from dead microbial mass (*Aspergillus oryzae* and baker's yeast). The maximum sorption capacity was 92.7%, and the purity of the obtained copper was 95.2%.

Zinc (Zn)

Zinc is a moderately reactive chalcophile that normally occurs in oxidation state II. It is an abundant element (75 ppm) and has five isotopes (^{64}Zn, ^{66}Zn, ^{67}Zn, ^{68}Zn, and ^{70}Zn). Zinc metal is generally extracted from zinc blende minerals. It has a wide range of industrial applications in sectors such as energy, alloy productions (zinc carbonate, zinc chloride), organic chemical treatments, and electrochemistry (due to its reducing properties). However, the extensive industrial applications, particularly by electroplating industries, have led to the disposal of huge volumes of zinc in the environment. The zinc concentration in this waste is as much as 16 times greater than expected. This has resulted in environmental damage affecting terrestrial as well as aquatic life.

The acceptable exposure limit according to the US Environmental Protection Agency EPA is 4.2 mg/L. Remediation of zinc-affected effluents has been attempted through several techniques such as solvent extraction, precipitation, electrocoagulation, and filtration. Currently, remediation with biosorption is preferred over conventional techniques due to its advantages, including cost-effective remediation and recovery of zinc metal from waste/effluents. Recovered zinc can be reused by manufacturing industries, thereby reducing zinc demand. For instance, (Aftab et al., 2017) sequestered Zn (II) ions from an aqueous solution using fungal biomass (*Aspergillus oryzae*). After optimization, the recovery efficacy was 85.5%; this efficacy was maintained up to five sorption-desorption cycles. Additionally, the sorption activity was endothermic. Recently, Pan et al. (2019) employed a stable carboxymethylated biosorbent for remediation of copper from synthetic wastewater. The maximum sorption capacity was 76.92 mg/g. Furthermore, the adsorption of copper to the biosorbent was selective even in the presence of K^+ and Na^+. In addition, recovery efficiency was maintained at 92% even after eight adsorption-desorption cycles. This sorption performance was primarily attributed to the chemical adsorption and ion-exchange process between Cu^{2+} and -COONa. The sorption activity was also attributed to chelated oxygen atoms, which involved hydroxy and carboxyl groups.

Zirconium (Zr)

Zirconium, a corrosion-resistant d-block metal, naturally occurs in five isotopic forms (^{90}Zr, ^{91}Zr, ^{92}Zr, ^{94}Zr, and ^{9}Zr). With a half-life (time required to lose half of its initial mass) of 1.53×10^6 years, Zr^{90} has applications in nuclear technologies (such as fuel cladding) because neutrons can pass through it. In addition, zirconium is used in chemical industries, tanning industries, and so on (Zhang and Fan, 2020; Zoli et al., 2021). However, like other heavy metals, zirconium has also been reported to affect the environment.

Ruthenium (Ru)

Ruthenium is a relatively rare platinum group (ruthenium, rhodium, palladium, osmium, iridium, platinum) transition metal. It is often found in electronics (supercapacitors), solar energy (dye-sensitized solar cells [DSSCs]), aerospace vehicles (turbine blades), and textile industries (nozzles) due to its outstanding physical, mechanical, chemical, and electronic properties (Aguirre-Araque et al., 2020; B. et al., 2020; Mevada and Mukhopadhyay, 2020). It is also used as a catalyst in acetic acid chemistry, thereby generating effluents containing several heavy metals, including ruthenium. These effluents can be treated for recovery of ruthenium through hydrometallurgical and pyrometallurgical techniques, adsorption, and so on. (Kwak et al., 2013) used polyethylenimine (PEI)-modified bacterial as a biosorbent fiber for ruthenium recovery from acid wastewater. The modified biosorbent fiber exhibited a sorption rate of 110.5 mg/g (16 times higher than ion-exchange resin). Furthermore, the binding of negatively charged ruthenium was enabled by protonated amine groups in the biosorbent.

Palladium (Pd)

Palladium is a rare, lustrous platinum group metal. It is deployed in several manufacturing industries (such as oil industries) due to its catalytic nature. It is also deployed in catalytic converters in automobiles, fuel cells, water treatment solutions, dye remediation, and the purification of toxic gases due to its redox nature (Bhatt et al., 2020; Ghavidel Kalishomi et al., 2020; Mari et al., 2020; Freitas et al., 2021). Thus, the limited availability of palladium and its wide range of applications make recovery of the metal very important.

Cho et al. (2016) reported chemically modified biomass fiber (PEI-coated polysulfone/*Escherichia coli*) as a biosorbent for Pd (II) recovery (216.9 mg/g) from acidic effluents. The authors attributed the faster sorption rates (seven times higher than native biosorbent) to PEI modification. In addition, electrostatic interactions resulted in the formation of stable complexes.

Cadmium (Cd)

Cadmium is a rare (0.1 ppm), heavy (8.65 g/cm^3), highly toxic metal. It can exist in two oxidation states (I and II), of which (II) is stable and naturally occurring. Its physical properties enable cadmium to resist corrosion as well as act as a flame retardant. Cadmium is deployed in a wide range of applications, including solar energy, electrochemistry, nuclear technologies, televisions, and research laboratories. In addition, it is used in sensors for sensitive detection of cadmium in water (Al-Kuhaili 2021; Borella et al., 2021; Rabai et al., 2021).

A Food and Agriculture Organization (FAO) / WHO expert committee on food additives limited cadmium tolerance to 7 μg/kg body weight. Its ability to accumulate in the food chain makes cadmium toxicity a significant concern. This toxicity can trigger physical problems (cancer, weight loss, hypertension, etc.) as well as mental issues in humans (Alkharashi et al., 2017). Several conventional techniques (precipitation, adsorption) to remediate cadmium toxicity have been implemented. However, higher implementation costs and inefficient treatment techniques mean cadmium contamination has been only partially alleviated, and the generation of secondary waste results in additional environmental distress. This has led to a search for a sustainable material to remediate, recover, and permit the reuse of cadmium.

For example, (Gupta and Nayak 2012) modified agricultural orange peel powder with a magnetic nano-adsorbent for cadmium ion removal and recovery from aqueous solution. The maximum removal efficacy by the biosorbent was 76.92 mg/g (82%). So far, the morphological properties of biosorbents are controlled by variables like contact time, metal concentration, pH, ionic strength, and organic matter. The biosorbent exhibited an optimized contact time of 40 minutes (90 minutes for unmodified orange peels), attributed to surface adsorption due to impermeable adsorption. Cadmium sorption was found to increase with the initial cadmium concentration in the effluent due to the abundance of cadmium ions. In addition, the modified biosorbent had a faster sorption rate when the pH was increased and leveled off (2–7) due to the presence of hydroxyl groups, resulting in enhanced cadmium adsorption. A decrease in electrolytic concentration also facilitated the complexation tendencies of cadmium ions with hydroxyl functional groups due to its ionic nature. In terms of recovery, the adsorbent was regenerated using 0.1 M HNO3 and resulted in 98% cadmium recovery.

In a similar instance, Bulgariu and Bulgariu (2016) deployed activated microbial biomass (algae waste activated by alkaline treatment) as a biosorbent for recovery and removal of cadmium (II) from an industrial effluent. The biosorbent exhibited a sorption capacity of 53.19 mg/g, which was 53.68% higher than the native microbial biomass. This was due to increased porosity and surface area, which increased not only the functional groups but also the number of available sites for sorption activity.

Lanthanum (La)

Lanthanum is a highly reactive, ductile, transition rare earth metal in the lanthanide series. It usually occurs in the (III) oxidation state and has strong basic activity. Its wide range of industrial applications includes batteries, hydrogen storage, and cathode materials. In addition, lanthanum has experienced increased industrial demand due to applications such as electronic gadgets (computers, smartphones, etc.), semiconductors, lasers, and catalysts. Because lanthanum is rare, it is imperative to recover it from effluent/waste using a cost-effective and environmentally friendly technique.

Biosorption facilitates remediation as well as recovery, and several biosorbents have been deployed. For instance, Birungi and Chirwa (2014) investigated a cost-effective method by deploying a micro-algal sorbent (*Desmodesmus multivariabilis*) for recovering lanthanum from a standard stock solution. The biosorbent showed a maximum sorption capacity of 100 mg/g. The desorption efficiency at the beginning of the sorption activity was 97.17% and was maintained at 77.2% for subsequent cycles. In another instance, chemically modified carboxylated cellulose fibers were deployed to remove and recover lanthanum (III) with maximum adsorption activity of 33.7 mg/g. This was attributed to the introduction of carboxylate groups through TEMPO-mediated oxidation, which facilitated electrostatic interactions between carboxylate groups and La (III) ions (Birungi and Chirwa 2014).

Neodymium (Nd)

Neodymium is a rare earth element from the lanthanide series. It has several industrial applications, particularly in the manufacturing of magnets, lasers, and glass, which makes the metal even more precious. In addition, neodymium is deployed in biomedical products, electrode materials in supercapacitors, and fuel cells (Rana et al., 2015; Hu et al., 2017a; Neelima et al., 2019; Ansari et al., 2020; Kubra et al., 2020). Consequently, discharged effluents from these industries contain a considerable amount of heavy metals, including neodymium. Even at low concentrations, its non-biodegradability makes neodymium even more toxic, mainly affecting aquatic life.

Rhenium (Re)

Rhenium is one of the rarest transition elements on the planet (1 ppb in the earth's crust). Its physicochemical characteristics have led to its application in advanced technologies such as reactors, smelters, solar energy, and oxygen sensing (Komreddy et al., 2020; Nogami et al., 2020; Valdés et al., 2020; Zhang et al., 2020). Further, it is used as a catalyst (catalytic reforming) for obtaining high-octane gasoline products in petrochemical industries. Rhenium is derived through pyrometallurgical processes, particularly from copper sulfide ores. However, low concentration of Re, ~ 30 mg/L effluent (perrhenate salts, Re(VII), ReO_4^-), are also found in the primary effluent (acid wastewater), thereby providing a secondary source for retrieval. Several biosorbents (such as biochar) have been reported for wastewater treatment; they can also be used to recover Re (VII) from acid wastewater.

Platinum (Pt)

Platinum is a precious, rare (0.005 ppm), noble transition metal belonging to the platinum group of metals. Its intriguing physical (such as corrosion resistance), chemical, and electrical properties have led to a wide range of applications including sensors (thermometers, research-grade metrological devices, etc.), catalytic converters in automobiles, and fuel cells (Mróz et al., 2019; Wang et al., 2019b). Consequently, these industries dispose of solid waste and wastewater containing a significant amount of deactivated platinum-based catalysts. Although pyrometallurgical and hydrometallurgical techniques are often used to treat these wastes before their disposal, they are expensive in terms of energy and capital and generate secondary effluent. Functionalized biosorbents, on the other hand, are selective to adsorbates and can recover platinum from the primary effluent. Moreover, biosorbents can be reused over and over with similar recovery efficiencies.

Mao et al. (2010) chemically modified microbial biomass (*E. coli*) using PAA/HCl to recover hexachloroplatinate. Further sorption activity (4.36 times higher than native microbial biomass) was attributed to the electrostatic interaction between the biomass surface and cationic polymer. Regeneration of the biosorbent was observed for four cycles under optimized conditions (pH 10.02). Later, the authors also observed the sorption behavior of PAA/HCl-modified *E. coli* chitosan fiber for the recovery of Pt from an aqueous solution. The biosorbent exhibited a desorption efficiency of 99.7% initially, followed by maintained desorption efficacy at 90.2% for the subsequent five cycles. The sorption activity was attributed to amine groups in the chitosan matrix, enabling electrostatic attraction for [PtCl6]2. Additionally, maximum sorption uptake was recorded at 421.1 mg/g with 77.42% purity. PEI-modified nano-celluloses were used as biosorbents to recover platinum from simulated automobile catalyst leachate. The maximum sorption capacity of 600 mg/g was due to an enhanced specific area of sorption sites. The authors also found that sorption capacity is proportional to the PEI density (Hong et al., 2019).

Gold (Au)

Gold is a noble, rare transition metal. Its fascinating physical, electrical, thermal, and optical properties have led to its deployment in a wide range of applications. For instance, its minimal electrical resistivity has led to its extensive use in the computer industry, especially through electroplating. In addition, it is used in biomedical applications, catalysts, biosensors, and electrodes for sensors due to its physicochemical and optical properties

(Mie et al., 2020; Nekounam et al., 2020; Gonzalez and Sequí-Castellano, 2021; Li et al., 2021). Gold has also been reported to exhibit antibacterial properties (Zhang et al., 2021).

Many of these applications, such as electroplating, involve acids (such as EDTA), alkalines (KCN), and neutral electrolytes (for example, sodium phosphates). Specific gold concentrations are introduced into electrolyte baths. Although these baths are chemically treated for pH neutrality before disposal into the environment, they still contain a reasonable gold concentration, which can be lost if not recovered. Hence, given the rarity of gold and its wide range of applications, recovery of gold from bath effluents becomes a necessary source of obtaining gold in metal ion (Au III) and zero-valent (Au I) forms. Unlike ion exchange or activated charcoal, biosorption allows the recovery of gold metal through passive physicochemical interactions at the biosorbent-metal interface in an ionic solution. Furthermore, biosorbents are unaffected by heavy metals and thus (for example, lignin, microbial biomass, etc.) can be reused, thereby reducing recovery costs (handling, maintenance, etc.) costs.

For instance, Ishikawa et al. (2002) used eggshell membrane as a biosorbent for recovery of gold ions [dicyanoaurate(I) and tetrachloroaurate(III)] from several sources (aqueous solution, cyanide bath). The maximum sorption capacities were 147 mg Au(I)/g (dry weight) and 618 mg Au(III)/g (dry weight). Interestingly, no changes in sorption activity of gold uptake were observed for five successive sorption-desorption cycles. Kwak et al. (2010) comparatively studied the sorption behavior of chemically modified microbial biomass (PEI-modified *Corynebacterium glutamicum*) with ion-exchange resin and activated carbon (granular) for the recovery of gold from a cyanide bath. The purity of the recovered gold was highest for microbial biomass (97.72%) in comparison to ion-exchange resin (96.56%) and activated carbon (94.57%). Similarly, Park et al. (2012) reported the recovery of gold in porous gold fibers from aqueous solutions through PEI-modified bacterial (*Corynebacterium glutamicum*) biosorbent. The fiber form was achieved through incineration. The maximum sorption uptake was recorded at 421.1 mg/g with 77.42% purity.

Lead (Pb)

Lead is an unreactive, chalcophile, post-transitional heavy metal (10.66 g/cm^3). It is also one of the most abundant elements on the planet (14 ppm) and generally exists within zinc ores. It is primarily used in energy storage devices (lead-acid batteries), dyes, optical displays, solar cells, and seismic sensors (Basavapoornima et al., 2020; Luo et al., 2020; Shaafi et al., 2020; Shi et al., 2020). Industries involved in manufacturing lead-based products often treat their effluents with conventional processes that are energy intensive and only partially effective in remediation, resulting in significant environmental damage.

Advances in Multi-Metal Recovery with Used Biosorbents

Saeed et al. (2005a) studied the sorption activity of papaya wood against copper (II), cadmium (II), and zinc (II). They attributed the metal recovery efficacy to the quantity of biomass used as the result of increased sorption sites. The efficacy remained unchanged even after five successive sorption-desorption cycles. In another instance, Farnane et al. (2017) investigated the removal of cadmium (II) and cobalt (II) from an aqueous solution by chemically pretreating raw carob shells with NaOH. They also optimized the functional variables (pH, adsorbent dosage, contact time, initial adsorbate concentration, and temperature) and enhanced and increased the selectivity of biosorbent toward the removal of cadmium (II) and cobalt (II). Recently,

Taksitta et al. (2020) improved the adsorption capacity of cellulose fiber by using epichlorohydrin as a cross-linking agent in the treatment of wastewater for Cu(II), Cd(II), and Pb(II). The maximum sorption capacities were 72.99 mg/g, 59.52 mg/g, and 38.02 mg/g, respectively. They attributed the reduced sorption capacities after five cycles to the inability of metal ions to vacate sorption sites following desorption. Likewise, Wang et al. (2021) used *Nephrolepis cordifolia* leaves along with cake-like biosorbents from tea waste to recover heavy metals Pb (II), Cu (II), and Cd (II), with maximum adsorption capacities of 247.20, 85.80, and 265.31 mg/g. Selective sequestration was attributed to ion-exchange, complexation, and microprecipitation mechanisms.

Herrera-Barros et al. (2020) recently reported 87 and 81% selective removal efficacies for cadmium and nickel, respectively, with oil palm bagasse and dimethyl sulfoxide (DMSO) modified alumina nanoparticles from industrial effluents. This removal was accounted for by the physicochemical properties of cadmium and nickel and chemical modification of the biosorbent. Dan et al. (2021) investigated sodium hydroxide-modified rice hulls to recover a wide range of heavy metals: Be^{2+}, Al^{3+}, Cr^{3+}, Co^{2+}, Ni^{2+}, Cu^{2+}, Zn^{2+}, Sr^{2+}, Ag^{+}, Cd^{2+}, Ba^{2+}, and Pb^{2+}. The authors attributed the removal and recovery of heavy metals to the electrostatic interaction between the rice hulls (negatively charged) and metal ions (positively charged). They also found that the pH of the effluent had a significant impact on sorption activity.

Feisther et al. (2019) reported the recovery of Cd^{2+} (15 mg/g), Pb^{2+} (45 mg/g), and Ni^{2+} (mg/g) from an aqueous solution (pH 3.8) by using acid/alkaline-treated *Eucalyptus globulus* residue biomass. Furthermore, they observed reduced efficacies after three sorption-desorption cycles due to an incomplete desorption process. Varala et al. (2017) reported the recovery of radionucleotides (thorium and zirconium) by using de-oiled Karanja seed cake. They also optimized desorption (96%, Thorium; 67%, zirconium) efficiency, attributed to its dependency on the eluent concentration. The occurrence of desorption was accounted for by h ion-exchange of metal ions with anion or cations in the eluent.

Basu et al. (2019) reported humic acid-modified rice husk (cellulose) as a biosorbent to remove nickel and chromium from an aquatic medium (potable water). Optimized (pH 5) sorption capacities (12.41 mg/g, nickel; 19.39 mg/g, chromium) were mostly attributed to humic acid modification due to the attraction of opposite charges possessed by the functional groups (carboxylic, ketonic, alcoholic, etc.). Ajao et al. (2020) obtained extracellular polymeric substance (EPS) biosorbent from nitrogen-limited glycerol/ethanol-rich wastewater and immobilized it to recover Cu^{2+} and Pb^{2+} from aqueous solutions. The EPS adsorbed 99.9% of influent metal (Cu^{2+} and Pb^{2+}). Tecovery efficacies were seen at 86 and 90% for Cu^{2+} and Pb^{2+}, respectively. These sorption-desorption activities were primarily attributed to the interaction between PEI and EPS and many available sites for sorption.

Saeed et al. (2005b) obtained a biosorbent by milling black gram for multi-metal recovery (Pb, Cd, Zn, Cu, and Ni). The maximum sorption capacity of the optimized (ph 5) biosorbent for recovery of Pb, Cd, Zn, Cu, and Ni was 49.97, 39.99, 33.81, 25.73, and 19.56 mg/g respectively. In addition, the authors found that the sorption of metal ions increased with initial metal concentration in the effluent. Lapo et al. (2020) used lignocellulosic biomass (banana rachis, banana pseudo-stem, and banana peel) as an adsorbent for recovery of rare earth metals (Nd^{3+}, Eu^{3+}, Y^{3+}, Dy^{3+}, and Tb^{3+}). The sorption capacity was around 100 mg/g for Nd^{3+}, Eu^{3+}, Y^{3+}, Dy^{3+}, and Tb^{3+} and was desorbed using EDTA. The desorption efficacy was maintained at 97%. Recent advances in metal recovery with used biosorbents are shown in Table 12.4.

Table 12.4 Advances in multi-metal removal by various biosorbents.

Biosorbent	Recovered metal	q_m, (mg/g) or percentage recovery efficiency (%)	References
Ca-alginate immobilized *Providencia vermicola*	Palladium (II)	197.23 mg/g	(Xie et al., 1996)
Soluble tannins, cross-linked *Lagerstroemia speciosa* leaf powder	Palladium (II)	46.3 mg/g, 99%	(Tsezos et al., 1989)
Cross-linked waste biomass (*Escherichia coli*) with polyallylamine hydrochloride (PAH)	Palladium (II), zero-valent palladium	265.3 mg/g	(Park et al., 2010)
Pd(II)-sorbed poly-ethylenimine-modified corynebacterium glutamicum	Palladium (II)	136.9 mg/g	(Won et al., 2013)
Functionalized wood pulp	Chromium (VI)	24.56 mg/g	(Varshney et al., 2019)
Black tea	Chromium (VI), Aluminium (III)	97%	(Bhatti and Yasir, 2016)
Laminaria seaweed	Chromium (III)	42 mg/g	(Dittert et al., 2012)
Saccharomyces cerevisiae	Chromium (III)	77%	(Şahin and Öztürk, 2005)
Coffee husks	Aroma compounds	Eugenol, 79.44%; eugenol acetate compound, 54.05%	(Maimulyanti et al., 2019)
T-fonticaldi AT-A2 proteins	Au (III)	1.45 mg Au/mg, 71%	(Han et al., 2017)
Lignin-based composite	Au (III) ions	4193 mg/g, 70.9%	(He et al., 2013)
Surface-functionalized biomass (PEI-modified *Lagerstromeia speciosa* leaf powder)	Au	>95%	(Choudhary et al., 2018)
Cladosporium Cladosporioides biomass beads	Au	36 mg/g, 55%	(Pethkar and Paknikar 1998)

(*Continued*)

Table 12.4 (Continued)

Biosorbent	Recovered metal	q_m, (mg/g) or percentage recovery efficiency (%)	References
Decarboxylated (*Corynebacterium glutamicum*) biosorbent	Zero-valent gold	86.16 mg/g, 93%	(Kwak and Yun 2010)
Nano-zno functionalized biochar	Re (VII)	24.5 mg/g	(Hu et al., 2019)
Agricultural waste	Zr	107.4 mg/g	(Kausar et al., 2016)
Candida tropicalis	Zr	179 mg/g	(Akhtar et al., 2008)
Polyethylenimine (PEI)-modified bacterial biosorbent fiber	Ru	110 mg/g	(Won et al., 2015)
Rhodopseudomonas palustris strains	Ru	40 mg/g, 72%	(Colica et al., 2012)
Amberlite ion-exchange resin	Nickel (II)	30.48 mg/g	(Kwak et al., 2011)
Fungal-based biosorbent	Nickel (II)	214 mg/g	(Suhasini et al., 1999)
Agarose gel	Pb (II)	115 mg/g	(Pandey et al., 2009)
Marine algae (*Sargassum*)	Pb (II)	95%	(Jalali et al., 2002)
Fungal-based (*Cladosporium cladosporioides*) biosorbent	Tetracyanocuprate(II), tetracyanonickelate(II)	95%	(Patil and Paknikar 1999a)
G-PGA and its sodium salt (g-pgana)	Neodymium	305 mg-Nd/(g-g-PGANa)$^{-1}$	(Hisada and Kawase 2018)
Lagerstroemia speciosa biomass modified by polyethylenimine (PEI-LS)	Platinum	122-154 mg/g, >90%	(Garole et al., 2018)
Polyethylenimine (PEI)-modified (inactive *Escherichia coli*) biomass	Platinum	108.8 mg/g, 98.7%	(Won et al., 2010)
Neem (*Azadirachta indica*)	Trace metal isotopes: hard, borderline, soft metals	99.999%	(Nahar and Zhang, 2013)
Fungal based *Ganoderma lucidum*	U(VI)	8.98 mg/g, 100%	(Kulshrestha and Venkobachar 2008)
Cladosporium cladosporioides	Copper cyanide, nickel cyanide	99.9%	(Patil and Paknikar 1999b)
Fsh scales, neem sawdust	Lanthanum (III)	84.6%	(Das et al., 2014)

Adsorption Kinetics

Adsorption kinetics enable the computation of several adsorption-related quantities such as maximum quantity adsorbed, equilibrium rate constants, and so on. The curves present sorption behavior with time. The methodology involves the description of experimental data according to suitable kinetic models. The best fit is usually determined by the highest correlation coefficient (R) through kinetic models. Kinetic models are shown in Table 12.5, and adsorption isotherm models are listed in Table 12.6.

The adsorption rate is given by

$$R\% = \frac{C_0 - C}{C_0} \times 100 \tag{12.1}$$

where C_0 = initial concentration, (mg/L) and C = equilibrium concentration (mg/L).

The adsorption capacity is calculated using

$$q_t = \frac{(C_0 - C)V}{M} \tag{12.2}$$

Table 12.5 Kinetic models and their expressions.

Kinetic model	Expressions, q_t	Usual terms/nomenclature	References
Pseudo-first-order	$q_t = q_e\left(1 - e^{-k_1 t}\right)$	q_t: Sorption quantity (metal) at time t, mg/g q_e: Sorption quantity (metal) at equilibrium state, mg/g k_1: First-order equilibrium rate constant, min^{-1}	(He et al., 2013)
Pseudo-second-order	$q_t = \frac{q_e^2 k_2 t}{1 + q_e k_2 t}$	k_2: Second-order equilibrium rate constant, g/mg/min	(He et al., 2013)
Elovich	$q_t = \frac{1}{\beta}\ln\left(1 + \alpha\beta t\right)$	β: Elovich constants t: Time, in hours	(Ritchie, 1977)
Two-compartment first-order	$\frac{q_t}{q_e} = F_{fast}\left(1 - \exp\left(-F_{fast}t\right)\right) + F_{slow}\left(1 - \exp\left(-K_{slow}t\right)\right) F_{fast} + F_{slow} = 1$	F_{fast} : Mass fraction rate, fast F_{slow} : Mass fraction rate, slow K_{slow} : Tirst-order rate constant into slow	(Hu et al., 2017b)
Intra-particle diffusion	$q_t = C + k_i\sqrt{t}$	C: Interparticle diffusion constant, mg/g k_i: Interparticle diffusion constant, mg $h^{0.5}$/g	(Pholosi et al., 2020)
Liquid film diffusion	$\ln(1 - F) = K_f t$; $F = \frac{q_t}{q_e}$	K_f: Adsorption rate constant	(Salam et al., 2014)

Table 12.6 Adsorption isotherms and their expressions.

Model	Expression	Usual terms/nomenclature	References
Langmuir	$q_e = \frac{q_m K_L C_r}{1 + k_L C_r}$ $R_L = \frac{1}{1 + K_L C_m}$	q_e: Quantity adsorbed, mg/g q_m: Maximum quantity adsorbed, mg/g b: Langmuir constant, L/mg R_L: Separation factor $R_L > 1$: Unfavorable $R_L = 0$: Linear $0 < R_L < 1$: Favorable $R_L < 0$: Irreversible	(Langmuir, 1916) (Ramanaiah et al., 2007)
Freundlich	$q = K_f C_r^{1/n}$	n: Heterogeneity factor, ($n < 1$: unfavorable, $n >$: favorable, $n = 1$: linear reaction) K_f: Freundlich constant	(Freundlich, 1907)
Dubinin-Radushkevich (D–R)	$\ln q_e = \ln q_m - kp^2$ $E = \frac{1}{(2k)^{-0.5}} p$ $= RT \ln\left(1 + \frac{1}{C_e}\right)$	E: Free energy, kJ/mol k: Dubinin-Radushkevich constant p: Polanyi's adsorption free energy, J/mol R: Universal gas constant T: Temperature, K	(Dubinin, 1947)
Redlich-Peterson	$q_e = \frac{K_{RP} c_r}{1 + a_{RP} C_r^g}$	K_{RP}, A_{RP}, g: Redlich-Peterson constants Cr: Concentration, mg/L	(Redlich and Peterson, 1959)
Temkin	$q_e = \frac{RT}{bT} \ln(k_T c_r)$	K_T: Equilibrium bond constant associated with maximum energy of the bond	(Temkin et al., 1940)

The desorption rate is calculated using

$$D\% = \frac{Desorbed\ quantity\,(mg)}{Initial\ adsorbed\ quantity\,(mg)} * 100 \tag{12.3}$$

The nature of sorption activity can be determined thermodynamically through the parameters listed in Table 12.7.

Current Challenges

In practice, selective adsorption of metal ions by biosorbents is not significant, given the quantum and diversity of metals present in industrial waste. The adsorbing capacity per biosorbent is limited, making them a less-than-ideal choice for industrial

Table 12.7 Thermodynamic expressions for Gibbs free energy change.

Thermodynamic parameter	Thermodynamic expression	Usual notations	References
Gibbs free energy change: ΔG^0 (KJ/mol)	$\Delta G^0 = -RT \ln K$ $\Delta G^0 = \Delta H^0 - T\Delta S^0$	K: Thermodynamic equilibrium	(Xie et al., 2020)
Enthalpy change: ΔH^0 (KJ/mol)		T: Absolute temperature (kelvin)	(Xie et al., 2020)
Entropy change: ΔS^0 (J/mol/K)		$\Delta G°$: Gibbs free energy change	(Xie et al., 2020)
		R: Gas constant (8.314 J/mol/K)	
		$\Delta H°$: Enthalpy change	
		$S°$: Entropy change	

applications. This means industrial effluents having lower metal concentrations would be less efficient for metal recovery. The expenditure incurred when recovering metals would also be much higher in such cases. When sorbents are deployed with microbial mass, separating it from industrial effluent is difficult and energy sensitive. This is usually due to the physicochemical properties of the microbial physique, such as smaller particle dimensions. Unless it is chemically modified, raw biomass is reported to have lower adsorption capacity as well as reusability. Although this augmentation improves sorption characteristics, these also result in the formation of secondary waste, which again increases the environmental burden.

Conclusion

The development of sustainable approaches involving used biosorbents and their deployment for metal recovery could present a source in addition to raw material. This would reduce the environmental impact caused by the continuous exploitation of natural resources by industries. Consequently, such an approach would result in economic benefits by conserving the environment and its resources. Biosorption studies reveal that untreated biomaterials exhibit good sorption activity. However, recent studies also report the impact of chemical modification in treating wastewater/industrial effluents. The studies show that chemically modified biosorbents exhibit enhanced sorption rates compared to native or naturally occurring biosorbents.

This chapter has described several biomaterial/microbial biomass sorption activities. Sorption studies indicate that the participation of specific functional group groups (hydroxyl, carboxyl, amines, etc.) leads to higher metal adsorption rates due to the interaction between metal and the biosorbent surface. The effectiveness of biosorbents is also affected by the pH profile of the effluents. From a thermodynamic point of view, adsorption is a continuous, endothermic process. Based on recent advances, it can be concluded that biosorbents are a cost-effective, convenient, and efficient way to remediate and recover metals from waste.

Summary

Biosorption is a promising technique to deploy used biosorbents in the recovery of heavy metals. This chapter has explored several microbial biomasses and biomaterials have been explored to recover various metals from industrial effluents and wastewater. Chemically modified biomass exhibits better sorption than native or natural forms in terms of sorption activity. The chapter has also discussed the mechanisms involved in biosorption, factors affecting desorption activity, and recent advances in bioremediation of heavy metals. It can be concluded that biosorption is a viable alternative to conventional remediation processes.

References

Aftab, K., Akhtar, K., Kausar, A. et al. (2017). Fungal strains isolation, identification and application for the recovery of Zn(II) ions. *J Photochem Photobiol B Biol* 175: 282–290. doi:10.1016/j.jphotobiol.2017.08.028.

Aguirre-Araque, J.S., Guimaraes, R.R., and Toma, H.E. (2020). Chemistry of ternary monocarboxyterpyridine-bipyridine-trimercaptotriazine ruthenium complexes and application in dye sensitized solar cells. *Polyhedron* 182: 114513. doi:10.1016/j.poly.2020.114513.

Ahmad, M., Yousaf, M., Nasir, A. et al. (2019). Porous Eleocharis@MnPE Layered Hybrid for Synergistic Adsorption and Catalytic Biodegradation of Toxic Azo Dyes from Industrial Wastewater. *Environ Sci Technol* 53: 2161–2170. doi:10.1021/acs.est.8b05866.

Ahmad, M.F. and Haydar, S. (2016). Evaluation of a newly developed biosorbent using packed bed column for possible application in the treatment of industrial effluents for removal of cadmium ions. *J Taiwan Inst Chem Eng* 62: 122–131. doi:10.1016/j.jtice.2015.12.032.

Ajao, V., Nam, K., Chatzopoulos, P. et al. (2020). Regeneration and reuse of microbial extracellular polymers immobilised on a bed column for heavy metal recovery. *Water Res* 171: 115472. doi:10.1016/j.watres.2020.115472.

Ajeya K V, Sadhasivam, T., Kurkuri, M.D. et al. (2020). Recovery of spent VOSO4 using an organic ligand for vanadium redox flow battery applications. *J Hazard Mater* 399: 123047. doi:10.1016/j.jhazmat.2020.123047.

Akhtar, K., Akhtar, M.W., and Khalid, A.M. (2008). Removal and recovery of zirconium from its aqueous solution by Candida tropicalis. *J Hazard Mater* 156: 108–117. doi:10.1016/j.jhazmat.2007.12.002.

Al-Kuhaili, M.F. (2021). Photoelectric properties of highly conductive samarium-doped cadmium telluride thin films for photovoltaic applications. *Sol Energy* 213: 163–171. doi:10.1016/j.solener.2020.11.028.

Alkharashi, N.A.O., Periasamy, V.S., Athinarayanan, J., and Alshatwi, A.A. (2017). Sulforaphane mitigates cadmium-induced toxicity pattern in human peripheral blood lymphocytes and monocytes. *Environ Toxicol Pharmacol* 55: 223–239. doi:10.1016/j.etap.2017.08.026.

Amphlett, J.T.M., Ogden, M.D., Foster, R.I. et al. (2018). Polyamine functionalised ion exchange resins: Synthesis, characterisation and uranyl uptake. *Chem Eng J* 334: 1361–1370. doi:10.1016/j.cej.2017.11.040.

Ansari, A.A., Khan, A., Alam, M. et al. (2020). Optically active neodymium hydroxide surface-functionalized mesoporous silica micro-cocoons for biomedical applications. *Colloids Surfaces B Biointerfaces* 189: 110877. doi:10.1016/j.colsurfb.2020.110877.

Ayub, A., Raza, Z.A., Majeed, M.I. et al. (2020). Development of sustainable magnetic chitosan biosorbent beads for kinetic remediation of arsenic contaminated water. *Int J Biol Macromol* 163: 603–617. doi:10.1016/j.ijbiomac.2020.06.287.

Bajwa, F., Ahmed, N., Khan, M.A. et al. (2020). Evaluating the ion exchange chromatography for matrix-assisted PEGylation and purification of consensus interferon. *Biotechnol Appl Biochem* 67: 196–205. doi:10.1002/bab.1832.

Ball, V., Winterhalter, M., Schwinte, P. et al. (2002). Complexation mechanism of bovine serum albumin and poly(allylamine hydrochloride). *J Phys Chem B* 106: 2357–2364. doi:10.1021/jp012522m

Basavapoornima, C., Kesavulu, C.R., Maheswari, T. et al. (2020). Spectral characteristics of Pr3+-doped lead based phosphate glasses for optical display device applications. *J Lumin* 228: 117585. doi:10.1016/j.jlumin.2020.117585.

Basu, H., Saha, S., Mahadevan, I.A. et al. (2019). Humic acid coated cellulose derived from rice husk: A novel biosorbent for the removal of Ni and Cr. *J Water Process Eng* 32: 100892. doi:10.1016/j.jwpe.2019.100892.

Bhatt, R., Sharma, A.K., Himanshi et al. (2020). Palladium complexes of chalcogenoethanamine (S/Se) bidentate ligands: Applications in catalytic arylation of CH and OH bonds. *Polyhedron* 185: 114597. doi:10.1016/j.poly.2020.114597.

Bhatti, H.N. and Yasir, M. (2016). Removal and recovery of Al(III) and Cr(VI) from aqueous solution by waste black tea. *Environ Eng Manag J* 15: 809–816. doi:10.30638/eemj.2016.087.

Birungi, Z.S. and Chirwa, E.M.N. (2014). The kinetics of uptake and recovery of lanthanum using freshwater algae as biosorbents: Comparative analysis. *Bioresour Technol* 160: 43–51. doi:10.1016/j.biortech.2014.01.033.

Borella, A., Bruggeman, M., Rossa, R., and Schillebeeckx, P. (2021). Peak shape calibration of a Cadmium Zinc Telluride detector and its application for the determination of uranium enrichment. *Nucl Instruments Methods Phys Res Sect A Accel Spectrometers, Detect Assoc Equip* 986: 164718. doi:10.1016/j.nima.2020.164718.

Brady, D. and Duncan, J.R. (1994). Bioaccumulation of metal cations by Saccharomyces cerevisiae. *Appl Microbiol Biotechnol* 41: 149–154. doi:10.1007/BF00166098.

Bulgariu, D. and Bulgariu, L. (2016). Potential use of alkaline treated algae waste biomass as sustainable biosorbent for clean recovery of cadmium(II) from aqueous media: batch and column studies. *J Clean Prod* 112: 4525–4533. doi:10.1016/j.jclepro.2015.05.124.

Burakov, A.E., Galunin E V., Burakova I V. et al. (2018). Adsorption of heavy metals on conventional and nanostructured materials for wastewater treatment purposes: A review. *Ecotoxicol Environ Saf* 148: 702–712. doi:10.1016/j.ecoenv.2017.11.034.

Chen, H., Xie, A., and You, S. (2018). A review: advances on absorption of heavy metals in the waste water by biochar. *IOP Conf Ser Mater Sci Eng* 301: 012160. doi:10.1088/1757-899X/301/1/012160.

Chen, P., Yan, L., Leng, F. et al. (2011). Bioleaching of realgar by Acidithiobacillus ferrooxidans using ferrous iron and elemental sulfur as the sole and mixed energy sources. *Bioresour Technol* 102: 3260–3267. doi:10.1016/j.biortech.2010.11.059.

Cho, C.-W., Kang S Bin, Kim, S. et al. (2016). Reusable polyethylenimine-coated polysulfone/bacterial biomass composite fiber biosorbent for recovery of Pd(II) from acidic solutions. *Chem Eng J* 302: 545–551. doi:10.1016/j.cej.2016.05.091.

Choi, J.-W., Song, M.-H., Bediako, J.K., and Yun, Y.-S. (2020). Sequential recovery of gold and copper from bioleached wastewater using ion exchange resins. *Environ Pollut* 266: 115167. doi:10.1016/j.envpol.2020.115167.

Chojnacka, K. (2010). Biosorption and bioaccumulation – the prospects for practical applications. *Environment International* 36 (3): 299–307. doi:10.1016/j.envint.2009.12.001.

Choudhary, B.C., Paul, D., Borse, A.U., and Garole, D.J. (2018). Surface functionalized biomass for adsorption and recovery of gold from electronic scrap and refinery wastewater. *Sep Purif Technol* 195: 260–270. doi:10.1016/j.seppur.2017.12.024.

Chua, H., Wong, P.K., Yu, P.H.F., and Li, X.Z. (1998). The removal and recovery of copper (II) ions from wastewater by magnetite immobilized cells of Pseudomonas putida 5-X. *Water Sci Technol* 38: 315–322. doi:10.1016/S0273-1223(98)00514-9.

Colica, G., Caparrotta, S., and De Philippis, R. (2012). Selective biosorption and recovery of Ruthenium from industrial effluents with Rhodopseudomonas palustris strains. *Appl Microbiol Biotechnol* 95: 381–387. doi:10.1007/s00253-012-4053-9.

Crini, G., Lichtfouse, E., Wilson, L.D., and Morin-Crini, N. (2019). Conventional and non-conventional adsorbents for wastewater treatment. *Environ Chem Lett* 17: 195–213. doi:10.1007/s10311-018-0786-8.

Sriharsha, D.V., Lokesh, K.R., and Savitha, J. (2017). Immobilized fungi on Luffa cylindrica: An effective biosorbent for the removal of lead. *J Taiwan Inst Chem Eng* 80: 589–595. doi:10.1016/j.jtice.2017.08.032.

Dai, Q., Zhang, T., Zhao, Y. et al. (2020). Potentiality of living Bacillus pumilus SWU7-1 in biosorption of strontium radionuclide. *Chemosphere* 260: 127559. doi:10.1016/j.chemosphere.2020.127559.

Dan, Y., Xu, L., Qiang, Z. et al. (2021). Preparation of green biosorbent using rice hull to preconcentrate, remove and recover heavy metal and other metal elements from water. *Chemosphere* 262: 127940. doi:10.1016/j.chemosphere.2020.127940.

Das, D., Jaya Sre Varshini, C., and Das, N. (2014). Recovery of lanthanum(III) from aqueous solution using biosorbents of plant and animal origin: Batch and column studies. *Miner Eng* 69: 40–56. doi:10.1016/j.mineng.2014.06.013.

Das, N. and Das, D. (2013). Recovery of rare earth metals through biosorption: An overview. *J Rare Earths* 31: 933–943. doi:10.1016/S1002-0721(13)60009-5.

Debnath, B., Chowdhury, R., and Ghosh, S.K. (2018). Sustainability of metal recovery from E-waste. *Front Environ Sci Eng* 12: 2. doi:10.1007/s11783-018-1044-9.

Deliormanlı, A.M., Oguzlar, S., and Ertekin, K. (2020). Photoluminescence and decay characteristics of cerium, gallium and vanadium-containing borate-based bioactive glass powders for bioimaging applications. *Ceram Int* doi:10.1016/j.ceramint.2020.09.237.

Dittert, I.M., Vilar, V.J.P., da Silva, E.A.B. et al. (2012). Adding value to marine macro-algae Laminaria digitata through its use in the separation and recovery of trivalent chromium ions from aqueous solution. *Chem Eng J* 193–194: 348–357. doi:10.1016/j.cej.2012.04.048.

Duarte, G., Ciminelli, V.S.T., Dantas, M.S.S. et al. (2012). As(III) immobilization on gibbsite: Investigation of the complexation mechanism by combining EXAFS analyses and DFT calculations. *Geochim Cosmochim Acta* 83: 205–216. doi:10.1016/j.gca.2011.12.019.

Dubinin, M.M (1947). The equation of the characteristic curve of activated charcoal. *Dokl Akad Nauk SSSR* 55: 327–329.

Emenike, C.U., Jayanthi, B., Agamuthu, P., and Fauziah, S.H. (2018). Biotransformation and removal of heavy metals: a review of phytoremediation and microbial remediation assessment on contaminated soil. *Environ Rev* 26: 156–168. doi:10.1139/er-2017-0045.

Erüst, C., Akcil, A., Gahan, C.S. et al. (2013). Biohydrometallurgy of secondary metal resources: a potential alternative approach for metal recovery. *J Chem Technol Biotechnol* 88: 2115–2132. doi:10.1002/jctb.4164.

Farnane, M., Tounsadi, H., Elmoubarki, R. et al. (2017). Alkaline treated carob shells as sustainable biosorbent for clean recovery of heavy metals: Kinetics, equilibrium, ions interference and process optimisation. *Ecol Eng* 101: 9–20. doi:10.1016/j.ecoleng.2017.01.012.

Feisther, V.A., Filho, J.S., Hackbarth, F.V. et al. (2019). Raw leaves and leaf residues from the extraction of essential oils as biosorbents for metal removal. *J Environ Chem Eng* 7: 103047. doi:10.1016/j.jece.2019.103047.

Freitas, N.S., Alzamora, M., Sánchez, D.R. et al. (2021). Green palladium nanoparticles prepared with glycerol and supported on maghemite for dye removal application. *J Environ Chem Eng* 9: 104856. doi:10.1016/j.jece.2020.104856.

Freundlich, H.M.F. (1906). Uber die adsorption in Losungen. *Zeitschrift für Physikalische Chemie* 57: 385–470.

Garole, D.J., Choudhary, B.C., Paul, D., and Borse, A.U. (2018). Sorption and recovery of platinum from simulated spent catalyst solution and refinery wastewater using chemically modified biomass as a novel sorbent. *Environ Sci Pollut Res* 25: 10911–10925. doi:10.1007/s11356-018-1351-5.

Gedgagov, E.I., Zakhar'yan S V., Sinyanskaya, O.M., and Zakhar'yan D V. (2018). Removal of impurities from saturated ion-exchange resins by frontal-gradient purification in schemes for recovery of nonferrous, rare, and rare earth metals. *Theor Found Chem Eng* 52: 920–927. doi:10.1134/S0040579518050111.

Ghanim, B., O'Dwyer, T.F., Leahy, J.J. et al. (2020). Application of KOH modified seaweed hydrochar as a biosorbent of Vanadium from aqueous solution: Characterisations, mechanisms and regeneration capacity. *J Environ Chem Eng* 8: 104176. doi:10.1016/j.jece.2020.104176.

Ghavidel Kalishomi, R., Rostamizadeh, S., Nouri, F., and Khazaei, A. (2020). Synthesis of nano magnetic supported NHC-palladium and investigation of its applications as a catalyst in the Mizoroki-Heck cross-coupling reaction in H2O. *Heliyon* 6: e04946. doi:10.1016/j.heliyon.2020.e04946.

Gomes, H.I., Funari, V., Dinelli, E., and Soavi, F. (2020). Enhanced electrodialytic bioleaching of fly ashes of municipal solid waste incineration for metal recovery. *Electrochim Acta* 345: 136188. doi:10.1016/j.electacta.2020.136188.

Gomes, H.I., Funari, V., Mayes, W.M. et al. (2018). Recovery of Al, Cr and V from steel slag by bioleaching: Batch and column experiments. *J Environ Manage* 222: 30–36. doi:10.1016/j.jenvman.2018.05.056.

Gonzalez, J. and Sequí-Castellano, J.A. (2021). Electrochemical determination of kinetic parameters of surface confined redox probes in presence of intermolecular interactions by means of cyclic voltammetry. Application to TEMPO monolayers in gold and platinum electrodes. *Electrochim Acta* 365: 137331. doi:10.1016/j.electacta.2020.137331.

Gül, Ü.D., İrdem, E., Yavuz, Ş.A., and İlhan, S. (2020). Determination of dye biosorption capacity of lichens and reusability of wastes as antimicrobial agents. *J Text Inst* 1–9. doi:10.1080/00405000.2020.1797263.

Gupta, V.K. and Nayak, A. (2012). Cadmium removal and recovery from aqueous solutions by novel adsorbents prepared from orange peel and Fe2O3 nanoparticles. *Chem Eng J* 180: 81–90. doi:10.1016/j.cej.2011.11.006.

Han, X., Wong, Y.S., and Tam, N.F.Y. (2006). Surface complexation mechanism and modeling in Cr(III) biosorption by a microalgal isolate, Chlorella miniata. *J Colloid Interface Sci* 303: 365–371. doi:10.1016/j.jcis.2006.08.028.

Han, Y.-L., Wu, J.-H., Cheng, C.-L. et al. (2017). Recovery of gold from industrial wastewater by extracellular proteins obtained from a thermophilic bacterium Tepidimonas fonticaldi AT-A2. *Bioresour Technol* 239: 160–170. doi:10.1016/j.biortech.2017.05.038.

He, Z.-W., He, L.-H., Yang, J., and Lü, Q.-F. (2013). Removal and recovery of Au(III) from aqueous solution using a low-cost lignin-based biosorbent. *Ind Eng Chem Res* 52: 4103–4108. doi:10.1021/ie303410g

Herrera-Barros, A., Tejada-Tovar, C., Villabona-Ortíz, A. et al. (2020). Cd (II) and Ni (II) uptake by novel biosorbent prepared from oil palm residual biomass and Al2O3 nanoparticles. *Sustain Chem Pharm* 15: 100216. doi:10.1016/j.scp.2020.100216.

Hima, K.A., Srinivasa, R.R., Vijaya, S.S. et al. (2007). Biosorption: An eco-friendly alternative for heavy metal removal. *African J Biotechnol* 6: 2924–2931. doi:10.5897/AJB2007.000-2461.

Hisada, M. and Kawase, Y. (2018). Recovery of rare-earth metal neodymium from aqueous solutions by poly-γ-glutamic acid and its sodium salt as biosorbents: Effects of solution pH on neodymium recovery mechanisms. *J Rare Earths* 36: 528–536. doi:10.1016/j.jre.2018.01.001.

Hoang, A.T., Bui, X.L., and Pham, X.D. (2018). A novel investigation of oil and heavy metal adsorption capacity from as-fabricated adsorbent based on agricultural by-product and porous polymer. *Energy Sources, Part A Recover Util Environ Eff* 40: 929–939. doi:10.1080/15567036.2018.1466008.

Hong, H.-J., Yu, H., Park, M., and Jeong, H.S. (2019). Recovery of platinum from waste effluent using polyethyleneimine-modified nanocelluloses: Effects of the cellulose source and type. *Carbohydr Polym* 210: 167–174. doi:10.1016/j.carbpol.2019.01.079.

Hu, H., Sun, L., Wang, T. et al. (2019). Nano-ZnO functionalized biochar as a superhydrophobic biosorbent for selective recovery of low-concentration Re(VII) from strong acidic solutions. *Miner Eng* 142:. doi:10.1016/j.mineng.2019.105885.

Hu, L., He, D., Chen, H. et al. (2017a). Research and development of neodymium phosphate laser glass for high power laser application. *Opt Mater (Amst)* 63: 213–220. doi:10.1016/j.optmat.2016.11.052.

Hu, X., Zhao, Y., Wang, H. et al. (2017b). Efficient removal of tetracycline from aqueous media with a fe3o4 nanoparticles@graphene oxide nanosheets assembly. *Int J Environ Res Public Health* 14: 1495. doi:10.3390/ijerph14121495.

Huang, H., Jia, Q., Jing, W. et al. (2020). Screening strains for microbial biosorption technology of cadmium. *Chemosphere* 251: 126428. doi:10.1016/j.chemosphere.2020.126428.

Isacfranklin, M., Deepika, C., Ravi, G. et al. (2020). Nickel, bismuth, and cobalt vanadium oxides for supercapacitor applications. *Ceram Int* 46: 28206–28210. doi:10.1016/j.ceramint.2020.07.320.

Ishigaki, T., Nakanishi, A., Tateda, M. et al. (2005). Bioleaching of metal from municipal waste incineration fly ash using a mixed culture of sulfur-oxidizing and iron-oxidizing bacteria. *Chemosphere* 60: 1087–1094. doi:10.1016/j.chemosphere.2004.12.060.

Ishikawa, S., Suyama, K., Arihara, K., and Itoh, M. (2002). Uptake and recovery of gold ions from electroplating wastes using eggshell membrane. *Bioresour Technol* 81: 201–206. doi:10.1016/S0960-8524(01)00134-1.

Işıldar, A., Rene, E.R., van Hullebusch, E.D., and Lens, P.N.L. (2018). Electronic waste as a secondary source of critical metals: Management and recovery technologies. *Resour Conserv Recycl* 135: 296–312. doi:10.1016/j.resconrec.2017.07.031.

Jadhav, U. and Hocheng, H. (2015). *Hydrometallurgical recovery of metals from large printed circuit board pieces. Sci Rep* 5. doi:10.1038/srep14574.

Jagaba, A.H., Kutty, S.R.M., Khaw, S.G. et al. (2020). Derived hybrid biosorbent for zinc(II) removal from aqueous solution by continuous-flow activated sludge system. *J Water Process Eng* 34: 101152. doi:10.1016/j.jwpe.2020.101152.

Jalali, R., Ghafourian, H., Asef, Y. et al. (2002). Removal and recovery of lead using nonliving biomass of marine algae. *J Hazard Mater* 92: 253–262. doi:10.1016/S0304-3894(02)00021-3.

Jobby, R., Jha, P., Yadav, A.K., and Desai, N. (2018). Biosorption and biotransformation of hexavalent chromium [Cr(VI)]: A comprehensive review. *Chemosphere* 207: 255–266. doi:10.1016/j.chemosphere.2018.05.050.

Kacholi, D.S. and Sahu, M. (2018). Levels and health risk assessment of heavy metals in soil, water, and vegetables of Dar es Salaam, Tanzania. *J Chem* 2018: 1–9. doi:10.1155/2018/1402674.

Kadivar, S., Pourhossein, F., and Mousavi, S.M. (2020). *Recovery of valuable metals from spent mobile phone printed circuit boards using biochar in indirect bioleaching. J Environ Manage* 111642. doi:10.1016/j.jenvman.2020.111642.

Kausar, A., Bhatti, H.N., and MacKinnon, G. (2016). Re-use of agricultural wastes for the removal and recovery of Zr(IV) from aqueous solutions. *J Taiwan Inst Chem Eng* 59: 330–340. doi:10.1016/j.jtice.2015.08.016.

Kim, J.H., Min, B.R., Won, J., and Kang, Y.S. (2002). Complexation mechanism of olefin with silver ions dissolved in a polymer matrix and its effect on facilitated olefin transport. *Chem - A Eur J* 8: 650–654.

Kinidi, L., Tan, I.A.W., Abdul Wahab, N.B. et al. (2018). Recent development in ammonia stripping process for industrial wastewater treatment. *Int J Chem Eng* 2018: 1–14. doi:10.1155/2018/3181087.

Kinoshita, T., Akita, S., Kobayashi, N. et al. (2003). Metal recovery from non-mounted printed wiring boards via hydrometallurgical processing. *Hydrometallurgy* 69: 73–79. doi:10.1016/S0304-386X(03)00031-8.

Kocaoba, S. and Arısoy, M. (2011). The use of a white rot fungi (Pleurotus ostreatus) immobilized on Amberlite, X.A.D.-4 as a new biosorbent in trace metal determination. *Bioresour Technol* 102: 8035–8039. doi:10.1016/j.biortech.2011.05.004.

Komreddy, V., Ensz, K., Nguyen, H., and Paul Rillema, D. (2020). Synthesis and characterization of rhenium(I) 4,4′-dicarboxy-2,2′-bipyridine tricarbonyl complexes for solar energy conversion. *Inorganica Chim Acta* 511: 119815. doi:10.1016/j.ica.2020.119815.

Kovacs, H. and Szemmelveisz, K. (2017). Disposal options for polluted plants grown on heavy metal contaminated brownfield lands – A review. *Chemosphere* 166: 8–20. doi:10.1016/j.chemosphere.2016.09.076.

Kubra, K.T., Sharif, R., Patil, B. et al. (2020). Hydrothermal synthesis of neodymium oxide nanoparticles and its nanocomposites with manganese oxide as electrode materials for supercapacitor application. *J Alloys Compd* 815: 152104. doi:10.1016/j.jallcom.2019.152104.

Kulshrestha, M. and Venkobachar, C. (2008). Removal and recovery of uranium (VI) using a fungal based low-cost biosorbent Ganoderma lucidum. *Int J Environ Pollut* 34: 83–96. doi:10.1504/ijep.2008.020784.

Kumar, M., Nandi, M., and Pakshirajan, K. (2021). Recent advances in heavy metal recovery from wastewater by biogenic sulfide precipitation. *J Environ Manage* 278: 111555. doi:10.1016/j.jenvman.2020.111555.

Kumar, R., Sharma, R.K., and Singh, A.P. (2017). Cellulose based grafted biosorbents - Journey from lignocellulose biomass to toxic metal ions sorption applications - A review. *J Mol Liq* 232: 62–93. doi:10.1016/j.molliq.2017.02.050.

Kuyucak, N. and Volesky, B. (1988). Biosorbents for recovery of metals from industrial solutions. *Biotechnol Lett* 10: 137–142. doi:10.1007/BF01024641.

Kwak, I.S., Bae, M.A., Won, S.W. et al. (2010). Sequential process of sorption and incineration for recovery of gold from cyanide solutions: Comparison of ion exchange resin, activated carbon and biosorbent. *Chem Eng J* 165: 440–446. doi:10.1016/j.cej.2010.09.027.

Kwak, I.S., Won, S.W., Choi, S.B. et al. (2011). Sorptive removal and recovery of nickel(II) from an actual effluent of electroplating industry: Comparison between Escherichia coli biosorbent and Amberlite ion exchange resin. *Korean J Chem Eng* 28: 927–932. doi:10.1007/s11814-010-0441-y

Kwak, I.S., Won, S.W., Chung, Y.S., and Yun, Y.-S. (2013). Ruthenium recovery from acetic acid waste water through sorption with bacterial biosorbent fibers. *Bioresour Technol* 128: 30–35. doi:10.1016/j.biortech.2012.10.146.

Kwak, I.S. and Yun, Y.-S. (2010). Recovery of zero-valent gold from cyanide solution by a combined method of biosorption and incineration. *Bioresour Technol* 101: 8587–8592. doi:10.1016/j.biortech.2010.06.080.

Langmuir, I. (1916). The constitution and fundamental properties of solids and liquids. *Part I. Solids. J Am Chem Soc* 38: 2221–2295. doi:10.1021/ja02268a002.

Lapo, B., Bou, J.J., Hoyo, J. et al. (2020). A potential lignocellulosic biomass based on banana waste for critical rare earths recovery from aqueous solutions. *Environ Pollut* 264: 114409. doi:10.1016/j.envpol.2020.114409.

Lee, M.-Y., Park, J.M., and Yang, J.-W. (1997). Micro precipitation of lead on the surface of crab shell particles. *Process Biochem* 32: 671–677. doi:10.1016/S0032-9592(97)00026-5.

Li, D., Li, R., Ding, Z. et al. (2020a). Discovery of a novel native bacterium of Providencia sp. with high biosorption and oxidation ability of manganese for bioleaching of heavy metal contaminated soils. *Chemosphere* 241: 125039. doi:10.1016/j.chemosphere.2019.125039.

Li, H., Li, Z., Liu, T. et al. (2008). A novel technology for biosorption and recovery hexavalent chromium in wastewater by bio-functional magnetic beads. *Bioresour Technol* 99: 6271–6279. doi:10.1016/j.biortech.2007.12.002.

Li, Q., Bai, A., Xue, Z. et al. (2020b). Nitrogen and sulfur co-doped graphene composite electrode with high electrocatalytic activity for vanadium redox flow battery application. *Electrochim Acta* 362: 137223. doi:10.1016/j.electacta.2020.137223.

Li, Y., Teng, S., Wang, M. et al. (2021). Molecular crowding-modulated fluorescence emission of gold nanoclusters: Ligand-dependent behaviors and application in improved biosensing. *Sensors Actuators B Chem* 330: 129290. doi:10.1016/j.snb.2020.129290.

Liaquat, F., Haroon, U., Munis, M.F.H. et al. (2020). Efficient recovery of metal tolerant fungi from the soil of industrial area and determination of their biosorption capacity. *Environ Technol Innov* 101237. doi:10.1016/j.eti.2020.101237.

Lo, Y.-C., Cheng, C.-L., Han, Y.-L. et al. (2014). Recovery of high-value metals from geothermal sites by biosorption and bioaccumulation. *Bioresour Technol* 160: 182–190. doi:10.1016/j.biortech.2014.02.008.

Luo, W., Li, H., Zhou, Y., and Zhou, H. (2020). Seismic performance of lead-filled steel tube damper: Laboratory test, parameter identification and application. *Eng Struct* 219: 110764. doi:10.1016/j.engstruct.2020.110764.

Mahmoud, M.E., Yakout, A.A., Abdel-Aal, H., and Osman, M.M. (2012). High performance SiO2-nanoparticles-immobilized-Penicillium funiculosum for bioaccumulation and solid phase extraction of lead. *Bioresour Technol* 106: 125–132. doi:10.1016/j.biortech.2011.11.081.

Maimulyanti, A., Prihadi, A.R., Rosita, T., and Safrudin, I. (2019). Adsorption and recovery of aroma compounds from wastewater of clove oil distillation using coffee husk biosorbent. *ScienceAsia* 45: 446–451. doi:10.2306/scienceasia1513-1874.2019.45.446.

Maleki, A., Hayati, B., Najafi, F. et al. (2016). Heavy metal adsorption from industrial wastewater by PAMAM/TiO2 nanohybrid: Preparation, characterization and adsorption studies. *J Mol Liq* 224: 95–104. doi:10.1016/j.molliq.2016.09.060.

Mangini, L.F.K., Valt, R.B.G., Ponte, M.J.J. de S, and Ponte H de A (2020). Vanadium removal from spent catalyst used in the manufacture of sulfuric acid by electrical potential application. *Sep Purif Technol* 246: 116854. doi:10.1016/j.seppur.2020.116854.

Mao, J., Kim, S., Wu, X.H. et al. (2015). A sustainable cationic chitosan/E. coli fiber biosorbent for Pt(IV) removal and recovery in batch and column systems. *Sep Purif Technol* 143: 32–39. doi:10.1016/j.seppur.2015.01.023.

Mao, J., Lee, S.Y., Won, S.W., and Yun, Y.-S. (2010). Surface modified bacterial biosorbent with poly(allylamine hydrochloride): Development using response surface methodology and use for recovery of hexachloroplatinate(IV) from aqueous solution. *Water Res* 44: 5919–5928. doi:10.1016/j.watres.2010.07.034.

Mari, E., Tsai, P.-C., Eswaran, M., and Ponnusamy, V.K. (2020). Efficient electro-catalytic oxidation of ethylene glycol using flower-like graphitic carbon nitride/iron oxide/palladium nanocomposite for fuel cell application. *Fuel* 280: 118646. doi:10.1016/j.fuel.2020.118646.

Mcafee, B.J., Gould, W.D., Nadeau, J.C. et al. (2001). Biosorption of metal ions using chitosan, chitin, and biomass of rhizopus oryzae. *Separation Science and Technology*, 36 (14): 3207–3222.

Mevada, C. and Mukhopadhyay, M. (2020). High mass loading tin oxide-ruthenium oxide-based nanocomposite electrode for supercapacitor application. *J Energy Storage* 31: 101587. doi:10.1016/j.est.2020.101587.

Mie, Y., Takayama, H., and Hirano, Y. (2020). Facile control of surface crystallographic orientation of anodized nanoporous gold catalyst and its application for highly efficient hydrogen evolution reaction. *J Catal* 389: 476–482. doi:10.1016/j.jcat.2020.06.023.

Mo, Y., Wang, S., Vincent, T. et al. (2019). New highly-percolating alginate-PEI membranes for efficient recovery of chromium from aqueous solutions. *Carbohydr Polym* 225: 115177. doi:10.1016/j.carbpol.2019.115177.

Mohammed, R.R. and Chong, M.F. (2014). Treatment and decolorization of biologically treated palm oil mill effluent (POME) using banana peel as novel biosorbent. *Journal of Environmental Management* 132: 237–249. doi:10.1016/j.jenvman.2013.11.031.

Mohd Udaiyappan, A.F., Abu Hasan, H., Takriff, M.S., and Sheikh Abdullah, S.R. (2017). A review of the potentials, challenges and current status of microalgae biomass applications in industrial wastewater treatment. *J Water Process Eng* 20: 8–21. doi:10.1016/j.jwpe.2017.09.006.

Mróz, W., Budner, B., Grigoriev, S.A. et al. (2019). Quantitative investigation using X-ray photoelectron spectroscopy of oxidation of platinum catalyst films deposited by sputtering and spraying for fuel cell applications. *Thin Solid Films* 683: 27–33. doi:10.1016/j.tsf.2019.05.015.

Murugesan, M.P., Kannan, K., and Selvaganapathy, T. (2020). Bioleaching recovery of copper from printed circuit boards and optimization of various parameters using response surface methodology (RSM). *Mater Today Proc* 26: 2720–2728. doi:10.1016/j.matpr.2020.02.571.

Nahar, M.S. and Zhang, J. (2013). Recovery of trace metal isotopes in seawater samples using multifunctional neem (azadirachta indica) biosorbent: A comparison with monofunctional NOBIASChelatePA1 resin. *ACS Sustain Chem Eng* 1: 488–495. doi:10.1021/sc3001433.

Naseri, T., Bahaloo-Horeh, N., and Mousavi, S.M. (2019). Environmentally friendly recovery of valuable metals from spent coin cells through two-step bioleaching using Acidithiobacillus thiooxidans. *J Environ Manage* 235: 357–367. doi:10.1016/j.jenvman.2019.01.086.

Neelima, G., Venkata Krishnaiah, K., Ravi, N. et al. (2019). Investigation of optical and spectroscopic properties of neodymium doped oxyfluoro-titania-phosphate glasses for laser applications. *Scr Mater* 162: 246–250. doi:10.1016/j.scriptamat.2018.11.018.

Nekouei, R.K., Pahlevani, F., Assefi, M. et al. (2019). Selective isolation of heavy metals from spent electronic waste solution by macroporous ion-exchange resins. *J Hazard Mater* 371: 389–396. doi:10.1016/j.jhazmat.2019.03.013.

Nekounam, H., Allahyari, Z., Gholizadeh, S. et al. (2020). Simple and robust fabrication and characterization of conductive carbonized nanofibers loaded with gold nanoparticles for bone tissue engineering applications. *Mater Sci Eng C* 117: 111226. doi:10.1016/j.msec.2020.111226.

Nisha, B., Vidyalakshmi, Y., and Razack, S.A. (2020). Enhanced formation of ruthenium oxide nanoparticles through green synthesis for highly efficient supercapacitor applications. *Adv Powder Technol* 31: 1001–1006. doi:10.1016/j.apt.2019.12.026.

Nogami, S., Hasegawa, A., Fukuda, M. et al. (2020). Tungsten modified by potassium doping and rhenium addition for fusion reactor applications. *Fusion Eng Des* 152: 111445. doi:10.1016/j.fusengdes.2019.111445.

Olguín, E.J. and Sánchez-Galván, G. (2012). Heavy metal removal in phytofiltration and phycoremediation: The need to differentiate between bioadsorption and bioaccumulation. *N Biotechnol* 30: 3–8. doi:10.1016/j.nbt.2012.05.020.

Özdemir, S., Kilinc, E., Poli, A. et al. (2009). Biosorption of Cd, Cu, Ni, Mn and Zn from aqueous solutions by thermophilic bacteria, Geobacillus toebii sub.sp. decanicus and Geobacillus thermoleovorans sub.sp. stromboliensis: Equilibrium, kinetic and thermodynamic studies. *Chem Eng J* 152: 195–206. doi:10.1016/j.cej.2009.04.041.

Özdemir, S., Oduncu, M.K., Kilinc, E., and Soylak, M. (2017). Tolerance and bioaccumulation of U(VI) by Bacillus mojavensis and its solid phase preconcentration by Bacillus mojavensis immobilized multiwalled carbon nanotube. *J Environ Manage* 187: 490–496. doi:10.1016/j.jenvman.2016.11.004.

Pal, A., Ghosh, S., and Paul, A.K. (2006). Biosorption of cobalt by fungi from serpentine soil of Andaman. *Bioresour Technol* 97: 1253–1258. doi:10.1016/j.biortech.2005.01.043.

Pan, J., Gao, Y., Gao, B. et al. (2019). One-step synthesis of easily-recoverable carboxylated biogas residues for efficient removal of heavy metal ions from synthetic wastewater. *J Clean Prod* 240: 118264. doi:10.1016/j.jclepro.2019.118264.

Panda, H., Tiadi, N., Mohanty, M., and Mohanty, C.R. (2017). Studies on adsorption behavior of an industrial waste for removal of chromium from aqueous solution. *South African J Chem Eng* 23: 132–138. doi:10.1016/j.sajce.2017.05.002.

Pandey, A., Shukla, A., and Ray, L. (2009). Uptake and recovery of lead by agarose gel polymers. *Am J Biochem Biotechnol* 5: 14–20. doi:10.3844/ajbb.2009.14.20.

Park, J., Won, S.W., Mao, J. et al. (2010). Recovery of Pd(II) from hydrochloric solution using polyallylamine hydrochloride-modified Escherichia coli biomass. *J Hazard Mater* 181: 794–800. doi:10.1016/j.jhazmat.2010.05.083.

Park, S.-I., Kwak, I.S., Bae, M.A. et al. (2012). Recovery of gold as a type of porous fiber by using biosorption followed by incineration. *Bioresour Technol* 104: 208–214. doi:10.1016/j.biortech.2011.11.018.

Patil, Y.B. and Paknikar, K.M. (1999a). Removal and recovery of metal-cyanides from industrial effluents. *Process Metall* 9: 707–715. doi:10.1016/S1572-4409(99)80161-8.

Patil, Y.B. and Paknikar, K.M. (1999b). Removal and recovery of metal cyanides using a combination of biosorption and biodegradation processes. *Biotechnology Letters* 21: 913–919.

Pethkar, A.V. and Paknikar, K.M. (1998). Recovery of gold from solutions using Cladosporium cladosporioides biomass beads. *J Biotechnol* 63: 121–136. doi:10.1016/S0168-1656(98)00078-9.

Pholosi, A., Naidoo, E.B., and Ofomaja, A.E. (2020). Intraparticle diffusion of Cr(VI) through biomass and magnetite coated biomass: A comparative kinetic and diffusion study. *South African J Chem Eng* 32: 39–55. doi:10.1016/j.sajce.2020.01.005.

Premkumar, M.P., Thiruvengadaravi K V., Senthil Kumar, P. et al. (2018). Eco-friendly treatment strategies for wastewater containing dyes and heavy metals. In: *Environmental Contaminants. Energy, Environment, and Sustainability* (ed. T. Gupta, A. Agarwal, R. Agarwal, and N. Labhsetwar), 317–360. Springer. doi:10.1007/978-981-10-7332-8_14.

Putro, J.N., Kurniawan, A., Ismadji, S., and Ju, Y.-H. (2017). Nanocellulose based biosorbents for wastewater treatment: Study of isotherm, kinetic, thermodynamic and reusability. *Environ Nanotechnology, Monit Manag* 8: 134–149. doi:10.1016/j.enmm.2017.07.002.

Rabai, S., Benounis, M., Catanante, G. et al. (2021). Development of a label-free electrochemical aptasensor based on diazonium electrodeposition: Application to cadmium detection in water. *Anal Biochem* 612: 113956. doi:10.1016/j.ab.2020.113956.

Ramanaiah S V, Venkata Mohan, S., and Sarma, P.N. (2007). Adsorptive removal of fluoride from aqueous phase using waste fungus (Pleurotus ostreatus 1804) biosorbent: Kinetics evaluation. *Ecol Eng* 31: 47–56. doi:10.1016/j.ecoleng.2007.05.006.

Rana, M.A., Akbar, M.N., Anwar, M. et al. (2015). Sol-gel fabrication of novel electrolytes based on neodymium doped ceria for application in low temperature solid oxide fuel cells. *Mater Today Proc* 2: 5649–5653. doi:10.1016/j.matpr.2015.11.099.

Redlich, O. and Peterson, D.L. (1959). A useful adsorption isotherm. *J Phys Chem* 63: 1024–1024. doi:10.1021/j150576a611.

Ritchie, A.G. (1977). Alternative to the Elovich equation for the kinetics of adsorption of gases on solids. *J Chem* Soc *Faraday Trans 1 Phys Chem Condens Phases* 73: 1650–1653. doi:10.1039/F19777301650.

Rocchetti, L., Vegliò, F., Kopacek, B., and Beolchini, F. (2013). Environmental impact assessment of hydrometallurgical processes for metal recovery from WEEE residues using a portable prototype plant. *Environ Sci Technol* 130125102742005. doi:10.1021/es302192t.

Saeed, A., Akhter, M.W., and Iqbal, M. (2005a). Removal and recovery of heavy metals from aqueous solution using papaya wood as a new biosorbent. *Sep Purif Technol* 45: 25–31. doi:10.1016/j.seppur.2005.02.004.

Saeed, A., Iqbal, M., and Akhtar, M.W. (2005). Removal and recovery of lead (II) from single and multiple (Cd, Ni, Cu, Zn) solutions by crop milling waste (black gram husk). *J Hazard Mater* 117: 65–73.

Şahin, Y. and Öztürk, A. (2005). Biosorption of chromium(VI) ions from aqueous solution by the bacterium Bacillus thuringiensis. *Process Biochem* 40: 1895–1901. doi:10.1016/j.procbio.2004.07.002.

Salam, M.A., El-Shishtawy, R.M., and Obaid, A.Y. (2014). Synthesis of magnetic multi-walled carbon nanotubes/magnetite/chitin magnetic nanocomposite for the removal of Rose Bengal from real and model solution. *J Ind Eng Chem* 20: 3559–3567. doi:10.1016/j.jiec.2013.12.049.

Sang, W., Xu, J., Bashir, M.H., and Ali, S. (2018). Developmental responses of Cryptolaemus montrouzieri to heavy metals transferred across multi-trophic food chain. *Chemosphere* 205: 690–697. doi:10.1016/j.chemosphere.2018.02.073.

Sar, P., K. Kazy, S., and D'Souza, S.F. (2004). Radionuclide remediation using a bacterial biosorbent. *Int Biodeterior Biodegradation* 54: 193–202. doi:10.1016/j.ibiod.2004.05.004.

Schippers, A., Hedrich, S., Vasters, J. et al. (2013). Biomining: metal recovery from ores with microorganisms. In: *Geobiotechnology I* (ed. A. Schippers, F. Glombitza, and W. Sand), 1–47. Springer Berlin Heidelberg. doi:10.1007/10_2013_216.

Sedlakova-Kadukova, J., Kopcakova, A., Gresakova, L. et al. (2019). Bioaccumulation and biosorption of zinc by a novel Streptomyces K11 strain isolated from highly alkaline aluminium brown mud disposal site. *Ecotoxicol Environ Saf* 167: 204–211. doi:10.1016/j.ecoenv.2018.09.123.

Shaafi, N.F., Muzakir, S.K., Aziz, S.B. et al. (2020). *The effect of activated carbon additives on lead sulphide thin film for solar cell applications. J Alloys Compd* 158117. doi:10.1016/j.jallcom.2020.158117.

Shamsuddin, M. (1986). Metal recovery from scrap and waste. *JOM* 38: 24–31. doi:10.1007/BF03257917.

Sharma, R., Talukdar, D., Bhardwaj, S. et al. (2020). Bioremediation potential of novel fungal species isolated from wastewater for the removal of lead from liquid medium. *Environ Technol Innov* 18: 100757. doi:10.1016/j.eti.2020.100757.

Shi, J., Lin, N., Wang, Y. et al. (2020). The application of rice husk-based porous carbon in positive electrodes of lead acid batteries. *J Energy Storage* 30: 101392. doi:10.1016/j.est.2020.101392.

Sinha, R., Chauhan, G., Singh, A. et al. (2018). A novel eco-friendly hybrid approach for recovery and reuse of copper from electronic waste. *J Environ Chem Eng* 6: 1053–1061. doi:10.1016/j.jece.2018.01.030.

Sobhanan, J., Jones, P., Kohara, R. et al. (2020). Toxicity of nanomaterials due to photochemical degradation and the release of heavy metal ions. *Nanoscale* 12: 22049–22058. doi:10.1039/D0NR03957H

Soliman, N.K. and Moustafa, A.F. (2020). Industrial solid waste for heavy metals adsorption features and challenges; a review. *J. Mater. Res. Technol.* 9: 10235–10253.

Soyluoglu, M., Ersan, M.S., Ateia, M., and Karanfil, T. (2020). Removal of bromide from natural waters: Bromide-selective vs. conventional ion exchange resins. *Chemosphere* 238: 124583. doi:10.1016/j.chemosphere.2019.124583.

Suganya, E., Saranya, N., Senthilkumar, S. et al. (2020). Experimentation on raw and phosphoric acid activated Eucalyptuscamadulensis seeds as novel biosorbents for hexavalent chromium removal from simulated and electroplating effluents. *Environ Technol Innov* 19: 100977. doi:10.1016/j.eti.2020.100977.

Suhasini, I.P., Sriram, G., Asolekar, S.R., and Sureshkumar, G.K. (1999). Nickel biosorption from aqueous systems: Studies on single and multimetal equilibria, kinetics, and recovery. *Sep Sci Technol* 34: 2761–2779. doi:10.1081/SS-100100803.

Taksitta, K., Sujarit, P., Ratanawimarnwong, N. et al. (2020). Development of tannin-immobilized cellulose fiber extracted from coconut husk and the application as a biosorbent to remove heavy metal ions. *Environ Nanotechnology, Monit Manag* 14: 100389. doi:10.1016/j.enmm.2020.100389.

Tkaczyk, A., Mitrowska, K., and Posyniak, A. (2020). Synthetic organic dyes as contaminants of the aquatic environment and their implications for ecosystems: A review. *Sci Total Environ* 717: 137222. doi:10.1016/j.scitotenv.2020.137222.

Tsezos, M., McCready, R.G.L., and Bell, J.P. (1989). The continuous recovery of uranium from biologically leached solutions using immobilized biomass. *Biotechnol Bioeng* 34: 10–17. doi:10.1002/bit.260340103.

Tuncuk, A., Stazi, V., Akcil, A. et al. (2012). Aqueous metal recovery techniques from e-scrap: Hydrometallurgy in recycling. *Miner Eng* 25: 28–37. doi:10.1016/j.mineng.2011.09.019.

Temkin, M.J., Pyzhev, V., Temkin, M. et al. (1940). Recent modifications to Langmuir isotherms, *Acta Physiochim*. https://www.scienceopen.com/document?vid=ef69d115-706c-4007-b3ce-2f45e972f91b.

Valdés, E., Cepeda-Plaza, M., Günther, G. et al. (2020). An amine linker group modulates luminescent properties in a Rhenium(I) tricarbonyl complex. How can it be applied for ratiometric oxygen sensing? *Dye Pigment* 172: 107787. doi:10.1016/j.dyepig.2019.107787.

Varala, S., Parthasarathy, R., Bhargava, S.K., and Bankupalli, S. (2017). Desorption studies for the recovery of radionuclides (Th and Zr) and optimization using Taguchi mixed design

L18(2132) – A regeneration step for loaded biosorbent, general mathematical model for multistage operation. *J Environ Chem Eng* 5: 5396–5405. doi:10.1016/j.jece.2017.09.046.

Varshney, S., Jain, P., and Srivastava, S. (2019). Removal and recovery of chromium from e-waste by functionalized wood pulp: a green bio-hydrometallurgical approach. *Natl Acad Sci Lett* 42: 99–103. doi:10.1007/s40009-018-0672-8.

Veglio, F. and Beolchini, F. (1997). Removal of metals by biosorption: A review. *Hydrometallurgy* 44: 301–316.

Vilardi, G., Ochando-Pulido, J.M., Verdone, N. et al. (2018). On the removal of hexavalent chromium by olive stones coated by iron-based nanoparticles: Equilibrium study and chromium recovery. *J Clean Prod* 190: 200–210. doi:10.1016/j.jclepro.2018.04.151.

Wang, C., Tan, H., Li, H. et al. (2020). Mechanism study of Chromium influenced soil remediated by an uptake-detoxification system using hyperaccumulator, resistant microbe consortium, and nano iron complex. *Environ Pollut* 257: 113558. doi:10.1016/j.envpol.2019.113558.

Wang, N., Qiu, Y., Hu, K. et al. (2021). One-step synthesis of cake-like biosorbents from plant biomass for the effective removal and recovery heavy metals: Effect of plant species and roles of xanthation. *Chemosphere* 266: 129129. doi:10.1016/j.chemosphere.2020.129129.

Wang, N., Qiu, Y., Xiao, T. et al. (2019a). Comparative studies on Pb(II) biosorption with three spongy microbe-based biosorbents: High performance, selectivity and application. *J Hazard Mater* 373: 39–49. doi:10.1016/j.jhazmat.2019.03.056.

Wang, Q., Yang, C., Yang, Q. et al. (2019b). Platinum-loaded mesoporous nickel phosphonate and its electrochemical application for sarcosine detection. *Anal Chim Acta* 1046: 93–98. doi:10.1016/j.aca.2018.09.027.

Wang, Q., Yang, Z. (2016). Industrial water pollution, water environment treatment, and health risks in China. *Environ Pollut* 218: 358–365. doi:10.1016/j.envpol.2016.07.011.

Won, S.W., Kwak, I.S., Mao, J., and Yun, Y.-S. (2015). Biosorption-incineration-leaching-smelting sequential process for Ru recovery from Ru-bearing acetic acid waste solution. *Ind Eng Chem Res* 54: 7148–7153. doi:10.1021/acs.iecr.5b01111.

Won, S.W., Lim, A., and Yun, Y.-S. (2013). Recovery of high-purity metallic Pd from Pd(II)-sorbed biosorbents by incineration. *Bioresour Technol* 137: 400–403. doi:10.1016/j.biortech.2013.03.143.

Won, S.W., Mao, J., Kwak I.-S. et al. (2010). Platinum recovery from ICP wastewater by a combined method of biosorption and incineration. *Bioresour Technol* 101: 1135–1140. doi:10.1016/j.biortech.2009.09.056.

Xie, J., Feng, N., Wang, R. et al. (2020). A reusable biosorbent using Ca-alginate immobilized providencia vermicola for Pd(II). Recovery from acidic solution. *Water, Air, Soil Pollut* 231: 36. doi:10.1007/s11270-020-4399-z.

Xie, J.Z., Chang, H.-L., and Kilbane, I.I. (1996). Removal and recovery of metal ions from wastewater using biosorbents and chemically modified biosorbents. *Bioresour Technol* 57: 127–136. doi:10.1016/0960-8524(96)00059-4.

Xu, X., Yang, Y., Zhao, X. et al. (2019). Recovery of gold from electronic wastewater by Phomopsis sp. XP-8 and its potential application in the degradation of toxic dyes. *Bioresour Technol* 288: 121610. doi:10.1016/j.biortech.2019.121610.

Yang, Y., Chen, S., Li, S. et al. (2014). Bioleaching waste printed circuit boards by Acidithiobacillus ferrooxidans and its kinetics aspect. *J Biotechnol* 173: 24–30. doi:10.1016/j.jbiotec.2014.01.008.

Zafar, M.N., Nadeem, R., and Hanif, M.A. (2007). Biosorption of nickel from protonated rice bran. *J Hazard Mater* 143: 478–485. doi:10.1016/j.jhazmat.2006.09.055.

Zagorodnyaya, A.N., Abisheva, Z.S., Sadykanova, S.E. et al. (2017). Purification of ammonium perrhenate solutions from potassium by ion exchange. *Miner Process Extr Metall Rev* 38: 284–291. doi:10.1080/08827508.2017.1292911.

Zeng, X., Xu, X., Qin, Q. et al. (2019). Heavy metal exposure has adverse effects on the growth and development of preschool children. *Environ Geochem Health* 41: 309–321. doi:10.1007/s10653-018-0114-z.

Zhang, B., Liu, H., Wang, W. et al. (2020). An innovative technique for recycling dissipated metal rhenium from the waste acid in copper smelting process. *China Geol* 3: 350–352. doi:10.31035/cg2020012.

Zhang, L., Su, J., Yang, S. et al. (2016). Extended X-ray absorption fine structure and density functional theory studies on the complexation mechanism of amidoximate ligand to uranyl carbonate. *Ind Eng Chem Res* 55: 4224–4230. doi:10.1021/acs.iecr.5b03217.

Zhang, Q., Li, M., Luo, B. et al. (2021). In situ growth gold nanoparticles in three-dimensional sugarcane membrane for flow catalytical and antibacterial application. *J Hazard Mater* 402: 123445. doi:10.1016/j.jhazmat.2020.123445.

Zhang, Z. and Fan, Z. (2020). *Morphological analysis of chromium in carbon quantum dots pairs Co-doped with zirconium and nitrogen and their applications in imaging of living cells. Spectrochim Acta Part A Mol Biomol Spectrosc* 119248. doi:10.1016/j.saa.2020.119248.

Zinicovscaia, I., Safonov, A., Zelenina, D. et al. (2020). Evaluation of biosorption and bioaccumulation capacity of cyanobacteria Arthrospira (spirulina) platensis for radionuclides. *Algal Res* 51: 102075. doi:10.1016/j.algal.2020.102075.

Zoli, L., Sciti, D., and Sani, E. (2021). Zirconium diboride-based nanofluids for solar energy applications. *J Mol Liq* 322: 114981. doi:10.1016/j.molliq.2020.114981.

13

Removal of Hexavalent Chromium from Aqueous Media Using Eco-Friendly and Cost-Effective Biological Methods

Veer Singh[1], Nidhi Singh[2], Priyanka Yadav[1], and Vishal Mishra[1]

[1]*School of Biochemical Engineering, IIT (BHU), Varanasi, Uttar Pradesh, India*
[2]*Centre of Bioinformatics, University of Allahabad, Prayagraj, Uttar Pradesh, India*

Introduction

In the environment, chromium originates in the form of Cr (III). It is found in ultramafic rocks at high concentrations and is a common cause of contamination in groundwater and the earth's surface. Cr (III) is less toxic and widely used in industrial processes (Saha et al., 2011). The key source of chromium is chromite ($FeCr_2O_4$), which is used at the commercial level and considered a primary product of mining (Jacobs et al., 2004). Hexavalent chromium, i.e. (Cr (VI)) in water and soil, is considered anthropogenic contamination. The industrial sources of Cr (VI) are leather industries, electroplating, chrome plating, coal mines, etc. (Owlad et al., 2009). Chromium is a widespread pollutant that enters the water supply when industrial effluents are released into bodies of water such as rivers and ponds. The highest allowable limit of hexavalent chromium, i.e. Cr (VI) in fresh and drinking water, is approximately 0.05 mg/L (Owlad et al., 2009; WHO, 2003).

Cr (VI) shows many toxic effects in humans and other animals. It causes sensitization in the mucosal layer of the respiratory tract of the human body. Cr (VI) enters cells and induces DNA mutations that can cause cancer. The probability of carcinogenicity of Cr (VI) depends on the type of chromium-containing compound such as strontium chromate, which is much more carcinogenic than other chromium compounds. Calcium chromate and zinc chromate are also responsible for the carcinogenicity effect. Some lead chromate is also carcinogenic, but it is much less potent than other chromate compounds (Aahluwalia et al., 2007).

Cr (III) is considered a nontoxic or less toxic metallic species. It is slightly soluble in water at near-neutral pH (Singh et al., 2021a). Cr (III) plays a key role in cellular metabolic processes like maintaining the metabolism of cholesterol, glucose, and fatty acids (Kimbrough et al., 1999). But Cr (III) also shows some harmful effects in high concentrations. On the other hand, Cr (VI) shows much more toxicity than trivalent chromium.

Biosorption for Wastewater Contaminants, First Edition.
Edited by Rangabhashiyam Selvasembian and Pardeep Singh.

It causes various types of genotoxic effects in humans. Hence, it is essential to minimize Cr (VI) concentrations in industrial effluents (Weikang et al., 2018; Venitt and Levy, 1974).

Today, various physicochemical techniques are used to eliminate Cr (VI) contamination from wastewater, including reduction, reverse osmosis, precipitation, coagulation, and ion-exchange resins (Singh et al., 2020). These techniques are very expensive and ineffective when the metal ions concentration is low in a large volume of wastewater. These conventional methods also have some other disadvantages, such as the requirement of a large quantity of chemicals, which generates a considerable amount of toxic sludge (Singh et al., 2021b). This toxic sludge contains many secondary pollutants (Bisht et al., 2017).

There is a need to develop an environmentally friendly process to remove hexavalent chromium contamination while generating no or minimal harmful waste materials. Biological methods are very effective compared to physicochemical processes because these methods are highly effective at low metal ion concentrations in large volumes of wastewater and less expensive than physicochemical methods (Song et al., 2016). This chapter focuses on sources, toxicity, and methods for the removal of hexavalent chromium, such as biosorption, bioaccumulation, and bioreduction of Cr (VI), and provides up-to-date information in these areas.

Sources of Hexavalent Chromium

Chromium metal ions mainly occur in the form of Cr (III), which comes from natural sources. Cr (VI) metal ions in wastewater and other environmental sources are derived from human activities (WHO, 2003). The primary sources of Cr (VI) emissions are industries such as rubber manufacturing, dyes for paints, metal finishing, plastic products manufacturing, chrome plating, leather processing, and manufacturers of pharmaceutical products (Owlad et al., 2009). The Cr (VI) concentration found in industrial effluents varies from 0.1 to 200 mg/l. Industries and Cr (VI) concentrations in their effluent are listed in Table 13.1.

Toxicity of Hexavalent Chromium

Hexavalent chromium causes many toxic effects in animals, plants, and microorganisms. The oxidation states of Cr-containing compounds define the degree of toxicity of Cr (VI). In general, a significant cause of chromium toxicity by Cr (VI) is the greater cellular uptake

Table 13.1 Industries and Cr (VI) concentrations in their effluents.

Industries	Cr (VI) concentration in mg/L	References
Leather processing	8.3	(Onyancha et al., 2008)
Chrome processing	3.7	(Gupta et al., 1999)
Electropolishing	42.8	(Davis et al., 1995)
Hardware	60.0	(Xu et al., 2005)
Electroplating	20.7	(Kiptoo et al., 2004)

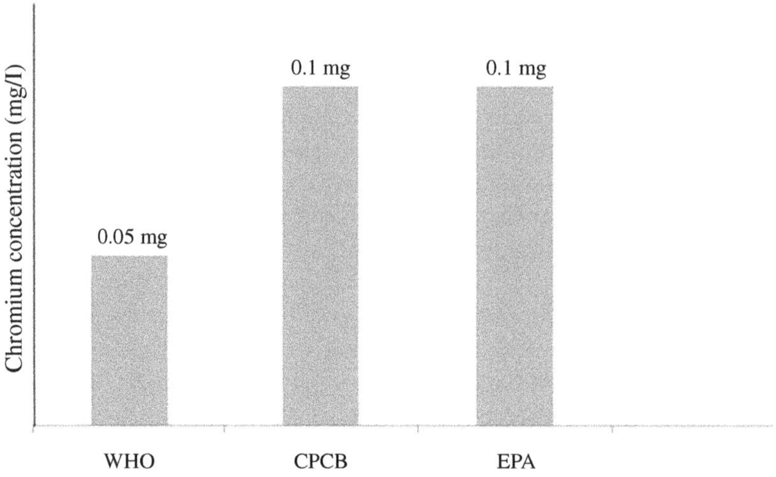

Figure 13.1 Schematic diagram of the permissible limit of Cr in drinking water as specified by Central Pollution Control Board (CPC), World Health Organization (WHO), and United States Environmental Protection Agency (US EPA).

of Cr (VI) compounds compared to Cr (III) (ATSDR, 2012). It is a very toxic state of chromium and causes many diseases such as respiratory disease and skin sensitization, and it has carcinogenic effects on humans and other animals (Saha et al., 2011). It is very dangerous for chromium plating workers and causes toxic effects. Cr (VI) is considered a priority pollutant in many countries because of its high solubility in aqueous solution and significant toxic effects. It is 100 times more toxic than trivalent chromium (Tchounwou et al., 2012; Koedrith et al., 2013). Environmental agencies including the World Health Organization (WHO), US Environmental Protection Agency (EPA), and India's Central Pollution Control Board (CPCB) have fixed the maximum limit of chromium contamination in drinking water. The maximum chromium (total chromium) according to the WHO, EPA, and CPCB standards are shown in Figure 13.1.

Cr (VI) easily enters the cell plasma membrane and is reduced into Cr (III) in the intracellular environment. Several species of chromium, including Cr (V), Cr (IV), and free radicals, are generated as reaction intermediates. These high-energy reaction intermediate products react with nucleic acid complexes and are responsible for DNA damage. Hexavalent chromium also causes many other health abnormalities in living organisms, such as kidney failure, cardiovascular effects, etc. (Xu et al., 2005; Xu et al., 2013; Ray, 2016; Cheung and Gu, 2007). Figure 13.2 summarizes chromium-mediated effects.

Removal of Hexavalent Chromium Ions

Various technologies are used to minimize hexavalent chromium concentration at contaminated sites. The most widely used technologies are ion exchange, coagulation, precipitation, adsorption, membrane technology, and reverse osmosis (Figure 13.3). Precipitation is a well-known technique used for biosorption of Cr (VI). The precipitation is done by

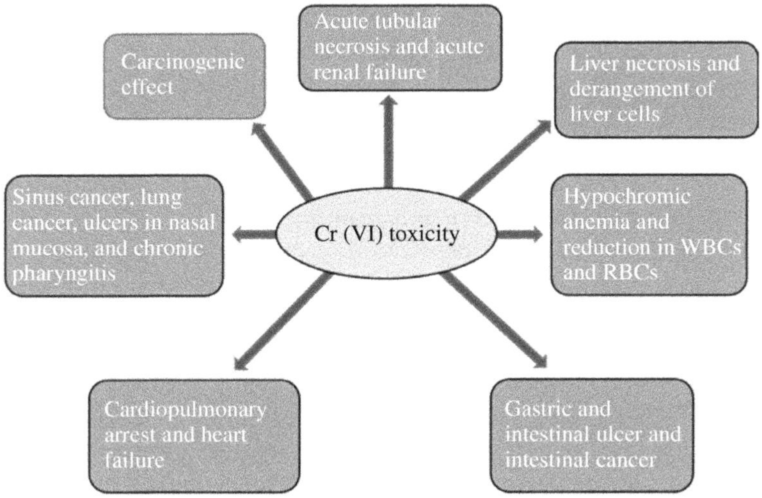

Figure 13.2 Schematic diagram of the toxic effect of hexavalent chromium ions on human health.

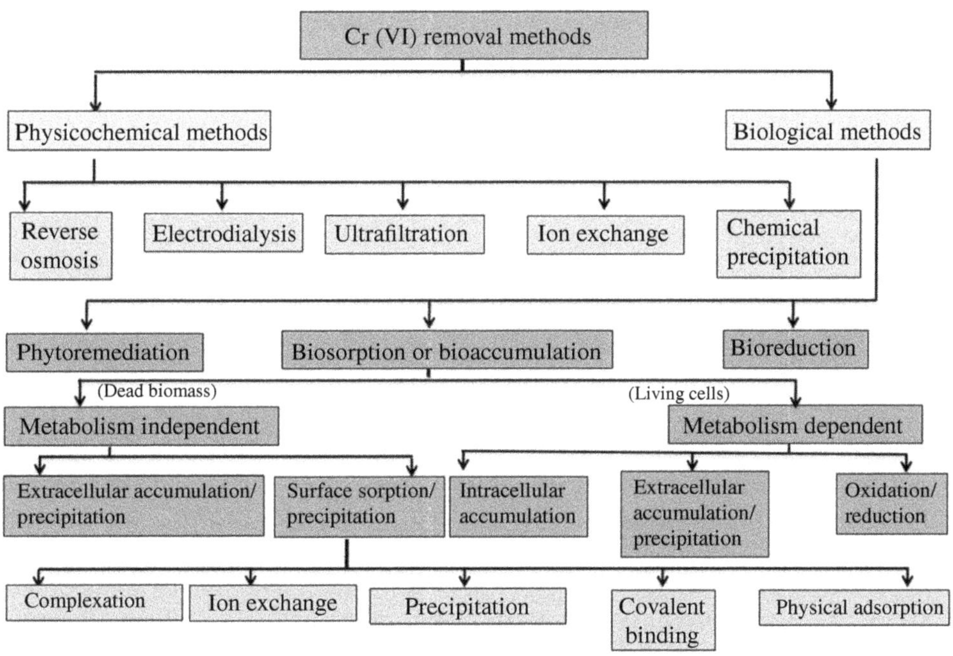

Figure 13.3 Physicochemical and biological methods applicable for decontamination of Cr (VI), including wastewater.

varying the pH with the help of reagents. This precipitation method is only effective when chromium concentration is above 2 mM (Devi et al., 2012).

Biological processes to remove Cr (VI) are very attractive compared to conventional methods because they are inexpensive and highly efficient at low concentrations of chromium in wastewater. Biological materials used to remove chromium contamination include plant biomass, agricultural waste, industrial byproducts, and microbial biomass. Living organisms also have an emerging role in the removal of chromium contamination. Microorganisms such as fungi, bacteria, and microalgae can uptake hexavalent chromium from the surrounding medium into their intracellular environment (Vieira et al., 2000). Recent research has shown that microbial species can decrease Cr (VI) to less-toxic Cr (III). Cr (III) ions are less mobile and less toxic and easily taken up by microorganisms.

Several methods have been reported for biosorption of chromium contamination, including biosorption, phytoremediation, bioaccumulation, and bioreduction (Mishra et al., 2015).

Biosorption

Biosorption is a metabolically independent phenomenon and generally performed using dead biomass. In this process, toxic metal ions bind or accumulate on the surface of biomass (Volesky and Holan, 1995; Gadd, 2007). However, it can also use living biomass; live cells can bind heavy metal ions on their surface by metabolically independent or passive adsorption of sorbate through surface complexation onto the cell surface (Malik, 2004; Peng, 2018). Preparation of biosorbent from biological waste is shown in Figure 13.4 (Saxena et al., 2017).

Biosorption is a biological ion-exchange process in which amino, amido, imidazole, sulfonate, and carboxyl functional groups bind with toxic metal ions. The pKa value affects the binding tendency of functional groups along with toxic metal ions (Volesky, 2007). One of the most important aspects of biosorption is selecting a suitable biosorbent. The raw

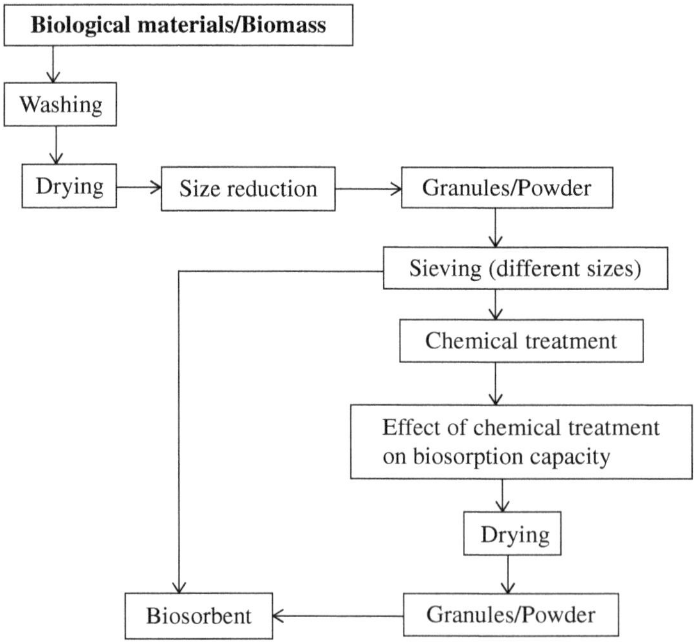

Figure 13.4 Schematic representation of biomass processing.

biosorbent material should be cheap and easily available in nature. The concentration of functional groups on the biosorbent surface is also determined by the biosorbent quality. The functional groups should be present in high concentrations. Techniques such as Fourier transform infrared spectroscopy (FTIR), scanning electron microscopy (SEM), energy dispersive X-ray (EDX), nuclear magnetic resonance (NMR) spectroscopy, and X-ray diffraction (XRD) analysis are used to characterize surface functional groups, surface morphology, elemental composition, molecular composition and crystalline structure of biosorbents (Chojnacka, 2010; Tsezos et al., 1997).

Many factors influence the biosorption process, such as the type of biomass, the occurrence of other metal ions (competing ions such as cations and anions), temperature, and pH. Generally, decreasing the pH value causes competition between positively charged metal ions, and increasing pH causes deprotonation and exposed surface binding groups. Biosorbents can also be regenerated using desorption. Metal ions can be recovered by varying the pH value of the medium (Naja and Volesky, 2010; Jobby et al., 2018).

Cr (VI) biosorption is a complex process in which anionic hexavalent chromium ions bind with positively charged groups and are reduced into Cr (III) chromium ions through various pathways. The biosorption mechanism of hexavalent chromium ions generally includes three steps, as shown in Figure 13.5. The first step of biosorption is binding negatively charged Cr (VI) ions with groups that have positive charges. The second step is a

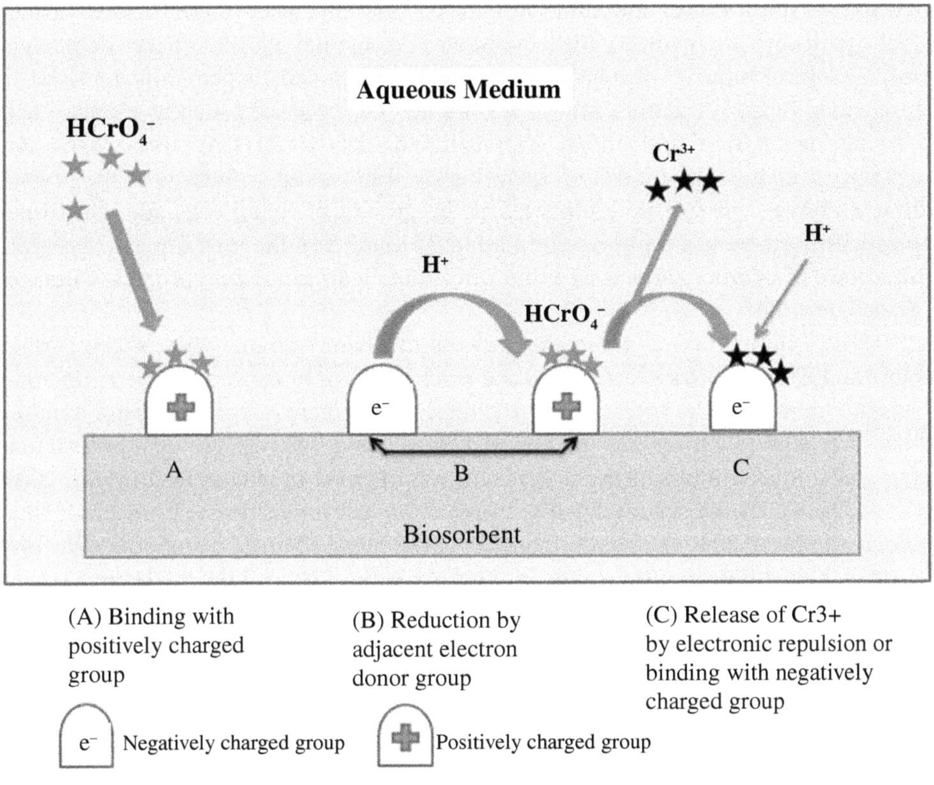

Figure 13.5 Mechanism of hexavalent chromium biosorption by non-living biological materials.

reduction process, where Cr (VI) ions are reduced to form Cr (III) ions with the help of electron donor groups. The third step releases Cr (III) ions into solution because of an electronic repulsion between the trivalent chromium – i.e. Cr (III) – with other positively charged groups or binding of the Cr (III) ions with other negative groups of the biosorbent (Park et al., 2005; Park et al., 2006; Deng et al., 2009).

Biosorption can be performed in a suspension medium onto immobilized biomass which enhances the heavy metal removal efficiency. Immobilization improves the rigidity and strength. Immobilized biomass is easily regenerated and can be used repeatedly which is responsible for long life of the biosorbent. Several types of matrices are used in immobilization methods, such as polysulfide, alginate, polyurethane, and polyacrylamide. Various plant and microbial biomasses are immobilized on these matrices, such as *Chlorella homosphaera* (biomass)-sodium alginate (matrices). Two types of bioreactors (continuous and batch) are generally applicable to eliminate Cr metal ions from aqueous solution (Chojnacka, 2010). Biosorption can be used with sources such as industrial biological wastes, agricultural wastes, and microbial biomass. We have summarized some examples of widely used biosorbents in Table 13.2.

Bioaccumulation

Bioaccumulation is defined as the accumulation of toxic metallic compounds into intracellular spaces. It is a slower and more complex process than biosorption because various metabolic pathways are involved. Bioaccumulation is a metabolic and energy-dependent process. Biological removal of heavy metal ions by living cells is generally achieved in batch reactors. Bioaccumulation minimizes many other processes such as biomass harvest, drying, preparation, and storage (Domezn and Aksu 2002). The process has a few disadvantages, such as the impact of operating conditions and pollutants in the growth medium. Different types of pollutants affect the growth of living cells and sometimes compete with toxic metals for bioaccumulation. Pollutants in the medium can aggregate on the surface of microorganisms and thus affect the bioaccumulation process (Yilmazer and Saracoglu, 2009).

Cr (VI) bioaccumulation is a complex process involving various types of cell surface receptors and channels. We summarized the mechanism of bioaccumulation in microorganisms in Figure 13.6. Cr (VI) ions enter the living cell and are bound with metal-binding proteins such as metallothionein, and ions are precipitated into the cell. Heavy metal ions incorporated into cells bind with the sulfhydryl group of enzymes and are toxic (Malik, 2004; Peng et al., 2018). The bioaccumulated harmful heavy metal ions promote the production of cellular products such as metallothionein. These metal-binding metallothionein proteins are rich in thiol groups, which interact with the intracellular toxic metal ions. Metallothionein proteins are less biologically active and not involved in the cell metabolic process, so binding toxic heavy metal ions with these proteins is non-toxic or less toxic for microorganisms. These proteins are frequently synthesized when toxic metal ions are present in the environment surrounding microorganisms. This is why heavy-metal-resistant microorganisms are more efficient than non-resistant microorganisms, and more bioaccumulation is achieved by microbial species isolated from chromium-contaminated sites and polluted environments.

Table 13.2 Biosorption capacity of biosorbents derived from algal, fungal, plant, and bacterial biomass.

Biosorbent	Biosorption capacity (mg/g)	pH	Temperature (°C)	Biosorbent dose (g/L)	Initial conc. (mg/L)	References
Bacillus salmalaya 139SI	20.35	3	25	1	120	(Dadrasnia et al., 2015)
Opuntia biomass	18.5	2	20	0.5	10	(Fernandez-Lopez et al., 2014)
Opuntia biomass	16.5	2	20	0.5	10	(Fernandez-Lopez et al., 2014)
Cotton waste	6.75	2	25	3	20	(Boosaeidi et al., 2017)
Barberry waste	7.89	2	25	5	40	(Boosaeidi et al., 2017)
Macadamia nutshell ground powder	45.23	2	—	0.5	100	(Pakade et al., 2017)
Trewia nudiflora fruit peel powder	294.12	1–2	20	0.75	22–248	(Bhattacharya et al., 2013)
Ceramium virgatum dry biomass	26.5	1.5	20	10	10	(Sari et al., 2008)
Groundnut hull modified biomass	131	2	28	4	8.3	(Owalude et al., 2016)
Chlorella pyrenoidosa	142.86	5	30	0.1	50	(Rezaei et al., 2011)
Chlorella minutissima (immobilized)	57.33	2	30	200	100	(Singh et al., 2012)
Rice husks	82.3%	2	—	1	5	(Mahdinia et al., 2014)
Oedogonium hatei biomass	31	2	45	0.8	50	(Gupta et al., 2009)
Oedogonium hatei biomass (acid-treated)	35.2	2	45	0.8	100	(Gupta et al., 2009)

(*Continued*)

Table 13.2 (Continued)

Biosorbent	Biosorption capacity (mg/g)	pH	Temperature (°C)	Biosorbent dose (g/L)	Initial conc. (mg/L)	References
Holly modified sawdust	18.86	7		6	60	(Siboni et al., 2011)
Acacia albida bark	2.98	2	37	20	7.5	(Gebrehawaria et al., 2015)
Eucleaschim peri leaves	3.94	2	37	20	7.5	(Gebrehawaria et al., 2015)
Pine needle powder	48	2–3	25	10	50	(Hadjmohammadi et al., 2010)
Macadamia nutshells (modified powder)	72.12	2	—	2.5	150	(Maremeni et al., 2018)
Aspergillus niger NU101	17.58	6	30	2	25	(Ghosh et al., 2015)
Dictyota dichotoma biomass	9.02	4	27	20	40	(Nandhagopal et al., 2018)
Cupressus lusitanica bark	305.4	1.5	28	10	100	(Netzahuatl-Munoz et al., 2015)
Cladophora albida	47.1	0.5–3	25	2	154	(Deng et al., 2009)
Candida lipolytica	6.35	1	—	1	30	(Ye et al., 2010)
Fusarium solani	60	4	30	4.5	500	(Sen and Dastidar,2011)

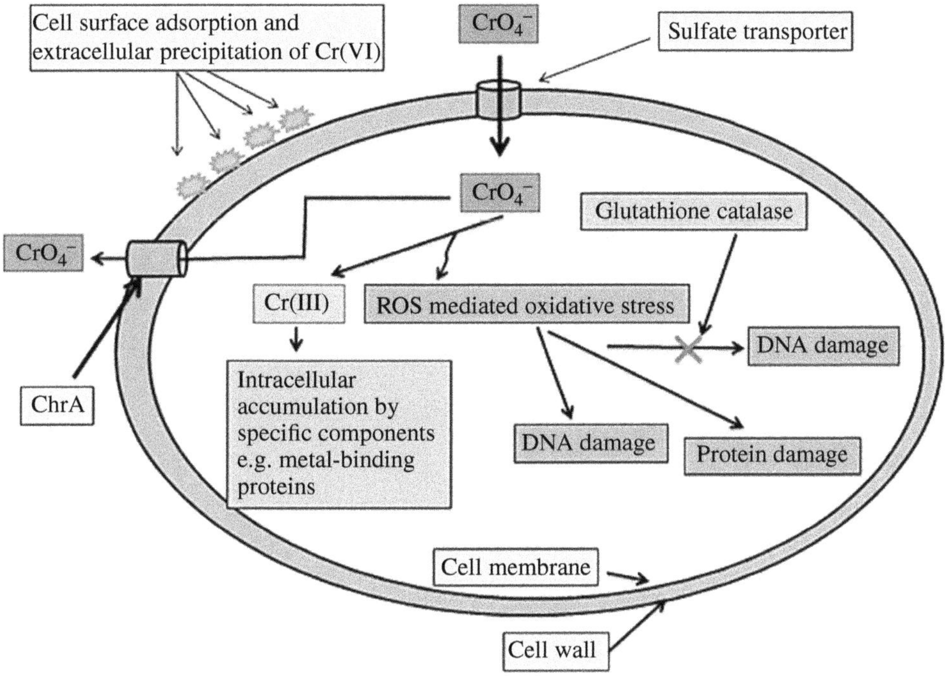

Figure 13.6 Schematic mechanism of Cr (VI) interaction with living microorganisms. Chromate enters into bacteria's cells through a sulfate transporter because sulfate has a structure similar to chromate. Plasmid encodes ChrA, which is responsible for expelling the intracellular chromate from the cell. Intracellular Cr (VI) is decremented into Cr (III) by various mechanisms. The reduced Cr (III) accumulates within the cell. During the reduction process, reactive oxygen species (ROS) mediate oxidative stress that can damage proteins and DNA molecules. Various intracellular proteins protect from oxidative stress, for example, glutathione.

Bioaccumulation is a very efficient process for removing metal ions, but there is a need to improve its efficiency. The expression of metal-binding proteins can be enhanced using genetic engineering techniques (Chojnacka, 2010). Potential microorganisms and their Cr (VI) removal capacity are listed in Table 13.3.

Biological Reduction of Hexavalent Chromium

Biological reduction is a significant method for minimizing Cr (VI) toxicity. Cr (VI) has a high level of toxicity due to being more soluble and permeable and having a higher oxidation state than Cr (III). In the reduction process, Cr (VI) is reduced into less-toxic and less-soluble Cr (III) (Baldiris et al., 2018). Microbial reduction of Cr (VI) was found to be the most useful and practical technique. These methods are inexpensive and environmentally friendly compared with other physicochemical methods (Raspor et al., 2000; Deng et al., 2010; Wang et al., 2012).

Reduction of Cr (VI) involves microbial substances/chemicals and enzymes secreted from several groups of anaerobic and aerobic bacteria (Donati et al., 2003). Anaerobic decrement or Cr (VI) reduction is a very slow process; hence it mainly depends on a membrane-bound

Table 13.3 Microorganisms and their Cr (VI) efficiency.

Microorganism	Removal efficiency (%)	Optimum pH	Optimum temperature	Initial chromium conc. (mg/L)	References
Bacillus subtilis (bacteria)	96	9	30	50	(Mangaiyarkarasi *et al.*, 2011)
Cellulosimicrobiumfunkei AR8 (bacteria)	100	7	35	100–250	(Karthik *et al.*, 2017)
Pseudomonas aeruginosa strain-1 (bacteria)	86	4–7	37	0.028	(Pandian et al., 2014)
Pseudomonas fluorescens (bacteria)	99	7	37	1000	(Devi et al., 2012)
Bacillus cereus IST105 (bacteria)	75	—	—	100	(Iyer et al., 2004)
Rhodococcus erythropolis (bacteria)	89.54	5	28	100	(Banerjee et al., 2017)
Staphylococcus capitis (bacteria)	89	7	37	—	(Zahoor and Rehman, 2009)
Bacillus sp. JDM-2-1 (bacteria)	86	6	37	—	(Zahoor and Rehman, 2009)
Pseudomonas GT7	40.8	7	30	—	(Zhang et al., 2016)
Pseudomonas putida (bacteria)	95.68	6.5	37	—	(Balamurugan et al., 2014)
Bacillus subtilis (bacteria)	95.19	7	37	—	(Balamurugan et al., 2014)
Scenedesmus incrassalulus (microalga)	52.7	—	25	1.2	(Penacastro et al., 2004)
Paecilomyces sp. (fungi)	100	4	28	50	(Juan et al., 2010)
Penicillium sp. N3 (fungi)	93	3	30	60	(Fukuda et al., 2008)
Aspergillus sp. N2 (fungi)	75	6	30	60	(Fukuda et al., 2008)

Rhizopus sp. CUC23 (fungi)	72.3	5.6	28	80	(Bibi et al., 2018)
Aspergillus fumigates ML43 (fungi)	72.9	5.6	28	80	(Bibi et al., 2018)
*Penicillium radicum*PL17 (fungi)	60.1	5.6	28	80	(Bibi et al., 2018)
*Fusarium proliferatum*FBL1 (fungi)	53.0	5.6	28	80	(Bibi et al., 2018)
Aspergillus niger (fungi)	96.3	4.5	30	20	(Shugaba et al., 2012)
Aspergillus parasiticus (fungi)	91.6	4.5	30	20	(Shugaba et al., 2012)
Paecilomyces sp. (fungi)	100	4	28	50	(Juan et al., 2010)
Trichoderma sp. BSCR02 (fungi)	77.36	5	35	200	(Smily et al., 2017)
Fusarium chlamydosporium SPFS2-g (fungi)	100	4.6	30	—	(Sharma et al., 2014)
Termitomycesclypeatus (fungi)	30	3	30	100	(Das et al., 2009)
*Acinetobacter*sp. (bacteria)	75	7	75	—	(Srivastava et al., 2007)

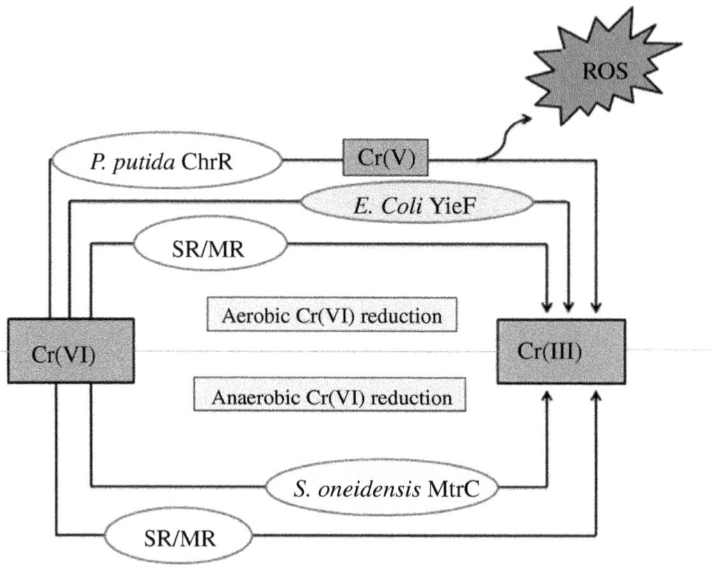

Figure 13.7 Microbial reduction mechanism of Cr (VI). Chromate reductase (ChrR) of *Pseudomonas putida* catalyzes the reduction of Cr (VI). It reduces Cr (VI) into Cr (V) and Cr (IV). The Cr (V) is finally reduced into Cr (III) and ROS. The chromate-reducing gene YieF of *E. coli* directly reduces Cr (VI) into Cr (III). MtrC is a terminal reductase of *Shewanella oneidensis* MR-1 that can reduce Cr (VI) to Cr (III). Moreover, soluble reductase (SR) and membrane-bound reductase (MR) decrement hexavalent chromium with the help of NADH in an anaerobic or aerobic environment.

enzymatic system. The enzymes participating in aerobic Cr (VI) reduction are primarily found in soluble form in cytosol. We have summarized the Cr (VI) reduction mechanism in Figure 13.7 (Somasundaram et al., 2009; Puzon et al., 2002).

Cr (VI) reduction efficiency is influenced by factors including the types of microorganisms and the hexavalent chromium concentration in the microbial growth medium. The aerobic or anaerobic reduction of hexavalent chromium also depends on temperature, medium components, and the pH of the medium (Alam and Malik, 2008; Xu et al., 2013). Three primary mechanisms are used for hexavalent chromium reduction (Ngwenya and Chirwa, 2011):

- *Anaerobic reduction of Cr (VI):* Under anaerobic conditions, cell protoplasm components are involved in decrementing Cr (VI). Cell components such as vitamins, heme-containing proteins, carbohydrates, amino acids, and glutathione act as electron donors during the reduction of Cr (VI) (Ahemad, 2014).
- *Aerobic reduction of Cr (VI):* The reduction of Cr (VI) ions depends on extracellular soluble reductase enzymes. The reduction of Cr (VI) depends on electron donors like nicotinamide adenine dinucleotide phosphate (NADPH), reduced glutathione (GSH), and d-glucose. NADPH donate electrons to Cr (VI) and reduces into less-toxic Cr (III). Chromate reduction enzymes are secreted by several groups of microbial species such as *Pseudomonas putida* PRS2000 and *Desulfovibrio vulgaris*. These chromate reductase enzymes utilize electron donors located inside or outside the bacterial cell (Loryuenyong

et al., 2014; Belchik et al., 2011). This is an energy-dependent and highly regulated process. Extracellular chromate reduction does not require the intake of chromium ions into the bacterial cell. The bacterial cell is protected from chromate-mediated DNA damage; this is the primary advantage of extracellular chromate reduction (Ahemad, 2014).

- *Membrane-associated reductase-mediated chromate reduction of Cr (VI)*: This process requires H_2 or glucose as an electron donor component. Cr (VI) accepts terminal electrons from the electron transport chain (Cheung et al., 2006; Baldiris et al., 2018). Many researchers report glucose as an external electron donor in this process. Glucose enhances the enzymatic activity of chromate reductase about 3.5-fold (Ibrahim et al., 2012).

Researchers have reported several bacteria species and their chromate reductases gene. The most-studied chromate reduction mechanisms were in *P. putida* (ChrR gene) and *E. coli* (YieF). ChrR and YieF flavoproteins of bacterial cytoplasm and reduce Cr (VI) completely into Cr (III). The chromate reductase-mediated reduction of Cr (VI) depends on factors such as temperature and availability of electron donor groups; for example, ChrR of *Pputida* reduces Cr (VI) at 70 °C, and YeiF of *E. coli* completely reduces Cr (VI) into Cr (III) at 35 °C (Ackerley et al., 2004).

Adsorption Kinetic Studies

Adsorption is greatly affected by the properties of the adsorbent materials and time. Adsorption kinetics play a vital role in adsorption and the design the treatment system. Generally, pseudo-first-order and pseudo-second-order kinetic models are helpful in kinetic studies.

Pseudo-First-Order Kinetics

The linear form of the pseudo-first-order equation is as follows

$$\log\left(q_e - q_t\right) = \log\left(q_e\right) - \frac{k_s}{2.303}\mathrm{t} \tag{13.1}$$

where k_s = equilibrium rate constant determined by the slope of log $(q_e - q_t)$ vs. time (min). q_t and q_e = adsorbate amount adsorbed per unit mass of an adsorbent at time *t* and equilibrium, respectively.

Pseudo-Second-Order Kinetics

The linear form of this model is as follows:

$$\frac{t}{q_t} = \frac{1}{k'_2 q_e} + \frac{1}{q_e}\mathrm{t} \tag{13.2}$$

Here, k'_2 is the equilibrium rate constant calculated using Eq. (13.3):

$$h = k'_2 q_e^2 \tag{13.3}$$

The rate constant k'_2 and h can be calculated using the t/q_t vs. t plot.

Adsorption Isotherm Studies

Isotherms consider the relationship between adsorption capacity and the remaining concentration of heavy metal ions at a constant temperature. Freundlich, Langmuir, Temkin, and Dubinin-Radushkevich (D-R) isotherm models are widely used to study adsorption isotherms.

Langmuir Isotherm

The basis of this isotherm is monolayer adsorption on a homogenous adsorbent by neglecting any surface interaction between two absorbed molecules. The linear form of the Langmuir equation is as follows

$$\frac{C_e}{q_e} = \frac{1}{Q^0 b} + \frac{C_e}{Q^0} \tag{13.4}$$

where Q^0 and b are isotherm constants that can be calculated from the intercept and slope of the plot of C_e/q_e vs. C_e.

Freundlich Isotherm

This adsorption isotherm model defines an adsorption phenomenon that occurs on a heterogeneous surface. The linear form of the Freundlich isotherm is as follows

$$\mathrm{Log}\, q_e = \log k_f + \frac{1}{n} \log C_e \tag{13.5}$$

where k_f and n are isotherm constants calculated from the intercept and slope of the plot of log C_e vs. log q_e.

Temkin Isotherm

This isotherm provides uniformly distributed binding energies over the population of surface binding adsorption. It predicts that the adsorption heat of all the molecules in a layer is decremented linearly with the coverage binding sites by the adsorbate. This model also indicates that adsorption is determined by the uniform distribution of the binding energies, up to some maximum binding energy (Temkin and Pyzhev, 1940; Gouamida et al., 2013). The linear form of the Temkin isotherm model is as follows

$$q_e = \frac{RT}{b_T} \ln A_T + \frac{RT}{b_T} \ln C_e \tag{13.6}$$

$$q_e = \beta \ln \alpha + \beta \ln C_e \tag{13.7}$$

where $\beta = \frac{RT}{b_T}$; $\alpha = A_T$; T = temperature (absolute) Kelvin; R = universal gas constant whose value is 8.314 J/mol K; and b_T = Temkin constant relating heat sorption (J/mg) and the equilibrium binding constant, i.e. A_T corresponding to the maximum binding energy (L/g). The Temkin constants A_T and b_T can be evaluated from the slope and intercept of C_e vs. InC_e.

D-R Isotherm

Radushkevich and Dubinin found that the nature of an adsorption phenomenon depended on the adsorbent structure, i.e. whether it was porous. The linear form of the D-R isotherm is as follows

$$\ln q_e = \ln X_m - \beta \mathrm{F}^2 \tag{13.8}$$

where X_m, and β are constants determined from the intercept and slope of the plot of lnq_e vs. F_2. F (Polany potential) is determined using Eq. (13.9)

$$F = RT \ln\left(1 + \frac{1}{C_e}\right) \tag{13.9}$$

where R = gas constant and T = adsorption equilibrium temperature.

The isotherm constant β relates the free energy of adsorption of 1 mole of heavy metal ions on the surface of the biosorbent. This energy can be calculated using Eq. (13.10):

$$E = \frac{1}{\sqrt{-2\beta}} \tag{13.10}$$

The classification of the current adsorption system as chemical or physical is based on the adsorption free energy *(E)* value.

Thermodynamics Studies

The parameters of thermodynamics, such as changes in enthalpy (ΔH^0), Gibbs free energy ($\Delta G°$), and entropy ($\Delta S°$) of adsorption, can be analyzed in various temperature ranges using these equations

$$k_c = \frac{C_{Ae}}{C_e} \tag{13.11}$$

$$\Delta G = -\mathrm{RT} \ln k_c \tag{13.12}$$

$$\ln k_c = \frac{\Delta S}{R} - \frac{\Delta H}{RT} \tag{13.13}$$

where C_{Ae} = equilibrium solute concentration of the adsorbent (μg/L); C_e = equilibrium solute concentration in a bulk solution (μg/L); T = absolute temperature (k); and R = universal gas constant (8.314 J/mol). The values of ($\Delta H°$) and ($\Delta S°$) are determined using the slope and intercept of the plot of ln k_c vs. $1000/T$.

A positive value of ΔH° means adsorption will be endothermic and hence can be increased by raising the temperature. A negative result or value of ΔG° suggests that adsorption is a spontaneous process and can be increased by incrementing the temperature.

Conclusion

Cr (VI) is applicable worldwide in many industrial processes that discharge chromium-containing effluents into the environment. Cr (VI) is non-biodegradable and persists in the environment for a long time; and Cr (VI) metal ions cause carcinogenic and mutagenic effects in humans and other animals. Various physical and chemical techniques are helpful for removing Cr (VI) contamination from industrial wastewater; each has disadvantages and advantages.

Biological methods are currently preferred to remove Cr (VI) contamination from industrial wastewaters. These methods are eco-friendly and inexpensive and are effective at lower concentrations of metal ions. Living microorganisms can accumulate heavy metal ions within their cells. The bioaccumulation process is very complex and involves intracellular pathways. Living organisms also reduce Cr (VI) into less-toxic trivalent chromium. The reduction process has the advantage of minimizing toxicity and improving bioaccumulation within cells.

Mathematical models such as isotherm modeling, kinetic modeling, and thermodynamic studies are used to understand the mechanisms of biosorption of hexavalent chromium. With these models, the heavy metal removal process can be optimized and biosorption of hexavalent chromium can be evaluated.

Acknowledgments

The authors of this chapter are grateful to IIT BHU, Varanasi, India, for instrumentation and financial support.

References

Ackerley, D.F., Gonzalez, C.F., Park, C.H. et al. (2004). Chromate-reducing properties of soluble flavoproteins from *pseudomonas putida* and *Escherichia coli*. *Applied and Environmental Microbiology* 70: 873–882.

Agency for Toxic Substances and Disease Registry (ASTDR) (2012). Toxicological PROFILE FOR CHROMIUM. CAS#: 7440-47-3. https://www.atsdr.cdc.gov/toxprofiles/tp.asp?id=62andtid=17.

Ahemad, M. (2014). Bacterial mechanisms for Cr (VI) resistance and reduction: An overview and recent advances. *Folia Microbiologica* 59: 1–12.

Ahluwalia, S.S. and Goyal, D. (2007). Microbial and plant derived biomass for removal of heavy metals from wastewater. *Bioresource Technology* 98 (12): 2243–2257. doi:10.1016/j.biortech.2005.12.006.

Alam, M.Z. and Malik, A. (2008). Chromate resistance, transport and bioreduction by *Exiguobacterium* sp. ZM-2 isolated from agricultural soil irrigated with tannery effluent. *J Basic Microbiol* 48: 416–420.

Balamurugan, D., Udayasooriyan, C., and Kamaladevi, B. (2014). Chromium (VI) reduction by Pseudomonas putida and *Bacillus subtilis* isolated from contaminated soils. *International Journal of Environmental Sciences* 5: 522–529.

Baldiris, R., Acosta-Tapia, N., Montes, A. et al. (2018). Reduction of hexavalent chromium and detection of chromate reductase (ChrR) in *Stenotrophomonas maltophilia*. *Molecules* 23: 406–445.

Banerjee, S., Joshi, S.R., Mandal, T., and Halder, G. (2017). Insight into Cr^{6+} reduction efficiency of *Rhodococcuserythropolis* isolated from coalmine waste water. *Chemosphere* 167: 269–281.

Belchik, S.M., Kennedy, D.W., Dohnalkova, A.C. et al. (2011). Extracellular reduction of hexavalent chromium by cytochromes MtrC and OmcA of *shewanella oneidensis* MR-1. *Applied and Environmental Microbiology* 2011: 4035–4041.

Bhattacharya, P., Banerjee, P., Mallick, K. et al. (2013). Potential of biosorbent developed from fruit peel of *Trewianudiflora* for removal of hexavalent chromium from synthetic and industrial effluent: Analyzing phytotoxicity in germinating Vigna seeds. *Journal of Environmental Science and Health, Part A* 48: 706–719.

Bibi, S., Hussain, A., Hamayun, M. et al. (2018). Bioremediation of hexavalent chromium by endophytic fungi; safe and improved production of *Lactuca sativa* L. *Chemosphere* 211: 653–663.

Boosaeidi, N., Pourkhabbaz, A., and Jahani, M. (2017). Biosorption of hexavalent chromium by the agricultural wastes of the cotton and barberry plants. *Advances in Environmental Technology* 3: 159–167.

Central Pollution Control Board (CPCB), New Delhi, India. (1993). Guidelines for the discharge of waste-water in natural water-bodies.

Cheung, K.H. and Gu, J.D. (2007). Mechanism of hexavalent chromium detoxification by microorganism and bioremediation application potential: a review. *Int Biodeterior Biodegrad* 59: 8–15.

Cheung, K.H., Lai, H.Y., and Gu, J.D. (2006). Membrane-associated hexavalent chromium reductase of *Bacillus megaterium* TKW3 with induced expression. *Journal of Microbiology and Biotechnology* 16: 855–862.

Chojnacka, K. (2010). Biosorption and bioaccumulation – the prospects for practical applications. *Environment International* 36 (3): 299–307. doi:10.1016/j.envint.2009.12.001.

Dadrasnia, A., Wei, K.S.C., Shahsavari, N. et al. (2015). Biosorption potential of *Bacillus salmalaya* strain 139SI for removal of Cr (VI) from aqueous solution. *Int J Environ Res Public Health* 12: 15321–15338.

Das, S.K. and Guha, A.K. (2009). Biosorption of hexavalent chromium by *Termitomyces clypeatus* biomass: Kinetics and transmission electron microscopic study. *J Hazard Mater* 167: 685–691.

Davis, A.P., Bernstein, C., and Gietka, P.M. (1995). Waste minimization in electropolishing: Process control. In: *Proceedings of the Twenty-Seventh Mid-Atlantic Industrial Waste Conference: Hazardous and Industrial Wastes* (ed A.K. Sengupta), 62–71. Technomic Publishing.

Deng, L., Zhang, Y., Qin, J., Wang, X., and Zhu, X. (2009). Biosorption of Cr (VI) from aqueous solutions by nonliving algae *Cladophora albida*. *Miner Eng* 22: 372–377.

Deng, S., Yu, Q., Huang, J., and Yu, G. (2010). Removal of perfluorooctane sulfonate from wastewater by anion exchange resins: Effects of resin properties and solution chemistry. *Water Res* 18: 5188–5195.

Devi, B.D., Thatheyus, A.J., and Ramya, D. (2012). Bioremeoval of hexavalent chromium, using *Pseudomonas fluorescens*. *J Microbiol Biotechnol Res* 2: 727–735.

Donati, E., Oliver, C., and Curutchet, G. (2003). Reduction of chromium (VI) by the indirect action of thiobacillusthioparus. *Braz J Chem Eng* 20: 69–73.

Donmez, G. and Aksu, Z. (2002). Removal of chromium (VI) from saline wastewaters by *Dunaliella* species. *Process Biochem* 38: 751–762.

Environmental Protection Act (2002). Standards for effluent discharge regulations.

Fernandez-Lopez, J.A., Angosto, J.M., and Aviles, M.D. (2014). Biosorption of hexavalent chromium from aqueous medium with opuntia biomass. *Scientific World Journal* 2014: 1–8.

Fukuda, T., Ishino, T., Ogawa, A. et al. (2008). Cr (VI) reduction from contaminated soils by *Aspergillus* sp. N2 and *Penicillium* sp. N3 isolated from chromium deposits. *J Gen Appl Microbiol* 54: 295–303.

Gadd, G.M. (2009). Biosorption: Critical review of scientific rationale, environmental importance and significance for pollution treatment. *Journal of Chemical Technology and Biotechnology* 84 (1): 13–28.

Gebrehawaria, G., Hussen, A., and Rao, V.M. (2015). Removal of hexavalent chromium from aqueous solutions using barks of *Acacia albida* and leaves of *Euclea schimperi*. *Int J Environ Sci Technol* 12: 1569–1580.

General Notice No. 44 of 2003. Environmental Protection Act (2002). http://faolex.fao.org/docs/texts/mat52519.doc.

Ghosh, S., Mondal, A., and Paul, A.K. (2015). Hexavalent chromium biosorption by dried biomass of *Aspergillus niger* NUA101 isolated from Indian ultramafic complex. *Afr J Microbiol Res* 9: 220–229.

Gupta, V.K., Morhan, D., Sharma, S., and Park, K.T. (1999). Removal of chromium (VI) from electroplating industry wastewater using bagasse fly ash a sugar industry waste material. *Environmentalist* 19: 129–136.

Gupta, V.K. and Rastogi, A. (2009). Biosorption of hexavalent chromium by raw and acid-treated green alga *Oedogoniumhatei* from aqueous solutions. *J Hazard Mater* 163: 396–402.

Hadjmohammadi, M.R., Salary, M., and Biparva, P. (2010). Removal of Cr (VI) from aqueous solution using pine needles powder as a biosorbent. *Journal of Applied Sciences in Environmental Sanitation* 6: 1–13.

Ibrahim, A.S.S., Elbadawi, Y.B., El-Tayeb, M.A., and Al-Salamah, A.A. (2012). Hexavalent chromium reduction by novel chromate resistant alkaliphilic *Bacillus* sp. strain KSUCr9a. *African Journal of Biotechnology* 11: 3832–3841.

Iyer, A., Mody, K., and Jha, B. (2004). Accumulation of hexavalent chromium by an exopolysaccharide producing marine *Enterobacter cloaceae*. *Marine Pollution Bulletin* 49: 974–977.

Jacobs, J.A. and Testa, S.M. (2004). Overview of chromium (VI) in the environment: background and history. In: *Chromium (VI) Handbook* (ed. J. Guertin, J.A. Jacobs, and C.P. Avakian), 1–21. CRC Press.

Jobby, R., Jha, P., Yadav, A.K., and Desai, N. (2018). Biosorption and biotransformation of hexavalent chromium [Cr(VI)]: A comprehensive review. *Chemosphere* 207: 255–266.

Juan, F., Cardenas-Gonzalez, and Acosta-Rodriguez, I. (2010). Hexavalent chromium removal by a *Paecilomyces* sp. fungal strain isolated from environment. *Bioinorganic Chemistry and Applications*doi:10.1155/2010/676243.

Karthik, C., Barathi, S., Pugazhendhi, A. et al. (2017). Evaluation of Cr (VI) reduction mechanism and removal by *Cellulosimicrobium funkei* strain AR8, a novel haloalkaliphilic bacterium. *Journal of Hazardous Materials* 333: 42–53.

Kiptoo, J.K., Ngila, J.C., and Sawula, G.M. (2004). Speciation studies of nickel and chromium in wastewater from an electroplating plant. *Talanta* 64: 54–59.

Koedrith, P., Kim, H., Weon, J.I., and Seo, Y.R. (2013). Toxicogenomic approaches for understanding molecular mechanisms of heavy metal mutagenicity and carcinogenicity. *Int J Hyg Environ Health* 216: 587–598.

Loryuenyong, V., Jarunsak, N., Chuangchai, T., and Buasri, A. (2014). The photocatalytic reduction of hexavalent chromium by controllable mesoporous anatase TiO2 nanoparticles. *Advances in Materials Science and Engineering* 2014: 1–8.

Mahdinia, S.M., Moenian, K., and Rastgoo, T. (2014). Rice husk silica adsorbent for removal of hexavalent chromium pollution from aquatic solution. *Iranica Journal of Energy and Environment* 5: 218–223.

Malik, A. (2004). Metal bioremediation through growing cells. *Environment International* 30: 261–278.

Mangaiyarkarasi, M.S.M., Vincent, S., Janarthanan, S. et al. (2011). Bioreduction of Cr (VI) by alkaliphilic Bacillus subtilis and interaction of the membrane groups. *Saudi Journal of Biological Sciences* 18: 157–167.

Maremeni, L.C., Modise, S.J., Mtunzi, F.M. et al. (2018). Adsorptive removal of hexavalent chromium by diphenylcarbazide-grafted macadamia nutshell powder. *Bioinorganic Chemistry and Applications* 2018: 1–14.

Mishra, A., Dubey, A., and Shinghal, S. (2015). Biosorption of chromium (VI) from aqueous solutions using waste plant biomass. *Int. J. Environ. Sci. Technol* 12: 1415–1426.

Naja, G.M. and Volesky, B. (2010). Treatment of metal-bearing effluents: removal and recovery. In: *Handbook on Heavy Metals in the Environment* (ed. L.K. Wang, J.P. Chen, Y.T. Hung, and N.K. Shammas). Taylor and Francis and CRC Press.

Nandhagopal, K., Munuswamy, E., and Krishnan, V. (2018). Biosorption of chromium vi by ubiquitous dictyota biomass. *International Journal of Pharmacy and Biological Sciences* 8: 127–131.

Netzahuatl-Munoz, A.R., Cristiani-Urbina, M.D.C., and Cristiani-Urbina, E. (2015). Chromium biosorption from Cr (VI) aqueous solutions by *Cupressus lusitanica* bark: kinetics, equilibrium and thermodynamic studies. *PLoS One* 10: 0137–086.

Ngwenya, N. and Chirwa, E.M.N. (2011). Biological removal of cationic fission products from nuclear wastewater. *Water Sci Technol* 63: 124–128.

Onyancha, D., Mavura, W., Ngila, J.C. et al. (2008). Studies of chromium removal from tannery wastewaters by algae biosorbents, *Spirogyra condensata* and *Rhizoclonium hieroglyphicum*. *Journal of Hazardous Materials* 158: 605–614.

Owalude, S.O. and Tella, A.C. (2016). Removal of hexavalent chromium from aqueous solutions adsorption on modified groundnut hull. *Beni-Suef University Journal of Basic Applied Sciences* 5: 377–388.

Owlad, M., Aroua, M.K., Daud, W.A.W., and Baroutian, S. (2009). Removal of hexavalent chromium-contaminated water and wastewater: A review. *Water Air and Soil Pollution* 200: 59–70.

Pakade, V.E., Ntuli, T.D., and Ofomaja, A.E. (2017). Biosorption of hexavalent chromium from aqueous solutions by *Macadamia* nutshell powder. *Appl Water Sci* 7: 3015–3030.

Pandian, K., Thatheyus, A.J., and Ramya, D. (2014). Bioremoval of chromium, nickel and zinc in electroplating effluent by *Pseudomonas aeruginosa*. *Open J Water Pollut Treat* 1: 75–82.

Park, D., Park, J.M., and Yun, Y.S. (2006). Mechanisms of the removal of hexavalent chromium by biomaterials or biomaterial-based activated carbons. *J Hazard Mater* 137: 1254–1257.

Park, D., Yun, Y.S., and Park, J.M. (2005). Studies on hexavalent chromium biosorption by chemically-treated biomass of *Eckloniasp.* *Chemosphere* 60: 1356–1364.

Penacastro, J.M., Martinez-Jeronimo, F., Esparza-Garcia, F., and Canizares-Villanueva, R.O. (2004). Heavy metals removal by the micro algae *Sceneunus in crassatulus* in continuous cultures. *Bioresource Technology* 94: 219–222.

Peng, S.H., Wang, R., Yang, L.Z. et al. (2018). Biosorption of copper, zinc, cadmium and chromium ions from aqueous solution by natural foxtail millet shell. *Ecotoxicol Environ Saf* 165: 61–69.

Puzon, G.J., Petersen, J.N., Roberts, A.G. et al. (2002). A bacterial flavin reductase system reduces chromates (III)–NAD+ complex. *Biochem Biophy Res* 294: 76–81.

Raspor, P., Batic, M., Jamnik, P. et al. (2000). The influence of chromium compounds on yeast physiology: (A review). *Acta Microbiol Inmunol Hung* 47: 143–173.

Ray, R.R. (2016). Adverse hematological effects of hexavalent chromium: an overview. *Interdiscip Toxicol* 9: 55–65.

Rezaei, H., Kulkarni, S., and Saptarshi, P.G. (2011). Study of physical chemistry on biosorption of hexavalent chromium by using *Chlorella pyrenoidosa*. Pelagia Research library. *Der Chemica Sinica* 2: 174–189.

Saha, R., Nandi, R., and Saha, B. (2011). Sources and toxicity of hexavalent chromium. *Journal of Coordination Chemistry* 64: 1782–1806.

Sari, A. and Tuzen, M. (2008). Biosorption of total chromium from aqueous solution by red algae (*Ceramiumvigatum*): Equilibrium, kinetics and thermodynamics studies. *J Hazard Mater* 160: 349–355.

Saxena A, Bhardwaj M, Allen T, Kumar S, Sahney R (2017). Adsorption of heavy metals from wastewater using agricultural–industrial wastes as biosorbents. *Water Science*. 31: 189–197, DOI: 10.1016/j.wsj.2017.09.002

Sen, M. and Dastidar, M.G. (2011). Biosorption of Cr (VI) by resting cells of *Fusarium solani*. *Iran J Environ Health Sci Eng* 8: 153–158.

Sharma, S. and Malaviya, P. (2014). Bioremediation of tannery wastewater by chromium resistant fungal isolate *Fusarium chlamydosporium* SPFS2-g. *Current World Environment* 9: 721–727.

Shugaba, A., Buba, F., Kolo, B.G. et al. (2012). Uptake and reduction of hexavalent chromium by *Aspergillus niger* and *Aspergillus parasiticus*. *J Pet Environ Biotechnol* 3: 1–8.

Siboni, M.S., Samarghandi, M.R., Azizian, S. et al. (2011). The removal of hexavalent chromium from aqueous solutions using modified holly sawdust: equilibrium and kinetics studies. *Environ Eng Res* 16: 55–60.

Singh, S.K., Bansal, A., Jha, M.K., and Dey, A. (2012). An integrated approach to remove Cr (VI) using immobilized *Chlorella minutissima* grown in nutrient rich sewage wastewater. *Bioresource Technology* 104: 257–265.

Singh, V., Singh, M.P., and Mishra, V. (2020). Bioremediation of toxic metal ions from coal washery effluent. *Desalination and Water Treatment* 197: 300–318.

Singh, V., Singh, J., and Mishra, V. (2021a). Sorption kinetics of an eco-friendly and sustainable Cr (VI) ion scavenger in a batch reactor. *Journal of Environmental Chemical Engineering* 9: 105125. doi:10.1016/j.jece.2021.105125.

Singh, V., Singh, J., and Mishra, V. (2021b). Development of a cost-effective, recyclable and viable metal ion doped adsorbent for simultaneous adsorption and reduction of toxic Cr (VI) ions. *Journal of Environmental Chemical Engineering* 9: 105124. doi:10.1016/j.jece.2021.105124.

Smily, J.R.M.B. and Sumithra, P.A. (2017). Optimization of chromium biosorption by fungal adsorbent, *Trichoderma* sp. BSCR02 and its desorption studies. *HAYATI Journal of Bioscience* 24: 65–71.

Somasundaram, V., Philip, L., and Bhallamudi, S.M. (2009). Experimental and mathematical modeling studies on Cr (VI) reduction by CRB, SRB and IRB, individually and in combination. *J Hazard Mater* 172: 606–617.

Song, D., Pan, K., Tariq, A. et al. (2016). Adsorptive removal of toxic chromium from wastewater using wheat straw and eupatorium adenophorum. *PLoS ONE* 11: e0167037. doi:10.1371/journal.pone.0167037.

Srivastava, S. and Thakur, I.S. (2007). Evaluation of biosorption potency of *Acinetobacter* sp. for removal of hexavalent chromium from tannery effluent. *Biodegradation* 18: 637–646.

Tchounwou, P.B., Yedjou, C.G., Patlolla, A.K. et al. (2012). Heavy metal toxicity and the environment. In: *Molecular, Clinical and Environmental Toxicology* (ed. A. Luch), 133–164. Springer Basel.

Tsezos, M., Remoudaki, E., and Angelatou, V. (1997). Biosorption sites of selected metals using Electron microscopy. *Comp Biochem Phys A* 118: 481–487.

Venitt, S. and Levy, L.S. (1974). Mutagenicity of chromates in bacteria and its relevance to chromate carcinogenesis. *Nature* 9: 493–495.

Vieira, R.H. and Volesky, B. (2000). Biosorption: A solution to pollution? *International Microbiology* 3: 17–24.

Volesky, B. (2007). Biosorption and me. *Water Res* 41: 4017–4029.

Volesky, B. and Holan, Z.R. (1990). Biosorption of heavy metals. *Biotechnol Prog* 11: 235–250.

Wang, S., Zhang, L., Xia, Z. et al. (2012). BCN graphene as efficient metal-free electrocatalyst for the oxygen reduction reaction. *Angew Chem Int Ed 5* 1: 4209–4212.

Weikang, L., Yang, Y., Xu, S. et al. (2018). Efficient removal of hexavalent chromium from water by an adsorption–reduction mechanism with sandwiched nanocomposites. *RSC Adv* 8: 15087–15093.

World Health Organization (WHO). (2003). Water sanitation hygiene Chromium in drinking-water Background document for development of WHO guidelines for drinking-water quality. www.who.int/water_sanitation_health/publications/chromium/en.

Xu, W.H., Liu, Y.G., Zeng, G.M. et al. (2013). Promoting influence of organic carbon source on chromate reduction by *Bacillus* sp. *Adv Mate Res* 610-613: 1789–1794.

Xu, Y., Xiao, H., and Sun, S. (2005). Study on anaerobic treatment of wastewater containing hexavalent chromium. *Journal of Zhejiang University Science* 6: 574–579.

Ye, J., Yin, H., Mai, B. et al. (2010). Biosorption chromium from aqueous solution and electroplating wastewater using mixture of *Candida lipolytica* and dewatered sewage sludge. *Bioresour Technol* 101: 3893–3902.

Yilmazer, P. and Saracoglu, N. (2009). Bioaccumulation and biosorption of copper (II) and chromium (III) from aqueous solutions by Pichia stipitis yeast. *Journal of Chemical Technology and Biotechnology* 84: 604–610.

Zahoor, A. and Rehman, A. (2009). Isolation of Cr (VI) reducing bacteria from industrial effluents and their potential use in bioremediation of chromium containing wastewater. *J Environ Sci* 21: 814–820.

Zhang, J.K., Wang, Z.H., and Ye, Y. (2016). Heavy metal resistances and chromium removal of a novel Cr (VI)-reducing *Pseudomonas* strain isolated from circulating cooling water of iron and steel plant. *Appl Biochem Biotechnol* 180: 1328–1344.

14

Biosorption of Arsenic from Wastewater

Bidhan Chandra Ruidas and Dan Bahadur Pal

Department of Chemical Engineering, Birla Institute of Technology, Mesra, Ranchi, Jharkhand, India

Introduction

The earth's surface consists of about 71% water, of which oceans contain about 96.5%. Water is also present in rivers and lakes, in ice caps and glaciers, in the ground as soil moisture, in aquifers and swamps, and in air as vapor. Only 2.5% of the total water is the fresh water required for life to survive. Only 1.2% of total water is surface water; the rest of the fresh water is found in glaciers and ice caps. This surface water is a key source of water for drinking, cooking, and household use. Groundwater is a crucial substance for all life on earth. In the twenty-first century, consistent access to affordable clean water is of prime importance and a significant challenge for all countries (Qu et al., 2013). Groundwater is seriously affected by increasing heavy metal content, which poses a threat to human health even at trace levels (Khan et al., 2010). Heavy metals, including arsenic, cause a long-term threat to the atmosphere and human beings when discharged in groundwater (Kumar et al., 2014, Pal et al., 2017). Chemical and tannery industries discharge various heavy metals, e.g., cadmium, iron, copper, chromium, arsenic, lead, etc. and their compounds (Prasad et al., 2020; Singh et al., 2015; Pal et al., 2021a; Pal et al., 2021b, Pal et al., accepted).

Arsenic is recognized as highly poisonous and carcinogenic worldwide because of the effects of mass poisoning from exposure to contaminated drinking water. Long-term arsenic exposure may cause severe health issues like bladder, lung, skin, and kidney cancer; skin lesions; respiratory disease; and liver and kidney disease. Countries like Taiwan, Bangladesh, India, Mexico, Hungary, Chile, Argentina, and the United States have high levels of arsenic in groundwater (Siddique et al., 2018). Most of the threatened regions are heavily populated; millions of people have become sick, and thousands of deaths occur each year.

There are two types of arsenic-derived compounds in natural water: inorganic and organic. Inorganic forms are present predominantly in groundwater and surface water (Pokhrel and Viraraghavan, 2008). Arsenic forms various oxidation states such as V, IV, III,

Biosorption for Wastewater Contaminants, First Edition.
Edited by Rangabhashiyam Selvasembian and Pardeep Singh.

II, I, 0, -I, -II, and -III, but the trivalent arsenic (III) and pentavalent arsenic (V) are predominantly found in natural water (Tuzen et al., 2009). As (III) usually appears in groundwater, while As (V) occurs in surface water (Pokhrel and Viraraghavan, 2006). According to Korte and Fernando (1991), arsenic (III) is 25 to 60 times more toxic than arsenic (V). A small dose of arsenic in drinking water can damage the skin and vital organs such as the central nervous system and liver, as well as cause cancer in various organs (Mohan et al. 2007; Kumar et al. 2014).

Filtration, reverse osmosis, flocculation, coagulation, electrocoagulation, precipitation, and adsorption are some methods used to remove heavy metals like arsenic (Siddiqui et al., 2018). Some of these are already used in affected areas. An area with arsenic contamination is generally identified by thorough monitoring of the ground and surface water. Some temporary solutions for arsenic-contaminated areas involve closing deep tube wells and installing activated carbon filters and other adsorber beds in the tap to eliminate arsenic from potable water (Pal et al., 2018; Singh et al., 2017; Singh et al., 2016).

Sources of Arsenic in Groundwater Pollution

Toxic arsenic is abundantly present in the earth's upper surface. Arsenic is a metalloid that can take on both metallic and nonmetallic forms and is found extensively in natural water, air, soil, mineral deposits, rocks, and biota (Miteva, 2005). It is found in over 200 minerals, including arsenopyrite, aealgar, orpiment, arsenolite, and others (Bissen and Frimmel, 2003). Natural processes such as desorption and degradation of arsenic-bearing rocks, as well as alluvial sediments, release arsenic, resulting in high groundwater arsenic content in deltas and alluvial plains (Polizzotto et al., 2006).

Arsenic enters organisms mainly through the drinking water, via either percolation from arsenic-rich rocks/soils or anthropogenic sources. Natural processes include volcanic emissions, biological activities, etc. Anthropogenic sources of groundwater pollution include mining and processing of sulfide ores; fly ash and combustion gas from fossils fuels; industrial effluent from electroplating, paint, and cement; waste from metallurgy, electronics, petroleum refining, and ceramic and glass processing; manufacturing and production of medicines, dyes, and wood preservatives; leaching of landfills; and the use of arsenic-containing herbicides, insecticides, and fungicides in agriculture and wood preservatives (Kord et al., 2013).

Effect of Arsenic on the Environment and Human Health

Arsenic alters the genetic material of fish and affects animals that eat affected fish. It impedes the production of numerous enzymes and antioxidants in plants. Inorganic arsenic is a known carcinogen in human beings. Its toxicity affects human health, causing skin lesions and cancer of the skin, brain, liver, kidney, and stomach (Murcott, 2012). Short-term arsenic exposure can cause nausea and vomiting, impair blood vessels, decrease red and white blood cell formation, cause irregular heart rhythms, and lead to a sensation of "pins and needles" on the palms and feet. Long-term arsenic exposure can lead to darkening of the skin and the appearance of small "corns" or "warts" on the palms. Tens of millions of people are exposed to excessive concentrations of arsenic through drinking water (Ng 2005). Death may be caused by the inhibition of enzyme metabolic activities by arsenic. Exposure to arsenic above 10 ppb may cause rhagades; impaired mucous membranes; irregularity in the respiratory, digestive, circulatory, and nervous systems; and skin, lung, and liver cancers (Wang and Wai, 2004).

Arsenic (III) is more toxic than arsenic (V) due to greater cellular uptake. The maximum contamination level (MCL) of total arsenic in drinking water has been decreased by regulatory authorities such as the World Health Organization (WHO) and the United States Environmental Protection Agency (US EPA) from 50 to 10 μg/L. However, arsenic concentrations exceed the permissible amount in many parts of the world, including Argentina, Bangladesh, India, Pakistan, Mexico, Mongolia, Germany, Thailand, China, Chile, the United States, Canada, Hungary, Romania, Vietnam, Nepal, Myanmar, and Cambodia (Chakraborti et al., 2002).

Methods for Removing Arsenic from Wastewater

There are various methods of eliminating arsenic from groundwater. These include oxidation followed by adsorption or precipitation, coagulation-flocculation, membrane filtration, reverse osmosis, adsorption, and biosorption.

Oxidation

The primary goal of the oxidation process is to convert soluble arsenic (III) to arsenic (V), which is then precipitated or absorbed. As (V) can be adsorbed onto solid surfaces more easily than arsenic (III). As a result, oxidation followed by adsorption is an efficient method for removing arsenic (Leupin and Hug, 2005). A number of oxidants like O_3, H_2O_2, Cl_2, permanganate, and ferrate have been utilized for oxidation, but oxidation with permanganate and ozone was faster than with hydrogen peroxide (Leupin and Hug 2005; Dodd et al., 2006).

Coagulation and Flocculation

Coagulation and flocculation are potential methods for removing arsenic. This approach involves adding a coagulant accompanied by the creation of a floc to extract arsenic from groundwater as a precipitate. Positively charged cationic coagulants decrease the colloid's negative charge, facilitating particle accumulation and the formation of larger particles (Choong et al., 2007). Flocs are produced in flocculation due to the polymeric bridge between the flocculent particles that ultimately agglomerate to create larger particles. Dissolved arsenic is precipitated/co-precipitated by the flocs and hence eliminated.

Coagulants based on iron (Fe) and aluminum (Al) are widely used as chemical coagulants (McNeill and Edwards, 1995). Pallier et al. (2010) confirmed removing more than 90 and 77% of arsenic (V) and arsenic (III), respectively, using kaolinite and $FeCl_3$ as a coagulant/flocculent. Hu et al. (2012) studied the coagulation-precipitation experiment on low-concentration arsenic solution at 280 ppb. They employed aluminum chloride and other two poly aluminum chloride coagulants and found arsenic concentration levels reduced below the MCL. Ravenscroft and Richards (2009) implied that Fe-based coagulants are more efficient at removing arsenic from water than Al-based coagulants.

Adsorption

Adsorption needs an activated surface to remove arsenic. Adsorption occurs onto activated/coated solid surfaces, and the process is free of sludge. After removing sequestered metal ions from the adsorbent surface, the surfaces can be reused, which is an additional benefit of this technology. More than 500 studies from the literature were examined by Mohan and Pittman (2007), and 40 different types of adsorbents were identified under experimentation (Prasad et al., 2020; Pal et al., 2017). They observed that a few adsorbents such as activated alumina and granulated ferric hydroxide are used successfully in removing As.Arsenic removal with adsorption depends on pH. A low pH (< 7) performs better for arsenic (V) removal than arsenic (III) (Wilkie and Hering, 1996; Giri et al., 2021). Ions such as HCO_3^-, Ca^{2+} silicate, and phosphate compete with arsenic for adsorption sites, decreasing the rate of arsenic removal (Lin and Wu, 2001).

Membrane Filtration

Membrane filtration is a modern technology for water treatment because of its eco-friendly nature, non-phase-transition procedure with arsenic, and ability to remove heavy metals with high efficiency. Various membrane separation processes are reported for arsenic removal in the literature, such as ultra-filtration (Amana et al., 2008), reverse osmosis (Kaczala et al., 2009) nanofiltration (Saeed et al., 2005), and electrodialysis (Park et al., 2007). Most of these processes have drawbacks, including high costs, incomplete elimination, poor selectivity, high energy consumption, and the release of significant volumes of toxic sludge (Alimohammadi et al., 2017). So, alternative low-cost technologies like biosorption are gaining in importance.

Biosorption

Biosorption is a low cost, eco-friendly approach for removing metal ions from aqueous solution. The process involves passive sequestration of toxic materials by dead or inactive biomass (Volesky, 1990), which preferably originates from sources such as industrial wastes from food, pharmaceutical industries, and wastewater treatment (Zouboulis, 1997). Investigators have reported a number of naturally occurring biomasses for arsenic removal from water, such as fungal biomass (Loukidou et al., 2003), coconut coir pith (Anirudhan and Unnithan, 2007), sea nodules, Lessonia nigrescens (Hansen et al., 2006), orange waste (Ghimire et al., 2003), chitosan (Mcafee, 2001), coconut husk carbon (Manju et al., 1998), bone char (Sneddon and Valsami-Jones,2005), crab shells (Niu et al., 2007), powdered eggshell (Oke and Adewusi, 2008), oat hull activated carbon (Chuang et al., 2005), activated carbon made from lignite, peat char (Mohan and Chander, 2006), and water hyacinth (Chigbo and Shore, 1982). Chowdhury and Mulligan (2011) investigated the biosorption of arsenic by anaerobic sludge obtained from a cheese manufacturer's anaerobic effluent treatment plant. They also investigated the kinetics of sorption, validated the experimental data with Langmuir and Freundlich isotherms, and studied the impacts of initial arsenic concentration, pH, and contact time on the arsenic sorption of anaerobic granules.

Principles of Biosorption

The main objective of biosorption is to selectively sequester metal ions by microbial cells. It is the physicochemical mechanism of inactive (i.e. non-metabolic) metal uptake by microbial biomass (Eccles and Hunt, 1986). Biosorption can sequester dissolved metals effectively with high efficiency even in a very low-concentration solution. The advantage of biosorption over other processes is in removing ions from low-concentration, complex wastewaters. Several mechanisms exist for sequestering metal ions by various biomass components: complexation, coordination, chelation, ion exchange, precipitation, and reduction (Wang and Chen, 2006). More than one process, such as metal complexation accompanied by metal reduction or metal precipitation, is involved in metal ion immobilization. Biosorption may occur by both metabolically active organisms and inactive cells. Dormant microbial cells can immobilize metal ions by adsorption on biomass, whereas active microbial cells can immobilize metal ions through biosorption and other mechanisms, including microbial metabolism (Hatzikioseyian and Tsezos, 2001).

Biosorption Sites

The microbial cell consists of the cell wall, plasma, and nucleus. The dissolved metal species first comes into contact with the cell wall in the extracellular environment. The cell walls of biomass play a vital role in biosorption. A bacterial cell barrier is a polymeric medium made up of polyoses, proteins, or a mixture of these complexes, found just outside a cell's plasma membrane. The cell wall provides mechanical strength to the cell plasma and protects the intracellular cytoplasm. The cell barriers of eubacteria, archaebacteria, fungi, and floras are collections of polysaccharides and lipids. Practically all eubacterial cell walls are made up of peptidoglycan, a polysaccharide. Due to their distinctive cell wall composition, eubacteria are categorized as gram +ve and gram -ve microorganisms. Different polysaccharides other than peptidoglycan and additional protein are found in the cell wall of archaebacteria, which made them distinctive from eubacteria. Some archaebacteria cell walls are also made of the polysaccharide pseudomurein. Typically, the cell wall of fungi consists of chitin and fiber polyoses. The cell barrier of algae and floras is made up primarily of cellulose polyose.

The three crucial polymeric groups – proteins, polysaccharides, and nucleic acids – are readily found in animal, plant, protozoan, and microorganism groups (Kapoor and Viraraghavan, 1997). These biopolymers, the critical components of the plasma, nucleus, and cell wall, have useful clusters that show significant metal ion binding capability (Kapoor and Viraraghavan, 1997). Metal ion fixation is also assisted by intracellular biopolymers such as proteins and DNA. Exo-Polysaccharides (EPSs), an extracellular polymeric substance, can also participate in metal immobilization (Diels et al., 1995b). The biopolymers bind with metal ions primarily through two mechanisms: ion exchange and the formation of complexes (coordination compounds) like chelates. It is obvious that more than one process takes place simultaneously during biosorption. A cation ion binds with negatively charged functional groups like C=O groups, organics of PO_4 groups, and SO_4 groups in biopolymers. The ligands possess more than one electron lone pair in their outer orbit. The electron-deficient metal ion accepts the lone electron pair from a ligand and forms a coordination bond. Primarily, nitrogen, oxygen, and sulfur atoms in biopolymers act as electron-pair donors.

Biosorption Mechanisms

The biosorption of metal ions by the metabolic mechanism is not well understood. On the other hand, non-metabolic mechanisms of biosorption occur through biosorption, including chelation coordination, complexation, ion exchange, reduction, and precipitation. The biosorption process depends on the following factors:

- The type of ligands in the bioorganism: chelating or single-pair donor
- The biomass state: alive or dead
- The electronic charge, stereochemical orientation, and organization features of the chosen metal ions
- The pH of the feed solution and the presence of competing co-ions (Esposito et al., 2002)

Complexation

Positively charged metal ions generally form covalent and coordination bonds with an organic functional group and ligand molecules. A covalent bond is a bond formed by two electrons, shared an electron from both sides: metal and atom in the functional group. Complex compounds are generally formed by the donation of a lone electron pair from the ligands to an electron-deficient metal ion. Complexation may occur by the formation of a coordination bond with mono-dentate (RNH_2) or poly-dentate (EDTA) ligands. According to Pearson classification, hard acids like Na, K, Ca, and Mg, the prerequisite nutrients for microorganisms, bind with oxygen-containing "hard bases" like OH^-, HPO_4^{2-}, and CO_3^{2-}. Soft acids, precious metals like Ag, Au, Pt, and Pd, form covalent bonds with "soft bases": ligands containing nitrogen or sulfur, present at the cell walls. Precious metal biosorption also involves in situ reduction after covalent bond formation (Esposito et al., 2002). Borderline metal ions between the hard and soft classifications, such as the transition metal ions $Mn^{2+,}$ Zn^{2+}, Cd^{2+}, Cu^{2+}, and Pb^{2+}, display a remarkable amount of covalent bonding with "soft" ligands containing nitrogen or sulfur (Brady and Tobin, 1995).

Chelation

Organic molecules occasionally contain multiple functional groups with electron pairs, which can be concurrently donated to the metal ion. Multiple donations of lone pairs from one organic ligand result in the formation of a loop structure involving the metal ion, referred to as *chelation*. The chelated complex with organic multi-dentate ligands is ethylenediamine tetracarboxylic acid (EDTA). Chelated compounds are very stable compared to complexes formed with mono-dentate ligands because bonding occurs at more than one position in a metal ion.

Ion Exchange

Gram-positive bacteria including *Salmonella* and other *Helicobacter*, acetic acid microorganisms, *Legionella*, etc. – primarily members of the genre *Bacillus* – have shown increased metal-binding ability due to their significant negative charge density (Wase and Forster, 1997). Because teichoic and teichuronic acids are involved in the peptidoglycan system in the cell wall, a negative charge arises. The phosphodiesters in teichoic acid and the carboxylate groups in teichuronic acid contribute to the cell wall with their ion-exchange ability.

Cation-binding ionizable groups are also present in biopolymers: for example, C=O groups, organics of PO_4 groups, and SO_4 groups. Sulfate is esterified into carbohydrate hydroxyl groups found in connective tissues and polysaccharides of algae, providing a negative charge density at very low pKa. Groupings of ester sulfate and phosphomonoester also occur in proteins. The phenolic hydroxyl group also provides a weak negative charge and binding potential with metal ions. In proteins, a positive charge is generated in protein components such as amino, imidazole, and guanidine groupings. The protonation of amino groups on purine, pyramiding rings, and heterocyclic nitrogen atoms gives rise to the positive charge in nucleic acids. These groups take part in anion exchange. A number of researchers have demonstrated the mechanism of ion exchange in removing metal ions by biosorption (Davis et al., 2003), but ion exchange is not the sole or key metal biosorption process (Brady and Tobin, 1995)

Precipitation

In metal biosorption by microbial cells, insoluble inorganic metal precipitates may be formed. Organic precipitates of metal may also be formed. This occurs when metal ions are bonded to extracellular polymeric materials discharged by procaryotic and eukaryotic microorganisms (Remoundaki et al., 2003).

Biosorption Isotherms

Three classical adsorption isotherms are used to validate experimental adsorption data. The Langmuir model is represented by the following equation (Langmuir et al., 1916)

$$q_e = \frac{q_m K C_e}{1 + K C_e} \tag{14.1}$$

where q_e and q_m = number of ions adsorbed at equilibrium and maximum adsorption capacity (mg of ions/g of adsorbent), respectively; K = equilibrium constant (L/mg) in the Langmuir isotherm; and C_e = optimum composition of heavy metals in aqueous solution (mg/L). The linearized form of the Langmuir isotherm model is:

$$\frac{C_e}{q_e} = \frac{1}{q_m K} + \frac{C_e}{q_m} \tag{14.2}$$

The Freundlich isotherm is presented in Eq. (14.3). This empirical isotherm shows that the mass or mole of adsorbate adsorbed at equilibrium follows a power law dependent on the adsorbate concentration (Freundlich et al., 1906). The isotherm can be represented as follows:

$$q_e = K_f C_e^{1/n} \tag{14.3}$$

where K_f = coefficient associated with the attachment energy; and n = heterogeneousness factor. Freundlich optimum coefficients may be calculated from this equation:

$$\ln q_e = \ln K_f + \frac{1}{n}\ln C_e \tag{14.4}$$

The Freundlich isotherm can be applied to adsorption on a surface that is energetically non-uniform. The slope (value of $1/n$) measures the adsorption intensity or surface heterogeneity, which ranges from 0 to 1. An angle above the charge of one specifies cooperative adsorption (Foo and Hameed, 2010).

Temkin et al. (1940) presented a model generalized and linearized equation for a biosorption isotherm, which is given by the following equation:

$$q_e = \frac{RT}{b_T}\ln\left(K_T C_e\right) \tag{14.5}$$

Biosorption Kinetics Model Analysis

For sorption arrangements and the response frequency monitoring phase, estimating kinetics is essential. The strategy of the adsorption-desorption process depends on the adsorbent's physiochemical characteristics and system conditions, such as temperature (Nadeem et al., 2006). The quantity of metal ions adsorbed, q_t, and contact time t were determined using this equation (Coskun et al., 2006):

$$\mathrm{q_{As}} = \frac{\left(\mathrm{C_{0}}\text{–}\mathrm{C_e}\right)\mathrm{V}}{\mathrm{W}} \tag{14.6}$$

The pseudo-first-order Lagrange model is generally expressed as follows (Farhan et al., 2012):

$$\frac{\mathrm{dq_t}}{\mathrm{dt}} = \mathrm{K}\left(\mathrm{q_e} - \mathrm{q_t}\right) \tag{14.7}$$

After integration and applying the boundary conditions $q_t = 0$ to $q_t = q_t$ at $T = 0$ to $t = t$, the integrating form of the equation becomes

$$\log\left(\mathrm{q_e} - \mathrm{q_t}\right) = \log \mathrm{q_e} + \frac{\mathrm{k_1}}{2.303}\mathrm{t} \tag{14.8}$$

$$\mathrm{q_t} = \mathrm{q_e}\left(1 - \left(\exp \mathrm{k_1 t}\right)\right) \tag{14.9}$$

The pseudo-second-order method is given by

$$\frac{\mathrm{dq_t}}{\mathrm{dt}} = \mathrm{k_2}\left(\mathrm{q_e} - \mathrm{q_t}\right)^2 \tag{14.10}$$

After including all limits, we found:

$$\frac{\mathrm{t}}{\mathrm{q_t}} = \frac{1}{\mathrm{q_e^2 k_2}} + \frac{\mathrm{t}}{\mathrm{q_e}} \tag{14.11}$$

A scheme of t/q_t v/s t shows the line equation from which k_2 and q_e may be calculated:

$$q_t = \frac{q_e^2\, k_2 t}{1 + q_e\, k_2 t} \tag{14.12}$$

The diffusion model can be written as an equation (Furusawa et al., 1974):

$$q_t = t^{0.5}\, k_{ipd} + C \tag{14.13}$$

The Elovich model equation is as follows (Zhang et al., 2019):

$$q_t = \frac{1}{\beta}\log(\alpha\beta) + \frac{1}{\beta}\log t \tag{14.14}$$

Biosorption of Arsenic from Wastewater

Several techniques and biomasses have been developed for sequestering arsenic from groundwater. Precipitation by treating with sulfide or hydroxide produces sludge that is difficult to manage. In addition, this technique consumes large quantities of reagents to precipitate, coagulate, and flocculate the pollutants (Gulledge and O'Connor, 1973). The advantages of biosorption with high adsorption capacities, low cost, and renderability of natural biosorption materials have been shown in advanced research over the past two decades (Gavrilescu, 2004).

Initially, biomass from acid-washed crab shells (AWCS) was developed to bind anionic metal species such as gold, cyanide, chromate, and vanadate (Niu and Volesky, 2003a). The carboxylic group present in yeast biomass was methylated and applied for adsorption of chromium (VI) and arsenic (V) (Seki et al., 2005). With a growing degree of methylation, adsorbed amounts of chromium (VI) and arsenic (V) increased significantly. Niu et al. (2007) further examined this crab shell to sequester arsenic (V). They found that the uptake of arsenic depends on the pH and presence of competitive ions. They further reported that arsenic uptake was primarily due to the arsenate bound on the amide groups of AWCS.

Baig et al. (2010) investigated the removal of arsenic using biomass derived from the stem of *Acacia nilotica* and found high efficiency (>95%) at a low concentration (200 μg/L). Baker's yeast (*Saccharomyces cerevisiae*) biomass obtained as a byproduct of fermentation processes has been studied for arsenic removal at ppm and ppb level with satisfactory results (Hadiania et al., 2019). Simultaneous biosorption of arsenic (III) and arsenic (V) by sea algae (*Sargassum glaucescens*) was studied by Tabaraki and Heidarizadi (2018). They fitted the biosorption equilibrium data with different adsorption isotherms and found that the Freundlich isotherm was the most suitable model.

Biosorption conditions such as pH, biosorbent dosage, and initial concentration of arsenic ions affect the uptake of arsenic. Biosorption of total arsenic and inorganic arsenic by alga *Sarcodia suiae* was investigated at different temperatures below 25 °C and light intensities. Maximum adsorption was obtained at the lowest temperature of 15 °C for arsenite, whereas no effect of light intensity was reported (Libatique et al., 2020).

Summary

In this chapter, we discussed the various health issues caused by arsenic uptake in the human body. Methodologies of arsenic sequestration were discussed from a rigorous literature review. Among the various methodologies, biosorption was found to be economical and safe to conduct and to have minimal toxic byproducts. Different adsorption isotherms were presented, which may be fitted with the biosorption equilibrium data obtained from various experiments. Among them, we found that the Freundlich isotherm is the most suitable.

Acknowledgments

The authors acknowledge the Birla Institute of Technology, Mesra, Ranchi, Jharkhand. DBP is also thankful to NPIU (TEQIP-III), Delhi, Govt. of India, for financial support.

References

Alimohammadi, M., Saeedi, Z., Akbarpour, B. et al. (2017). Adsorptive removal of arsenic and mercury from aqueous solutions by eucalyptus leaves. *Water, Air, and Soil Pollution* 228 (11): 429. doi:10.1007/s11270-017-3607-y.

Aman, T., Kazi, A.A., Sabri, M.U. et al. (2008). Potato peels as solid waste for the removal of heavy metal copper(II) from waste water/industrial effluent. *Colloids and Surfaces B: Biointerfaces* 63 (1): 116–121. doi:10.1016/j.colsurfb.2007.11.013.

Anirudhan, T.S. and Unnithan, M.R. Arsenic (V) removal from aqueous solutions using an anion exchanger derived from coconut coir pith and its recovery, *Chemosphere* 2007, 66, 60–66.

Baig, J.A., Kazi, T.G., Shah, A.Q. et al. (2010). Biosorption studies on powder of stem of Acacia nilotica: Removal of arsenic from surface water. *Journal of Hazardous Materials* 178 (1–3): 941–948. doi:10.1016/j.jhazmat.2010.02.028.

Bissen, M. and Frimmel, F.H. (2003). Arsenic – a review. Part I: occurrence, toxicity, speciation, mobility. *Acta hydrochimica et hydrobiologica* 31 (1): 9–18. doi:10.1002/aheh.200390025.

Brady, J.M. and Tobin, J.M. (1995). Binding of hard and soft metal ions to Rhizopus arrhizus biomass. *Enzyme and Microbial Technology* 17 (9): 791–796. doi:10.1016/0141-0229(95)00142-R.

Chakraborti, D. (2002). Arsenic calamity in the Indian subcontinent. What lessons have been learned? *Talanta* 58 (1): 3–22. doi:10.1016/S0039-9140(02)00270-9.

Chigbo, F.E., Smith, R.W., and Shore, F.L. (1982). Uptake of arsenic, cadmium, lead and mercury from polluted waters by the water hyacinth Eichornia crassipes. *Environmental Pollution Series, A., Ecological and Biological* 27 (1): 31–36. doi:10.1016/0143-1471(82)90060-5.

Choong, T.S.Y., Chuah, T.G., Robiah, Y. et al. (2007). Arsenic toxicity, health hazards and removal techniques from water: an overview. *Desalination* 217 (1–3): 139–166. doi:10.1016/j.desal.2007.01.015.

Chowdhury, M.D.R.I. and Mulligan, C.N. (2011). Biosorption of arsenic from contaminated water by anaerobic biomass. *Journal of Hazardous Materials* 190 (1–3): 486–492. doi:10.1016/j.jhazmat.2011.03.070.

Chuang, C.L., Fan, M., Xu, M. et al. (2005). Adsorption of arsenic(V) by activated carbon prepared from oat hulls. *Chemosphere* 61 (4): 478–483. doi:10.1016/j.chemosphere.2005.03.012.

CoŞKun, M., Steinnes, E., Frontasyeva, M.V. et al. (2006). Heavy metal pollution of surface soil in the Thrace region, Turkey. *Environmental Monitoring and Assessment* 119 (1–3): 545–556. doi:10.1007/s10661-005-9042-3.

Davis, T.A., Volesky, B., and Mucci, A. (2003). A review of the biochemistry of heavy metal biosorption by brown algae. *Water Research* 37 (18): 4311–4330. doi:10.1016/S0043-1354(03)00293-8.

Diels, L., Corbisier P., Hooyberghs L. et al. (1995). Heavy metal resistance and biosorption in Alcaligenes Eutrophus, E.R. 121. In: Proceedings of the International Biohydrometallurgy Symposium IBS 1995.

Dodd, M.C., Vu, N.D., Ammann, A. et al. (2006). Kinetics and mechanistic aspects of As(III) oxidation by aqueous chlorine, chloramines, and ozone: relevance to drinking water treatment. *Environmental Science and Technology* 40 (10): 3285–3292. doi:10.1021/es0524999.

Eccles, H. and Hunt, S. (1986). *Immobilisation of Ions by Bio-Sorption*. Ellis Horwood Ltd.

Esposito, A., Pagnanelli, F., and Vegliò, F. (2002). pH-related equilibria models for biosorption in single metal systems. *Chemical Engineering Science* 57 (3): 307–313. doi:10.1016/S0009-2509(01)00399-2.

Farhan, A.M., Salem, N.M., Al-Dujaili, A.H. et al. (2012). Biosorption studies of Cr(VI) ions from electroplating wastewater by walnut shell powder. *American Journal of Environmental Engineering* 2 (6): 188–195.

Foo, K.Y. and Hameed, B.H. (2010). Insights into the modeling of adsorption isotherm systems. *Chemical Engineering Journal* 156 (1): 2–10. doi:10.1016/j.cej.2009.09.013.

Freundlich, H.M.F. (1906). Uber die adsorption in Losungen. *Zeitschrift für Physikalische Chemie* 57: 385–470.

Furusawa, T. and Smith, J.M. (1974). Intraparticle mass transport in slurries by dynamic adsorption studies. *AIChE Journal* 20 (1): 88–93. doi:10.1002/aic.690200111.

Gavrilescu, M. (2004). Removal of heavy metals from the environment by biosorption. *Engineering in Life Sciences* 4 (3): 219–232. doi:10.1002/elsc.200420026.

Ghimire, K. (2003). Adsorptive separation of arsenate and arsenite anions from aqueous medium by using orange waste. *Water Research* 37 (20): 4945–4953. doi:10.1016/j.watres.2003.08.029.

Giri, D.D., Jha, J.M., Tiwari, A.K. et al. (2021). Java plum and amaltash seed biomass based bio-adsorbents for synthetic wastewater treatment. *Environmental Pollution* 280: 116890. doi:10.1016/j.envpol.2021.116890.

Gulledge, J.H. and O'Connor, J.T. (1973). Removal of arsenic (V) from water by adsorption on aluminum and ferric hydroxides. *Journal - American Water Works Association* 65 (8): 548–552. doi:10.1002/j.1551-8833.1973.tb01893.x.

Hadiani, M.R., Khosravi-Darani, K., and Rahimifard, N. (2019). Optimization of As (III) and As (V) removal by Saccharomyces cerevisiae biomass for biosorption of critical levels in the food and water resources. *Journal of Environmental Chemical Engineering* 7 (2): 102949. doi:10.1016/j.jece.2019.102949.

Hansen, H.K., Ribeiro, A., and Mateus, E. (2006). Biosorption of arsenic(V) with Lessonia nigrescens. *Minerals Engineering* 19 (5): 486–490. doi:10.1016/j.mineng.2005.08.018.

Hatzikioseyian, A., Remoudaki, E., and Tsezos, M. (2001). Biosorption and biological reduction of chromium by microbial biomass in batch and pilot scale rotating biological contactor (RBC). In: *International Biohydrometallurgy Symposium IBS 2001: Biohydrometallurgy Fundamentals, Technology and Sustainable Development* (ed. V.S.T. Ciminelli, O. Garcia Jr.).

Hu, C., Liu, H., Chen, G. et al. (2012). Effect of aluminum speciation on arsenic removal during coagulation process. *Separation and Purification Technology* 86: 35–40.

Kaczala, F., Marques, M., and Hogland, W. (2009). Lead and vanadium removal from a real industrial wastewater by gravitational settling/sedimentation and sorption onto Pinus sylvestris sawdust. *Bioresource Technology* 100 (1): 235–243.

Kapoor, A. and Viraraghavan, T. (1997). Heavy metal biosorption sites in Aspergillus niger. *Bioresource Technology* 61 (3): 221–227.

Khan, M.J., Zia, M.S., and Qasim, M. (2010). Contamination of agro-ecosystem and human health hazards from wastewater used for irrigation. *J. Chem. Soc. Pak.* 32: 370–378.

Kord Mostafapour, F., Bazrafshan, E., Farzadkia, M. et al. (2012). Arsenic Removal from Aqueous Solutions by Salvadora persica Stem Ash. *Journal of Chemistry and Chemical Engineering*. doi:10.1155/2013/740847.

Korte, N.E. and Fernando, Q. (1991). A review of arsenic (III) in groundwater. *Critical Reviews in Environmental Control* 21 (1): 1–39.

Kumar, S., Bhanjana, G., Jangra, K. et al. (2014). Utilization of carbon nanotubes for the removal of rhodamine B dye from aqueous solutions. *Journal of Nanoscience and Nanotechnology* 14 (6): 4331–4336. doi:10.1166/jnn.2014.8077.

Langmuir, I. (1918). The adsorption of gases on plane surfaces of glass, mica and platinum. *Journal of the American Chemical Society* 40 (9): 1361–1403.

Leupin, O.X. and Hug, S.J. (2005). Oxidation and removal of arsenic (III) from aerated groundwater by filtration through sand and zero-valent iron. *Water Research* 39 (9): 1729–1740.

Libatique, M.J.H., Lee, M.-C., Yeh, H.-Y. et al. (2020). Total and inorganic arsenic biosorption by Sarcodia suiae (Rhodophyta), as affected by controlled environmental conditions. *Chemosphere* 248: 126084.

Lin, T.F. and Wu, J.K. (2001). Adsorption of arsenite and arsenate within activated alumina grains: equilibrium and kinetics. *Water Research* 35 (8): 2049–2057.

Loukidou, M.X., Matis, K.A., Zouboulis, A.I. et al. (2003). Removal of As(V) from wastewaters by chemically modified fungal biomass. *Water Research* 37 (18): 4544–4552.

Manju, G.N., Raji, C., and Anirudhan, T.S. (1998). Evaluation of coconut husk carbon for the removal of arsenic from water. *Water Research* 32 (10): 3062–3070.

Mcafee, B.J., Gould, W.D., Nadeau, J.C. et al. (2001). Biosorption of metal ions using chitosan, chitin, and biomass of rhizopus oryzae. *Separation Science and Technology* 36 (14): 3207–3222.

McNeill, L.S. and Edwards, M. (1995). Soluble arsenic removal at water treatment plants. *Journal of American Water Works Association* 87 (4): 105–113.

Miteva, E., Hristova, D., Nenova, V. et al. (2005). Arsenic as a factor affecting virus infection in tomato plants: changes in plant growth, peroxidase activity and chloroplast pigments. *Scientia Horticulturae* 105 (3): 343–358.

Mohan, D. and Chander, S. (2006). Removal and recovery of metal ions from acid mine drainage using lignite-a low-cost sorbent. *Journal of Hazardous Materials* 137: 1545–1553.

Mohan, D. and Pittman, C.U. (2007). Arsenic removal from water/wastewater using adsorbents—A critical review. *Journal of Hazardous Materials* 142 (1): 1–53.

Murcott, S. (2012). *Arsenic Contamination in the World: An International Sourcebook*. IWA Publishing.

Nadeem, T., Shankar, P., and Iftode, L. (2006). A comparative study of data dissemination models for VANETs. In: *2006 Third Annual International Conference on Mobile and Ubiquitous Systems: Networking Services*, 1–10.

Ng, J.C. (2005). Environmental contamination of arsenic and its toxicological impact on humans. *Huan jing hua xue = Environmental chemistry*. www.publish.csiro.au/EN/EN05062.

Niu, C.H., Volesky, B., and Cleiman, D. (2007). Biosorption of arsenic (V) with acid-washed crab shells. *Water Research* 41 (11): 2473–2478.

Niu, H. and Volesky, B. (2003). Characteristics of anionic metal species biosorption with waste crab shells. *Hydrometallurgy* 71 (1): 209–215.

Oke, I.A., Olarinoye, N.O., and Adewusi, S.R.A. (2008). Adsorption kinetics for arsenic removal from aqueous solutions by untreated powdered eggshell. *Adsorption* 14 (1): 73–83.

Pal, D.B., Alsukaibi, A.K.D., Singh, R. et al. (accepted for publication). Removal of cadmium ions from simulated wastewater using copper and cerium based adsorbent. *Frontiers in Bioscience-Landmark*.

Pal, D.B., Giri, D.D., Singh, R. et al. (2017). Arsenic removal from synthetic waste water by CuO nano-flakes synthesized by aqueous precipitation method. *Desalination and Water Treatment* 62: 355–359.

Pal, D.B., Lavania, R., Srivastava, P. et al. (2018). Photo-catalytic degradation of methyl tertiary butyl ether from wastewater using CuO/CeO2 composite nanofiber catalyst. *Journal of Environmental Chemical Engineering* 6 (2): 2577–2587.

Pal, D.B., Selvasembian, R., and Singh, P. (2021a). Cadmium removal by composite copper oxide/ceria adsorbent from synthetic wastewater. *Biomass Conversion and Biorefinery*. doi:10.1007/s13399-021-01534-6.

Pal, D.B., Singh, A., Jha, J.M. et al. (2021b). Low-cost biochar adsorbents prepared from date and delonix regia seeds for heavy metal sorption. *Bioresource Technology* 339: 125606.

Pallier, V., Feuillade-Cathalifaud, G., Serpaud, B. et al. (2010). Effect of organic matter on arsenic removal during coagulation/flocculation treatment. *Journal of Colloid and Interface Science* 342 (1): 26–32.

Park, H.J., Jeong, S.W., Yang, J.K. et al. (2007). Removal of heavy metals using waste eggshell. *Journal of Environmental Sciences* 19 (12): 1436–1441.

Pokhrel, D. and Viraraghavan, T. (2006). Arsenic removal from an aqueous solution by a modified fungal biomass. *Water Research* 40 (3): 549–552.

Pokhrel, D. and Viraraghavan, T. (2008). Organic arsenic removal from an aqueous solution by iron oxide-coated fungal biomass: An analysis of factors influencing adsorption. *Chemical Engineering Journal* 140 (1-3): 165–172. doi:10.1016/j.cej.2007.09.038.

Polizzotto, M.L., Harvey, C.F., Li, G. et al. (2006). Solid-phases and desorption processes of arsenic within Bangladesh sediments. *Chemical Geology* 228 (1-3): 97–111. doi:10.1016/j.chemgeo.2005.11.026.

Prasad, N., Kumar, P., and Pal, D.B. (2020). Cadmium removal from aqueous solution by jackfruit seed bio-adsorbent. *SN Applied Sciences* 2 (6). doi:10.1007/s42452-020-2750-z.

Qu, X., Alvarez, P.J.J., and Li, Q. (2013). Applications of nanotechnology in water and wastewater treatment. *Water Research* 47 (12): 3931–3946. doi:10.1016/j.watres.2012.09.058.

Ravenscroft, P., Brammer, H., and Richards, K. (2009). *Arsenic Pollution: A Global Synthesis*. John Wiley and Sons.

Remoudaki, E., Hatzikioseyian, A., Kousi, P. et al. (2003). The mechanism of metals precipitation by biologically generated alkalinity in biofilm reactors. *Water Research* 37 (16): 3843–3854. doi:10.1016/s0043-1354(03)00306-3.

Saeed, A., Iqbal, M., and Akhtar, M.W. (2005). Removal and recovery of lead (II) from single and multiple (Cd, Ni, Cu, Zn) solutions by crop milling waste (black gram husk). *J Hazard Mater* 117: 65–73.

Seki, H., Suzuki, A., and Maruyama, H. (2005). Biosorption of chromium(VI) and arsenic(V) onto methylated yeast biomass. *Journal of Colloid and Interface Science* 281 (2): 261–266. doi:10.1016/j.jcis.2004.08.167.

Siddiqui, S.I. and Chaudhry, S.A. (2018). A review on graphene oxide and its composites preparation and their use for the removal of As3 and As5 from water under the effect of various parameters: Application of isotherm, kinetic and thermodynamics. *Process Safety and Environmental Protection* 119: 138–163. doi:10.1016/j.psep.2018.07.020.

Singh, P., Chaturvedi, R.K., Mishra, A. et al. (2015). Assessment of ground and surface water quality along the river Varuna, Varanasi, India. *Environmental Monitoring and Assessment* 187 (4). doi:10.1007/s10661-015-4382-0.

Singh, P., Jain, R., Srivastava, N. et al. (2017). Current and emerging trends in bioremediation of petrochemical waste: A review. *Critical Reviews in Environmental Science and Technology* 47 (3): 155–201. doi:10.1080/10643389.2017.1318616.

Singh, P., Vishnu, M.C., Sharma, K.K. et al. (2016). Photocatalytic degradation of Acid Red dye stuff in the presence of activated carbon-TiO2 composite and its kinetic enumeration. *Journal of Water Process Engineering* 12: 20–31. doi:10.1016/j.jwpe.2016.04.007.

Sneddon, I.R., Garelick, H., and Valsami-Jones, E. (2005). An investigation into arsenic(V) removal from aqueous solutions by hydroxylapatite and bone-char. *Mineralogical Magazine* 69 (5): 769–780. doi:10.1180/0026461056950287.

Tabaraki, R. and Heidarizadi, E. (2018). Simultaneous biosorption of Arsenic (III) and Arsenic (V): Application of multiple response optimizations. *Ecotoxicology and Environmental Safety* 166: 35–41. doi:10.1016/j.ecoenv.2018.09.063.

Temkin, M.I. and Pyzhev, V. (1940). Kinetics of ammonia synthesis on promoted iron catalyst. *Acta Phys. Chim. USSR* 12: 327–356.

Tuzen, M., Sari, A., Mendil, D. et al. (2009). Characterization of biosorption process of As(III) on green algae Ulothrix cylindricum. *Journal of Hazardous Materials* 165 (1-3): 566–572.

Volesky, B. (1990). *Biosorption of Heavy Metals*. CRC Press.

Wang, J. and Chen, C. (2006). Biosorption of heavy metals by Saccharomyces cerevisiae: A review. *Biotechnology Advances* 24 (5): 427–451. doi:10.1016/j.biotechadv.2006.03.001.

Wang, J.S. and Wai, C.M. (2004). Arsenic in drinking water – a global environmental problem. *J. Chem. Educ.* 81: 207–213.
Wase, J. and Forster, C. (1997). *Biosorbents for Metal Ions*. Taylor and Francis Ltd.
Wilkie, J.A. and Hering, J.G. (1996). Adsorption of arsenic onto hydrous ferric oxide: effects of adsorbate/adsorbent ratios and co-occurring solutes. *Colloids and Surfaces A: Physicochemical and Engineering Aspects* 107: 97–110. doi:10.1016/0927-7757(95)03368-8.
Zhang, H., Omer, A.M., Hu, Z. et al. (2019). Fabrication of magnetic bentonite/carboxymethyl chitosan/sodium alginate hydrogel beads for Cu (II) adsorption. *International Journal of Biological Macromolecules* 135: 490–500. doi:10.1016/j.ijbiomac.2019.05.185.
Zouboulis, A.I., Matis, K.A., and Hancock, I.C. (1997). Biosorption of metals from dilute aqueous solutions. *Separation and Purification Methods* 26 (2): 255–295. doi:10.1080/03602549708014160.

Index

Note: Page numbers in *italic* refer to figures, page numbers in **bold** refer to tables.

Biosorption for Wastewater Contaminants, First Edition.
Edited by Rangabhashiyam Selvasembian and Pardeep Singh.

b

c

d

www.ingramcontent.com/pod-product-compliance
Ingram Content Group UK Ltd.
Pitfield, Milton Keynes, MK11 3LW, UK
UKHW050418051026
15470UKWH00022B/1621

9 781119 737599